Quantum Science and Technology

The book series Quantum Science and Technology is dedicated to one of today's most active and rapidly expanding fields of research and development. In particular, the series will be a showcase for the growing number of experimental implementations and practical applications of quantum systems. These will include, but are not restricted to: quantum information processing, quantum computing, and quantum simulation; quantum communication and quantum cryptography; entanglement and other quantum resources; quantum interfaces and hybrid quantum systems; quantum memories and quantum repeaters; measurement-based quantum control and quantum feedback; quantum nanomechanics, quantum optomechanics and quantum transducers; quantum sensing and quantum metrology; as well as quantum effects in biology. Last but not least, the series will include books on the theoretical and mathematical questions relevant to designing and understanding these systems and devices, as well as foundational issues concerning the quantum phenomena themselves. Written and edited by leading experts, the treatments will be designed for graduate students and other researchers already working in, or intending to enter the field of quantum science and technology.

More information about this series at http://www.springer.com/series/10039

Giuliano Gadioli La Guardia

Quantum Error Correction

Symmetric, Asymmetric, Synchronizable,
and Convolutional Codes

 Springer

Giuliano Gadioli La Guardia
Department of Mathematics and Statistics
Ponta Grossa State University
Ponta Grossa, Paraná, Brazil

ISSN 2364-9054 ISSN 2364-9062 (electronic)
Quantum Science and Technology
ISBN 978-3-030-48553-5 ISBN 978-3-030-48551-1 (eBook)
https://doi.org/10.1007/978-3-030-48551-1

This Springer imprint is published by the registered company Springer Nature Switzerland AG
The registered company address is: Gewerbestrasse 11, 6330 Cham, Switzerland

To my parents and my family

Preface

Theory of quantum information and computation has been extensively investigated in the last two decades (see the textbook [121] by Nielsen and Chuang and the references [25, 53, 144, 147, 148] for the first constructions or construction methods of quantum codes shown in the literature; see also the papers [1, 2, 20, 21, 82, 122, 146] concerned with constructions of topological quantum codes, which will not be investigated in this book). This book is written in order to familiarize graduate or postgraduate students with respect to several constructions or construction methods of quantum codes as well as techniques of constructions of quantum convolutional codes available in the literature. More precisely, we gathered great part of the most relevant papers that we have published concerning quantum coding theory to present such results here, in form of book. To this end, we utilize the well-known Calderbank–Shor–Steane (CSS) construction, the Hermitian and the Steane enlargement construction to certain classes of classical codes. These quantum codes have good parameters and they are introduced recently in the literature. Furthermore, the book also presents several constructions of families of asymmetric quantum codes with good parameters.

Keeping in mind a similar approach, the book also contains a careful description of the procedures adopted to construct families of quantum convolutional codes. To close the book, we introduce and construct families of asymmetric quantum convolutional codes (this concept was introduced in Ref. [102] (La Guardia, G.G.: Asymmetric quantum convolutional codes. Quantum Inform. Processing **15**, 167–183 (2016)).

Although the book does not bring new didactic approach nor new form of presentation, I tried to write it carefully, with accessible language and clear explanations, in order to improve the quality and the accessibility of it. In my point of view, the book has some advantages. The first one is to teach the reader certain algebraic techniques of code construction that could improve the capacity of abstraction of him/her. It is also an attempt to motivate the reader to perform their own contributions from this area of research. Another important contribution is that all constructions presented here are performed algebraically, i.e., the procedures adopted are capable of constructing families of codes, and not only codes with specific parameters.

Description of the Book

The book is organized in such a way that the reader can skip some introductory chapters without major problems.

Chapter 1 presents a review of some basic concepts on linear algebra and metric spaces, necessary to define the scenario and the structure of the quantum mechanics.

In Chap. 2, the postulates of quantum mechanics, the definition of single and multiple qubit gates and the most common types of quantum channels are reviewed.

Chapter 3 is concerned with the first constructions or construction methods of quantum codes shown in the literature. The well-known five qubit and the Steane code are examples of such codes. We also review the CSS construction and the stabilizer quantum code construction.

Chapter 4 is devoted to review some definitions and results on linear block codes. We recall the concept of Euclidean and Hermitian dual of a linear code as well as the techniques to obtain new codes from old. Additionally, the classes of cyclic and algebraic geometry codes are reviewed, since such classical linear codes are necessary for the quantum code constructions presented in this book.

Chapter 5 brings the most relevant constructions of quantum codes that we have published in the last ten years of research. They include several families of quantum codes derived from (classical) Bose–Chaudhuri–Hocquenghem (BCH) and from (classical) algebraic geometry codes. Moreover, constructions of quantum synchronizable codes derived from (classical) cyclic, BCH and product codes are also presented here.

In Chap. 6, as in Chap. 5, we present my most relevant contributions concerning asymmetric quantum code constructions, which were published in the last years. We construct several families of asymmetric quantum codes (AQQs) derived from (classical) Reed–Solomon and generalized Reed–Solomon codes, generalized Reed–Muller and BCH codes. Additionally, we generalize to AQQs the well-known methods which are valid to quantum codes, namely: puncturing, extending, expanding, direct sum and the $(u|u+v)$ construction.

In Chap. 7, we present my main contributions published in the last years concerning constructions of quantum convolutional codes. We explain how to construct families of quantum convolutional codes with good parameters derived from

(classical) convolutional codes. These classical convolutional codes were obtained from linear block codes: BCH, negacyclic and algebraic geometry codes. Moreover, we introduce a new class of codes: the asymmetric quantum convolutional codes.
Have a good read and enjoy the book!!

Ponta Grossa, Paraná, Brazil

Giuliano Gadioli La Guardia
gguardia@uepg.br

Contents

Chapter 1
Some Linear Algebra

1.1 Introduction

In this chapter we recall elementary concepts of linear algebra necessary for the development of the book. If the reader is interested in more details, we recommend the textbooks [64, 137]; see also [13] for Theory of Modules which are, roughly speaking, "vector spaces over rings". The reader can consult the Appendix to recall the concept of ring and field if necessary.

We begin by defining the concept of vector space, one of the main concepts of linear algebra.

Definition 1.1.1 A vector space over a field K is a ordered triple $(V, +, \cdot)$, where V is a set,

$$+ : V \times V \longrightarrow V$$
$$(u, v) \longrightarrow u + v$$

is an internal operation, called vector addition, and

$$\cdot : K \times V \longrightarrow V$$
$$(a, v) \longrightarrow av$$

is an external operation, called scalar multiplication, such that

(1) $(V, +)$ is an abelian group:
(1.1) There exists $0 \in V$ such that $v + 0 = 0 + v = v, \forall v \in V$ (zero or null vector);
(1.2) $\forall v \in V$, there exists $-v \in V$ such that $v + (-v) = (-v) + v = 0$ (symmetric);
(1.3) $\forall u, v, w \in V, u + (v + w) = (u + v) + w$ (associative);
(1.4) $\forall u, v \in V, u + v = v + u$ (commutative);

© Springer Nature Switzerland AG 2020
G. G. La Guardia, *Quantum Error Correction*, Quantum Science and Technology,
https://doi.org/10.1007/978-3-030-48551-1_1

(2) The operations $\cdot$ and $+$ satisfy

(2.1) $\forall\, a \in K, \forall\, v, w \in V, a(v + w) = av + aw$;
(2.2) $\forall\, a, b \in K$ and $\forall\, v \in V, (a + b)v = av + bv$;
(2.3) $\forall\, a, b \in K$ and $\forall\, v \in V, (ab)v = a(bv)$;
(2.4) $\forall\, v \in V, 1v = v$.

Remark 1.1.1 (1) In order to avoid stress of notation we will denote only by V the vector space $(V, +, \cdot)$ in the cases where there is no possibility of confusion.

(2) Throughout this chapter we always assume that all vector spaces are defined over the same field K.

Subspaces of a vector space V are subsets of V which are also vector spaces under the operations of V.

Definition 1.1.2 Let V be a vector space. A subset $W \subseteq V$ is a subspace of V if the following conditions hold:

(1) $0 \in W$;
(2) W is closed under vector addition in V, i.e., $\forall\, v, w \in W$, it follows that $v + w \in W$;
(3) W is closed under scalar multiplication, i.e., $\forall\, v \in W$ and $\forall\, a \in K$, it implies that $av \in W$.

Definition 1.1.3 Let V be a vector space and consider that $v, v_1, v_2, \ldots v_n \in V$. We say that v is a linear combination of the vectors $v_1, v_2, \ldots, v_n$ if v can be written as

$$v = \sum_{i=1}^{n} a_i v_i, \text{ where } a_i \in K, \text{ for } i = 1, 2, \ldots, n.$$

Definition 1.1.4 Let V be a vector space and let $S = \{v_1, v_2, \ldots, v_n\}$ be a set of vectors in V. Then the set $\langle v_1, v_2, \ldots, v_n \rangle$ is the set of all linear combinations of $v_1, v_2, \ldots, v_n$, called subspace spanned by S.

The following result characterizes subspaces spanned by a set.

Theorem 1.1.1 *Let V be a vector space and $S = \{v_1, v_2, \ldots, v_n\}$ be a set of vectors in V. Then the intersection of all the subspaces of V containing S is $\langle v_1, v_2, \ldots, v_n \rangle$, i.e.,*

$$\langle v_1, v_2, \ldots, v_n \rangle = \bigcap_{S \subseteq W \subseteq V} W,$$

where W runs through the subspaces of V containing S. Hence, $\langle v_1, v_2, \ldots, v_n \rangle$ is the smallest subspace of V which contains S.

Exercise 1.1.1 Show Theorem 1.1.1.

Definition 1.1.5 Let V be a vector space and $v_1, v_2, \ldots, v_n \in V$. We say that the set $S = \{v_1, v_2, \ldots, v_n\}$ is linearly independent (LI) (or that the vectors $v_1, v_2, \ldots, v_n$

are linearly independent) if the equation $\sum_{i=1}^{n} a_i v_i = 0$ implies that $a_i = 0$ for all $i = 1, 2, \ldots, n$. If there exists at least one $a_i \neq 0$ such that $\sum_{i=1}^{n} a_i v_i = 0$, then S (or that the vectors $v_1, v_2, \ldots, v_n$) is called linearly dependent (LD).

Definition 1.1.6 Let V be a vector space. A basis for V is a linearly independent set of vectors in V which spans V. We say that V is finite-dimensional if it has a finite basis.

Theorem 1.1.2 (Invariance of dimension) *Let V be a vector space with bases $S = \{v_1, v_2, \ldots, v_n\}$ and $T = \{w_1, w_2, \ldots, w_m\}$. Then it follows that $m = n$.*

Because of the invariance of dimension (Theorem 1.1.2), one can define the *dimension* of a vector space.

Definition 1.1.7 The dimension $\dim_K(V)$ of a finite-dimensional vector space V (over K) is the number of elements in a basis of V.

1.2 Linear Transformations

Here, we recall the concepts of linear transformation and linear operator.

Definition 1.2.1 Let V and W be two vectors spaces over the same field K. A function $T : V \longrightarrow W$ is a linear transformation from V into W if the following hold:

(T1) $\forall\, u, v \in V, T(u + v) = T(u) + T(v)$;
(T2) $\forall\, v \in V$, and $\forall\, a \in K, T(av) = aT(v)$.

In Linear Algebra, isomorphic vector spaces are essentially the same.

Definition 1.2.2 Let V and W be two vectors spaces. We say that V and W are isomorphic, written $V \cong W$, if there exists a bijective linear transformation $T : V \longrightarrow W$.

Exercise 1.2.1 Show that if V and W are finite-dimensional vector spaces then $V \cong W$ if and only if $\dim_K(V) = \dim_K(W)$.

Theorem 1.2.1 *Let V and W be vector spaces and let $v_1, v_2, \ldots, v_n$ be a basis of V. Given any list of vectors $w_1, w_2, \ldots, w_n$ in W, there exists a unique linear transformation $T : V \longrightarrow W$ such that $T(v_i) = w_i$ for all $i = 1, 2, \ldots, n$.*

Exercise 1.2.2 Show Theorem 1.2.1.

Definition 1.2.3 Let V and W be two vector spaces over the same field K. Consider that $B_V = \{v_1, v_2, \ldots, v_n\}$ is a basis of V and $B_W = \{w_1, w_2, \ldots, w_m\}$ is a basis of W. If $T : V \longrightarrow W$ is a linear transformation from V into W, then the matrix of T relative to B_V and B_W is the $m \times n$ matrix $M = [a_{ij}]$ such that $T(v_j) = \sum_{i=1}^{m} a_{ij} w_i$. The matrix M is written as $M = [T]_{B_V}^{B_W}$ and it is said to be a representation of the operator T.

The following theorem states that the study of linear transformations is, in essence, the same as the study of matrices.

Theorem 1.2.2 *Let V and W be (finite-dimensional) vector spaces over a field K with bases $B_V = \{v_1, v_2, \ldots, v_n\}$ and $B_W = \{w_1, w_2, \ldots, w_m\}$, respectively. Let us denote $\mathrm{HOM}_K(V, W)$ the set of all linear transformations from V into W and let $\mathrm{Mat}_{m \times n}(K)$ be the set of all $m \times n$ matrices with entries in K. Then there exists a bijection $F : \mathrm{HOM}_K(V, W) \longrightarrow \mathrm{Mat}_{m \times n}(K)$ given by $F(T) = [T]_{B_V}^{B_W}$.*

Exercise 1.2.3 Prove Theorem 1.2.2.

Definition 1.2.4 Given a linear transformation $T : V \longrightarrow W$, we define the kernel of T, written $\ker(T)$, as $\ker(T) = \{v \in V | T(v) = 0\}$. The image of T, $\mathrm{im}(T)$, is the set $\mathrm{im}(T) = \{w \in W | w = T(v), \text{ for some } v \in V\}$.

The kernel $\ker(T)$ is a subspace of V and $\mathrm{im}(T)$ is a subspace of W. Moreover, T is injective if and only if $\ker(T) = \{0\}$.

Theorem 1.2.3 *If V is a vector space over K with $\dim_K(V) = \dim(V) = n < \infty$, and if $T : V \longrightarrow V$ is a linear transformation, then the following statements are equivalent:*

  (i) T is an isomorphism;
 (ii) T is surjective;
(iii) T is injective.

1.3 Diagonalizable Operators

Let V be a vector space over a field K. Recall that a linear transformation T from V into V is called a linear operator on V.

Definition 1.3.1 Let $T : V \longrightarrow V$ be a linear operator on V. If there exists an $v \neq 0$, $v \in V$, and $\lambda \in K$ such that $T(v) = \lambda v$, then we say that λ is an eigenvalue of T and v is an eigenvector of T associated with λ.

The subspace $V_\lambda = \{v \in V | T(v) = \lambda v\}$ of V is called eigenspace associated with λ.

Theorem 1.3.1 *Let V be a n-dimensional vector space over K and $\lambda \in K$. Let $T : V \longrightarrow V$ be a linear operator on V and I_n be the identity matrix of order n. Then the following are equivalents:*

  (i) λ *is an eigenvalue of T;*
  (ii) *the operator $T - \lambda I_n$ is not invertible;*
 (iii) $\det(T - \lambda I_n) = 0$.

Exercise 1.3.1 Prove the equivalences given of Theorem 1.3.1.

From Theorem 1.2.2, there exists a bijection between the set of all linear transformations $T : V \longrightarrow W$ and the set of all matrices $m \times n$ with entries in K. In particular, if $V = W$ and if B_V is an basis of V, by considering $M = [T]_{B_V}^{B_V}$, it follows that $T - \lambda I_n$ is invertible if and only if $M - \lambda I_n$ is invertible. Thus, if A is a square matrix of order n over the field K, an eigenvalue of A in K is a scalar $\lambda \in K$ such that $\det(A - \lambda I_n) = 0$. The polynomial $p(\lambda) = \det(A - \lambda I_n)$ is called characteristic polynomial of A.

Definition 1.3.2 Let V be a finite-dimensional vector space and let $T : V \longrightarrow V$ be a linear operator on V. Then T is said to be diagonalizable if there exists a basis of eigenvectors for V.

The following theorem characterizes diagonalizable operators.

Theorem 1.3.2 *Let $T : V \longrightarrow V$ be a linear operator on V, where V is n-dimensional. Assume that $\lambda_1, \lambda_2, \ldots, \lambda_r$ are distinct eigenvalues of T and $W_i = \ker(T - \lambda_i I_n)$. Then the following conditions are equivalent:*

  (i) T *is diagonalizable;*
  (ii) *the characteristic polynomial for T is $p(\lambda) = (\lambda - \lambda_1)^{d_1} \cdot \ldots \cdot (\lambda - \lambda_r)^{d_r}$, where $\dim(W_i) = d_i$, for all $i = 1, 2, \ldots, r$;*
 (iii) $\dim(W_1) + \dim(W_2) + \cdots + \dim(W_r) = \dim(V)$.

Exercise 1.3.2 Show Theorem 1.3.2.

1.4 Tensor Products

Let us now consider that U, V, W are vector spaces over K. A function $B : U \times V \longrightarrow W$ is called *bilinear* if for satisfies two conditions:

(1) for each $u \in U$, the function $B_u : V \longrightarrow W$ defined as $B_u(v) = B(u, v)$ is a linear transformation;
(2) for each $v \in V$, the function $B_v : U \longrightarrow W$ defined as $B_v(u) = B(u, v)$ is a linear transformation.

Keeping this concept in mind, we can define *tensor product* of vector spaces.

Definition 1.4.1 Let U and V be two finite-dimensional vector spaces over the same field K. The tensor product of U and V is a vector space $U \otimes V$ over K together with a bilinear function $B_{U \otimes V} : U \times V \longrightarrow U \otimes V$ written as $B(u, v) = u \otimes v$, with the universal property: for every vector space W over K and any bilinear function $\widetilde{B} : U \times V \longrightarrow W$, there exists a unique linear transformation $T_w : U \otimes V \longrightarrow W$ such that $T_w \circ B = \widetilde{B}$. The elements $u \otimes v$ are said to be *tensor* of u and v.

Note that in Definition 1.4.1 we have written "**The** tensor product of U and V...”; this can be done because tensor products are unique up to isomorphisms, according to the next theorem.

Theorem 1.4.1 *If* (T_1, B_{T_1}) *and* (T_2, B_{T_2}) *are tensor products of the vector spaces* U *and* V, *then there exists a unique isomorphism* $f : T_1 \longrightarrow T_2$ *such that* $f \circ B_{T_1} = B_{T_2}$.

Exercise 1.4.1 Show the uniqueness of tensor products up to isomorphisms.

If the vector spaces are finite-dimensional then the tensor product between them exists.

Theorem 1.4.2 *Let* U *and* V *be finite-dimensional vector spaces over* K *with dimensions* m *and* n, *respectively. Then* $U \otimes V$ *exists: if* $\{u_1, u_2, \ldots, u_m\}$ *is a basis for* U *and* $\{v_1, v_2, \ldots, v_n\}$ *is a basis for* V *then the* mn *vectors* $u_i \otimes v_j$ *for all* $i = 1, 2, \ldots, m$ *and* $j = 1, 2, \ldots, n$, *form a basis for* $U \otimes V$, *i.e.,* $\dim(U \otimes V) = [\dim(U)][\dim(V)]$.

Exercise 1.4.2 Show the existence of tensor products.

Remark 1.4.1 (1) In Definition 1.4.1, we have defined the tensor product of two vector spaces. The generalization for a finite number of vector spaces is performed in the same way by applying the concept of multi-linear functions. More precisely, if W and $V_1, V_2, \ldots, V_n$ are vector spaces over the same field K, a function $M : V_1 \times V_2 \times \cdots \times V_n \longrightarrow W$ is said to be multi-linear if, fixing $n - 1$ variables (except the ith variable), M is a linear transformation in the ith variable. Thus, we can also consider a finite number of tensor products of vectors spaces $V_1 \otimes V_2 \otimes \cdots \otimes V_n$.

As a consequence of its definition, the tensor product satisfies the properties given below:
(i) $\forall z \in K$, $v \in U$ and $w \in V$ one has $z(v \otimes w) = (zv) \otimes w = v \otimes (zw)$;
(ii) $\forall v_1, v_2 \in U$ and $w \in V$, $(v_1 + v_2) \otimes w = (v_1 \otimes w) + (v_2 \otimes w)$;
(iii) $\forall v \in U$ and $w_1, w_2 \in V$, $v \otimes (w_1 + w_2) = (v \otimes w_1) + (v \otimes w_2)$.

1.5 Modules

In this subsection we discuss a little bit about vectorial modules. Roughly speaking, a module is a "vector space-like", where the scalars belong to a ring and not in a field (the reader can compare Definitions 1.1.1 and 1.5.1 to see the complete similarity of both definitions). We next give the formal definition of a module.

Definition 1.5.1 Assume that R is a commutative ring with unit. An R-module is a set M endowed with two operations, an (internal) operation $+ : M \times M \longrightarrow M$ (called addition), and an external operation $\cdot : R \times M \longrightarrow M$ (called scalar multiplication) such that $+$ satisfies

(A1) $\forall v \in M$, there exists $0 \in M$ such that $v + 0 = 0 + v = v$ (identity);
(A2) $\forall u, v, w \in M, u + (v + w) = (u + v) + w$ (associative);
(A3) $\forall v \in M$, there exists $-v \in M$ such that $v + (-v) = (-v) + v = 0$ (symmetric);
(A4) $\forall u, v \in M, u + v = v + u$ (commutative);

and the operations $\cdot$ and $+$ satisfy

(M1) $\forall u, v \in M$ and $\forall r \in R, r(u + v) = ru + rv$;
(M2) $\forall u \in M$ and $\forall r_1, r_2 \in R, (r_1 + r_2)u = r_1 u + r_2 u$;
(M3) $\forall u \in M$ and $\forall r_1, r_2 \in R, (r_1 r_2)u = r_1(r_2 u)$;
(M4) $\forall u \in M$ and $1 \in R, 1m = m$.

Remark 1.5.1 Note that an R-module is, in fact, an ordered triple of the form $(M, +, \cdot)$. However, to avoid stress of notation, we simply write M when the operations $+$ and $\cdot$ are known from the context.

The concepts of submodule and homomorphism of modules are also analogous to that of subspace and linear transformation, respectively, as the reader can see in the following definition.

Definition 1.5.2 Let R be a commutative ring with unit and let M be an R-module. A subset $N \subseteq M$ is a called an *R-submodule* of M if N is an additive subgroup of M closed under scalar multiplication.

Definition 1.5.3 Let R be a commutative ring with unit and let M and N be two R-modules. A function $h : M \longrightarrow N$ is called an *R-homomorphism* if

(H1) $\forall u, v \in M, h(u + v) = h(u) + h(v)$;
(H2) $\forall r \in R$ and $\forall u \in M, h(ru) = rh(u)$.

A special class of module is the class of free-modules, i.e., modules that have basis.

Definition 1.5.4 Let R be a commutative ring with unit. An R-module M is said to be *free R-module* if M is isomorphic to a direct sum of copies of R. In other words, there exists an index set $\mathcal{I}$ such that $M = \sum_{i \in \mathcal{I}} R_i$, where $R_i = \langle b_i \rangle \cong R$ ($\cong$ as R-modules) for every $i \in \mathcal{I}$. The index set $\mathcal{I}$ can be finite or infinite and $\mathcal{B} = \{b_i | i \in \mathcal{I}\}$ is said to be *a basis of* M.

The concept of free module will be utilized in the definition of convolutional codes (see Sect. 7.1).

1.6 Metric Spaces

In the following we recall the concept of *metric space*.

Definition 1.6.1 A metric space consists of an ordered pair (M, d), where M is a nonempty set and d is a function $d : M \times M \longrightarrow \mathbb{R}$ such that, $\forall \, x, y, z \in M$, the following conditions are satisfied:

(1) $d(x, y) \geq 0$;
(2) $d(x, y) = 0 \Longleftrightarrow x = y$;
(3) (Symmetry) $d(x, y) = d(y, x)$;
(4) (Triangle inequality) $d(x, z) \leq d(x, y) + d(y, z)$.

The function d is called *metric* (or distance function) on M.

A metric space is a particular case of a topological space. Since this is not the subject of this book, we will not describe the latter concept here. For the reader who is interested to learn about topological spaces, we suggest the textbooks [8, 118].

As examples, let us consider the n-dimensional Euclidean space $\mathbb{R}^n = \{(x_1, x_2, \ldots, x_n), x_i \in \mathbb{R}\}$ and the unitary (sometimes called complex Euclidean n-space) n-dimensional space $\mathbb{C}^n = \{(x_1, x_2, \ldots, x_n), x_i \in \mathbb{C}\}$. In the first case, if $x = (x_1, x_2, \ldots, x_n)$ and $y = (y_1, y_2, \ldots, y_n)$ are vectors in $\mathbb{R}^n$ then the Euclidean metric on $\mathbb{R}^n$ is defined as

$$d_E(x, y) = \sqrt{\sum_{i=1}^{n} (x_i - y_i)^2}.$$

In the second case, if $z = (z_1, z_2, \ldots, z_n)$ and $w = (w_1, w_2, \ldots, w_n) \in \mathbb{C}^n$, then we define a metric on $\mathbb{C}^n$ by

$$d(z, w) = \sqrt{\sum_{i=1}^{n} |z_i - w_i|^2},$$

where $| \cdot |$ denotes the norm of a complex number.

Exercise 1.6.1 Show that the two functions defined above satisfies the four conditions of Definition 1.6.1 proving, therefore, that these functions are, in fact, two metrics defined on $\mathbb{R}^n$ and on $\mathbb{C}^n$, respectively.

We next define the concept of *convergence* in metric spaces.

Let (M, d) be a metric space. A sequence in M is a function $x : \mathbb{N} \longrightarrow M$ which associates for every $n \in \mathbb{N}$ an element $x(n) := x_n \in M$. The image of the function x is denoted by $(x_n)_n$ or, simply, by (x_n). By abuse of notation we simply write (x_n) to denote a sequence.

Definition 1.6.2 Let (M, d) be a metric space and let (x_n) be a sequence in M. We say that (x_n) converges if there exists an element $x \in M$ such that

$$\lim_{n \to \infty} d(x_n, x) = 0.$$

The element x is said to be the limit of (x_n) and it is written

$$\lim_{n \to \infty} x_n = x.$$

One says that (x_n) *converges to* x or has limit x. On the other hand, if (x_n) does not converge, then it is said to be divergent.

Definition 1.6.3 Let (M, d) be a metric space and (x_n) be a sequence in M. We say that (x_n) is a Cauchy sequence if for every positive real number $\epsilon > 0$, there exists a $n_0 = n_0(\epsilon) \in \mathbb{N}$ such that, for every $m, n > n_0$ one has $d(x_m, x_n) < \epsilon$. The space M is called *complete* if every Cauchy sequence in M converges (to an element of M).

1.7 Inner Product Spaces

From now on, in order to fit our notation to quantum theory, we adapt the notation of a vector as follows. Since the scenario of quantum mechanics is the vector space $\mathbb{C}^n$ over the complex field $\mathbb{C}$ (or complex vector space $\mathbb{C}^n$), under usual addition and scalar multiplication, we denote a vector $v \in \mathbb{C}^n$ by $|v\rangle$, which is called *ket* (Dirac notation). This notation is the same utilized in [121] and is natural to quantum mechanics. If $|v\rangle \in \mathbb{C}^n$, then we represent $|v\rangle$ by

$$|v\rangle \equiv \begin{bmatrix} z_1 \\ z_2 \\ \vdots \\ z_n \end{bmatrix}.$$

As usual, given a (complex) scalar $z \in \mathbb{C}$, we define by z^* the complex conjugate of z. Given a matrix A with complex entries, we denote $A^\dagger$ (the Hermitian conjugate or adjoint of A) to be $A^\dagger = (A^T)^*$, where A^T is the transpose of the matrix A.

For example, if we consider the complex vector space $\mathbb{C}^2$, then the canonical basis for $\mathbb{C}^2$ is given by the vectors

$$|v_1\rangle \equiv \begin{bmatrix} 1 \\ 0 \end{bmatrix} \quad \text{and} \quad |v_2\rangle \equiv \begin{bmatrix} 0 \\ 1 \end{bmatrix}.$$

Another basis of $\mathbb{C}^2$ is the basis consisting of the vectors

$$|w_1\rangle \equiv 1/\sqrt{2} \begin{bmatrix} 1 \\ 1 \end{bmatrix} \quad \text{and} \quad |w_2\rangle \equiv 1/\sqrt{2} \begin{bmatrix} 1 \\ -1 \end{bmatrix}.$$

If

$$|v\rangle \equiv \begin{bmatrix} z_1 \\ z_2 \\ \vdots \\ z_n \end{bmatrix},$$

then its dual vector $\langle v|$ (known as *bra* in the Dirac notation) is the row vector $\langle v| = [z_1^* z_2^* \cdots z_n^*]$.

In the following, we define inner product on vector spaces over $\mathbb{C}$ (as mentioned above, this is the scenario of quantum mechanics).

Definition 1.7.1 Let V be a vector space over $\mathbb{C}$. An inner product on V is a function $(\cdot, \cdot) : V \times V \longrightarrow \mathbb{C}$ that satisfies the following conditions:

(i) $\forall\ |v\rangle, |w_i\rangle \in V$, and $\forall\ c_i \in \mathbb{C}$, $i = 1, 2, \ldots, n$, one has $\left(|v\rangle, \sum_{i=1}^{n} c_i |w_i\rangle \right) = \sum_{i=1}^{n} c_i(|v\rangle, |w_i\rangle)$;

(ii) $\forall\ |v\rangle, |w\rangle \in V$, $(|v\rangle, |w\rangle) = (|w\rangle, |v\rangle)^*$, where $*$ denotes the complex conjugate of the complex number $(|w\rangle, |v\rangle)$;

(iii) $\forall\ |v\rangle \in V$, $(|v\rangle, |v\rangle) \geq 0$ and $(|v\rangle, |v\rangle) = 0 \iff |v\rangle = 0$.

A vector space V endowed with an inner product is said to be an inner product space.

Definition 1.7.2 An inner product space V is said to be a Hilbert space if it is a complete (with respect to the metric defined by the inner product) inner product space. The metric defined by the inner product is given by $d(v, w) = \sqrt{\langle v - w, v - w \rangle}$ for all $v, w \in V$, where $\langle v - w, v - w \rangle = (|v - w\rangle, |v - w\rangle)$

Remark 1.7.1 (1) If $|v\rangle, |w\rangle$ are vectors in V then they are called orthogonal if $(|v\rangle, |w\rangle) = 0$. An orthogonal basis for V is a basis $B_V = \{|v_i\rangle | i \in S\}$ of V such that $|v_i\rangle$ is orthogonal to $|v_j\rangle$ for every $i \neq j, i, j \in S$. If, in addition, the vectors $|v_j\rangle$ are unit vectors (see the line after Definition 1.7.3 for the definition of unit vector) then B_V is said to be an orthonormal basis.

(2) If $(\cdot, \cdot) : V \times V \longrightarrow \mathbb{C}$ is an inner product on V, and $|v\rangle, |w\rangle \in V$ we denote $(|v\rangle, |w\rangle) \equiv \langle v|w \rangle$. This notation (the same as [121]) is convenient since $(|v\rangle, |w\rangle)$ can be viewed as the matrix product of the dual vector $\langle v|$ by the vector $|w\rangle$.

Definition 1.7.3 Given an inner product $(\cdot, \cdot)$ on the complex vector space V, the norm of a vector $|v\rangle \in V$ with respect to $(\cdot, \cdot)$ is defined by

$$\||v\rangle\| = \sqrt{(|v\rangle, |v\rangle)}.$$

If $|v\rangle \in V$ is such that $\||v\rangle\| = 1$ then $|v\rangle$ is called a *unit vector*.

Example 1.7.1 As an example, we can define the following inner product in $\mathbb{C}^n$: if

$$|z\rangle \equiv \begin{bmatrix} z_1 \\ z_2 \\ \vdots \\ z_n \end{bmatrix} \quad \text{and} \quad |w\rangle \equiv \begin{bmatrix} w_1 \\ w_2 \\ \vdots \\ w_n \end{bmatrix}$$

are vectors in $\mathbb{C}^n$, then $(z, w) := \sum_{i=1}^{n} z_i^* w_i$. It is easy to see that this definition satisfies the conditions of Definition 1.7.1.

In the sequence, we define the concept of quantum bit (qubit for short). This is a fundamental concept for quantum mechanics.

Definition 1.7.4 A quantum bit (qubit or qubit's state) is a unit vector in a two-dimensional complex vector space. Mathematically, a qubit is a vector $|\psi\rangle = \alpha|0\rangle + \beta|1\rangle$ with $|\alpha|^2 + |\beta|^2 = 1$. The states $|0\rangle$ and $|1\rangle$ are called computational basis states; they form an orthonormal basis for such space.

Since, from Theorem 1.2.2, the matrix representation $M = [T]_{B_V}^{B_W}$ (here with complex entries) is equivalent to the operator T, sometimes we utilize the matrix representation and sometimes we utilize the operator form.

There exist some matrices that play a fundamental role in quantum mechanics: the Pauli matrices.

Definition 1.7.5 The Pauli matrices are the following matrices:

$$\rho_0 \equiv I \equiv \begin{bmatrix} 1 & 0 \\ 0 & 1 \end{bmatrix}, \quad \rho_1 \equiv \rho_x \equiv X \equiv \begin{bmatrix} 0 & 1 \\ 1 & 0 \end{bmatrix},$$

$$\rho_2 \equiv \rho_y \equiv Y \equiv \begin{bmatrix} 0 & -i \\ i & 0 \end{bmatrix}, \quad \rho_3 \equiv \rho_z \equiv Z \equiv \begin{bmatrix} 1 & 0 \\ 0 & -1 \end{bmatrix}.$$

Definition 1.7.6 Let V be a complex Hilbert space (recall that we are dealing with quantum mechanics). Assume also that $T : V \longrightarrow V$ is a linear operator on V. Then there exists a unique linear operator $T^\dagger : V \longrightarrow V$ such that, for every $|v\rangle, |w\rangle \in V$, it follows that $(|v\rangle, T|w\rangle) = (T^\dagger|v\rangle, |w\rangle)$. The operator $T^\dagger$ is called adjoint or Hermitian conjugate of T.

Hermitian and normal operators play an important role in quantum computation and quantum information.

Definition 1.7.7 Let $T : V \longrightarrow V$ be a linear operator, where V is a Hilbert space. If $T^\dagger = T$ then T is called *Hermitian* or *self-adjoint* operator.

Definition 1.7.8 Let V be a complex Hilbert space. A positive operator $T : V \longrightarrow V$ is a linear operator such that for every $|v\rangle \in V$, $(|v\rangle, T|v\rangle)$ is a real, nonnegative number. We say that T is positive definite if $(|v\rangle, T|v\rangle)$ is a real, strictly greater than zero number for all $|v\rangle \neq 0$.

Definition 1.7.9 Let $T : V \longrightarrow V$ be a linear operator. Then T is called normal if $TT^\dagger = T^\dagger T$.

Remark 1.7.2 (1) There exists an important and well-known property of normal operators: $T : V \longrightarrow V$ is normal if and only if T is diagonalizable. This result is known as Theorem of Spectral Decomposition.

(2) If $T : V \longrightarrow V$ is Hermitian then T is normal; thus T is also diagonalizable.

Unitary operators are fundamental to quantum mechanics since they describe the evolution of closed quantum systems.

Definition 1.7.10 Let U be a matrix with entries in $\mathbb{C}$. We say that U is *unitary* if $U^\dagger U = I$. In terms of operators, we say that an operator $U : V \longrightarrow V$ is unitary if $U^\dagger U = I$.

Remark 1.7.3 (1) If $U : V \longrightarrow V$ is unitary then $UU^\dagger = I$, so U is normal.

(2) Unitary operators preserve inner products between vectors.

The next results states that Pauli matrices have nice properties.

Proposition 1.7.1 *The Pauli Matrices are Hermitian and unitary, so they are also normal operators. Therefore, the Pauli matrices have spectral decomposition.*

Exercise 1.7.1 Show Proposition 1.7.1.

We now return the attention to tensor products of vector spaces. We begin by observing that sometimes we use the notation $|vw\rangle = |v\rangle \otimes |w\rangle$ to denote the tensor of $|v\rangle$ and $|w\rangle$. Let V and W be two vector spaces of dimensions m and n, respectively. Let $T : V \longrightarrow V$ be a linear operator on V and $S : W \longrightarrow W$ a linear operator on W. The elements of $V \otimes W$ are linear combinations of tensors $|v_i\rangle \otimes |w_i\rangle$, where $|v_i\rangle \in V$ and $|w_i\rangle \in W$. Then the function $T \otimes S : V \otimes W \longrightarrow V \otimes W$ defined by

$$[T \otimes S](|v\rangle \otimes |w\rangle) = [T \otimes S] \left(\sum_{i=1}^{n} a_i |v_i\rangle \otimes |w_i\rangle \right) \equiv \sum_{i=1}^{n} a_i T |v_i\rangle \otimes S |w_i\rangle,$$

(where we have written $T|v_i\rangle$ meaning $T(|v_i\rangle)$) is a linear operator on $V \otimes W$.

More generally, assume that $R : V_1 \longrightarrow V_2$ and $S : W_1 \longrightarrow W_2$ are two linear transformations, where the vector spaces are distinct. Then any linear transformation $T : V_1 \otimes W_1 \longrightarrow V_2 \otimes W_2$ can be represented by linear combinations of tensor products of linear transformations $R_i : V_1 \longrightarrow V_2$ and $S_i : W_1 \longrightarrow W_2$, $i = 1, 2, \ldots n$, i.e., $T = \sum_{i=1}^{n} z_i R_i \otimes S_i$, where $z_i \in \mathbb{C}$. By definition, we have

$$T(|v\rangle \otimes |w\rangle) = \left(\sum_{i=1}^{n} z_i R_i \otimes S_i \right) (|v\rangle \otimes |w\rangle) = \sum_{i=1}^{n} (z_i R_i(|v\rangle)) \otimes S_i(|w\rangle).$$

Definition 1.7.11 Let V and W be two vector spaces (over the same field K) endowed with inner products, and consider the tensor product $V \otimes W$. Then, an inner product $(\cdot, \cdot)_{V \otimes W} : (V \otimes W) \times (V \otimes W) \longrightarrow \mathbb{K}$ can be defined naturally by

$$\left(\sum_{i=1}^{n} a_i |v_i\rangle \otimes |w_i\rangle, \sum_{j=1}^{m} a_j' |v_j'\rangle \otimes |w_j'\rangle \right) \equiv \sum_{i=1}^{n} \sum_{j=1}^{m} a_i^* b_j \langle v_i | v_j' \rangle \langle w_i | w_j' \rangle.$$

In the following we define the *outer product representation*.

Definition 1.7.12 Assume that V and W are two inner product spaces. For every $|v\rangle \in V$ and $|w\rangle \in W$, we define the linear transformation $|w\rangle\langle v| : V \longrightarrow W$ by

$$(|w\rangle\langle v|)(|v^\circ\rangle) = \langle v|v^\circ\rangle |w\rangle.$$

In this context, it is possible to define the concept of *projector* onto a subspace.

Definition 1.7.13 Suppose that $\{|1\rangle, |2\rangle, \ldots, |n\rangle\}$ is an orthonormal basis for the n-dimensional inner product space V such that $\{|1\rangle, |2\rangle, \ldots, |m\rangle\}$ is an orthonormal basis for a m-dimensional subspace W of V. Then we can define the projector $P \equiv \sum_{i=1}^{m} |i\rangle\langle i|$ onto the subspace W.

Exercise 1.7.2 Show that $P \equiv \sum_{i=1}^{m} |i\rangle\langle i|$ given in Definition 1.7.13 does not depend on the choice of the orthogonal basis for W.

Fortunately, the characteristic of certain types of operators are maintained by applying tensor products $V \otimes W$ of vectors spaces.

Proposition 1.7.2 *Let $T_1 : V \longrightarrow V$ and $T_2 : W \longrightarrow W$ be linear operators.*

(1) If T_1 and T_2 are unitary then also is $T_1 \otimes T_2$.
(2) If T_1 and T_2 are Hermitian then also is $T_1 \otimes T_2$.
(3) If T_1 and T_2 are projectors then $T_1 \otimes T_2$ is also a projector.
(4) If T_1 and T_2 are positive operators then also is $T_1 \otimes T_2$.

Exercise 1.7.3 Show Proposition 1.7.2.

We are now interested to define tensor products of matrices since, from Theorem 1.2.2, linear transformations and matrices are equivalent (fixing the bases).

Definition 1.7.14 Let K a field. Assume that $A = [a_{ij}]$ is an $m \times n$ and $B = [b_{ij}]$ is a $r \times s$ matrix both with entries in K. Then the *Kronecker product* $A \otimes B$ of A and B is defined as

$$A \otimes B = \begin{bmatrix} a_{11}B & a_{12}B & \cdots & a_{1n}B \\ a_{21}B & a_{22}B & \cdots & a_{2n}B \\ \vdots & \vdots & \vdots & \vdots \\ a_{m1}B & a_{m2}B & \cdots & a_{mn}B \end{bmatrix}.$$

Since vectors can be considered as column (or row) matrices we also have a tensor product of vectors. As an example, if $K = \mathbb{C}$ and we consider the vectors $v_1 \in \mathbb{C}^2$ and $v_2 \in \mathbb{C}^3$ given, respectively, by

$$v_1 = \begin{bmatrix} i \\ 2 \end{bmatrix}, \quad v_2 = \begin{bmatrix} 1 \\ 2i \\ 5 \end{bmatrix}, \text{ then } v_1 \otimes v_2 = \begin{bmatrix} i \\ -2 \\ 5i \\ 2 \\ 4i \\ 10 \end{bmatrix}.$$

The Kronecker product satisfies nice properties.

Proposition 1.7.3 *Let A and B be two matrices with complex entries. Then the following hold:*

(1) $(A \otimes B)^T = A^T \otimes B^T$;
(2) $(A \otimes B)^* = A^* \otimes B^*$;
(3) $(A \otimes B)^\dagger = A^\dagger \otimes B^\dagger$.

Exercise 1.7.4 Show Proposition 1.7.3.

1.8 Commutator

We now present the concept of commutators and anti-commutators of linear operators.

Definition 1.8.1 Let $T_1, T_2 : V \longrightarrow V$ be two linear operators on V. The *commutator* between T_1 and T_2 is defined by

$$[T_1, T_2] := T_1 T_2 - T_2 T_1.$$

If $[T_1, T_2] \equiv 0$, then T_1 *commutes* with T_2.

Note first that $[T_1, T_2]$ is a function from V into V. Since T_1 and T_2 are linear then also is $[T_1, T_2]$. Thus we can view the commutator as an operation on the vector space $\mathrm{Hom}(V, V)$ of all linear operators on V, that is, $[,]$ is a function $[,] : \mathrm{Hom}(V, V) \times \mathrm{Hom}(V, V) \longrightarrow \mathrm{Hom}(V, V)$ such that, for every pair (T_1, T_2) of linear operators assigns the linear operator $[T_1, T_2]$.

Analogously, we can define the *anti-commutator* between two operators.

Definition 1.8.2 Let $T_1, T_2 : V \longrightarrow V$ be two linear operators on V. The anti-commutator between T_1 and T_2 is defined as

$$\{T_1, T_2\} := T_1 T_2 + T_2 T_1.$$

If $[T_1, T_2] \equiv 0$, then T_1 *commutes* with T_2. When $\{T_1, T_2\} \equiv 0$ we say that T_1 anti-commutes with T_2.

As in the case of commutators, anti-commutators can be also viewed as an operation on the vector space $\mathrm{Hom}(V, V)$: $\{, \} : \mathrm{Hom}(V, V) \times \mathrm{Hom}(V, V) \longrightarrow \mathrm{Hom}(V, V)$ such that, for every pair (T_1, T_2) of linear operators assigns the linear operator $\{T_1, T_2\}$.

Theorem 1.8.1 (Simultaneous diagonalization theorem) *Assume that $T_1, T_2 : V \longrightarrow V$ are two Hermitian operators on V. Then $[T_1, T_2] \equiv 0$ if and only if there exists an orthonormal basis $\mathcal{B}$ such that both T_1 and T_2 are diagonal with respect to $\mathcal{B}$. If there exists such basis, the operators T_1 and T_2 are called* simultaneously diagonalizable.

In other words, two Hermitian operators commute themselves if and only if they are simultaneously diagonalizable.

Exercise 1.8.1 (a) Show that X and Y are not simultaneously diagonalizable.

(b) Show the same to Y and Z e for Z and X.

(c) Adopting the notation of Definition 1.7.5, show that $\{\rho_i, \rho_j\} \equiv 0$, for each $i \neq j, i, j = 1, 2, 3$.

An interesting case is when two operators commute and anti-commute themselves at the same time and one of them is invertible. Let us see what happens. Assume that $T_1, T_2 : V \longrightarrow V$ are linear operators such that $[T_1, T_2] \equiv 0$, $\{T_1, T_2\} \equiv 0$ and T_1 is invertible. We then have

$$T_1 T_2 = T_2 T_1 \qquad (1.1)$$

and

$$T_1 T_2 = -T_2 T_1. \tag{1.2}$$

Applying T_1^{-1} to the left in Eq. (1.1) it follows that

$$T_2 = T_1^{-1} T_2 T_1. \tag{1.3}$$

Replacing Eq. (1.2) in Eq. (1.3) it follows that T_2 must be the zero operator.

Exercise 1.8.2 Let $T_1, T_2 : V \longrightarrow V$ be linear operators. Show that
(a) $[T_1, T_2]^\dagger = [T_2^\dagger, T_1^\dagger]$;
(b) $[T_1, T_2] = -[T_2, T_1]$;
(c) $i[T_1, T_2]$ is Hermitian;
(d) Is it true that if T_1 and T_2 are unitary operators then so is $[T_1, T_2]$? What about this assertion with respect to anti-commutator?

Chapter 2
A Little Bit of Quantum Mechanics

In this chapter we present a brief review of quantum mechanics, necessary for the development of this book. We suggest the textbook [121] for more details. We start by the four postulates of quantum mechanics.

2.1 Postulates of Quantum Mechanics

Postulate 2.1.1 *There exists a complex vector space with inner product which is complete in the metric defined by the inner product (that is, a Hilbert space) associated with any isolated physical system. This Hilbert space is known as the state space of the system. The system is completely described by its state vector, which is a unit vector in system's state space.*

The second postulate of quantum mechanics tells us when the system evolves over time.

Postulate 2.1.2 *The evolution of a closed quantum system is described by a unitary transformation. More precisely, the state $|v\rangle$ of the system at time t_1 is related to the state $|w\rangle$ of the system at time t_2 by a unitary operator U which depends only on t_1 and t_2: $|w\rangle = U|v\rangle$.*

Postulate 2.1.3 *Quantum measurements are described by a collection $\{M_n\}$ of measurement operators. These are operators acting on the state space of the system being measured. The index n refers to the measurement outcomes that may occur in the experiment. If the state of the quantum system is $|v\rangle$ immediately before the measurement then the probability that result n occurs is given by $p(n) = \langle v|M_n^\dagger M_n|v\rangle$, and the state of the system after measuring is $M_n|v\rangle/\sqrt{p(n)}$. The measurement operators satisfy $\sum_n M_n^\dagger M_n = I$ (called the completeness equation), where I is the identity operator.*

© Springer Nature Switzerland AG 2020

G. G. La Guardia, *Quantum Error Correction*, Quantum Science and Technology,
https://doi.org/10.1007/978-3-030-48551-1_2

A special case of Postulate 2.1.3 is a suitable type of measurement, that is, the *projective measurements* [121, p. 87]: A projective measurement is described by an observable, M, which is a Hermitian operator on the state space of the system being observed. The observable has a spectral decomposition, $M = \sum_m m P_m$, where P_m is the projector onto the eigenspace of M with eigenvalue m. The possible outcomes of the measurement correspond to the eigenvalues, m, of the observable. After the measurement of the qubit $|v\rangle$, the probability to have m is equal to $p(m) = \langle v| P_m |v\rangle$. If m occurred, the state of the quantum system after the measurement is $\frac{P_m |v\rangle}{\sqrt{p(m)}}$.

Postulate 2.1.4 *The state space of a composite physical system is the tensor product of the state spaces of the component physical systems. Moreover, if the number of systems is finite, say $S_1, S_2, \ldots, S_r$, and S_i is prepared in the state $|v_i\rangle$, then the joint state of the total system is $|v_1\rangle \otimes |v_2\rangle \otimes \cdots \otimes |v_r\rangle$.*

2.2   Ensemble of Quantum States

We next introduce the concepts of density operator, after writing the postulates of quantum mechanics in terms of these operators.

There exist mainly two types of developments of quantum mechanics: based on state vectors or by means of *density operator* or *density matrix*. Until now we have formulated quantum mechanics using the language of state vectors. In the sequence, we will introduce the concept of density operator. The density operator formalism is convenient for a description of quantum systems whose state is not completely known.

Definition 2.2.1 Assume that a quantum system is in one of the states $|\phi_i\rangle$, with probability p_i, where i is an index. Then the set $\{p_i, |\phi_i\rangle\}$ is said to be an ensemble of pure states. The density operator (or density matrix) for the system is defined as

$$\rho \equiv \sum_i p_i |\phi_i\rangle \langle \phi_i|.$$

Assume that the evolution of a closed quantum system (see Postulate 2.1.2) is described by the unitary operator U. If the system is initially in the state $|\phi_i\rangle$ with probability p_i, then, after the evolution by means of the operator U, the system will be in the state $U|\phi_i\rangle$ with probability p_i as shown in the sequence:

$$\rho = \sum_i p_i |\phi_i\rangle \langle \phi_i| \xrightarrow{U} \sum_i p_i U |\phi_i\rangle \langle \phi_i| U^\dagger = U \rho U^\dagger.$$

Assume that a measurement described by measurement operators M_m is done. Thus, if the initial state is $|\phi_i\rangle$, the probability of having m is given by

$$p(m|i) = \langle\phi_i|M_m^\dagger M_m|\phi_i\rangle = \text{Tr}(M_m^\dagger M_m|\phi_i\rangle\langle\phi_i|).$$

The density operator after obtaining the result m is

$$\rho_m = \sum_i p(i|m)|\phi_i^m\rangle\langle\phi_i^m| = \sum_i p(i|m)\frac{M_m|\phi_i\rangle\langle\phi_i|}{\langle\phi_i|M_m^\dagger M_m|\phi_i\rangle} = \frac{M_m\rho M_m^\dagger}{\text{Tr}(M_m^\dagger M_m\rho)},$$

where

$$|\phi_i^m\rangle = \frac{M_m|\phi_i\rangle}{\sqrt{\langle\phi_i|M_m^\dagger M_m|\phi_i\rangle}}$$

is the state after getting the result m, generating therefore an ensemble of states $|\phi_i^m\rangle$ with respective probabilities $p(m|i)$.

Definition 2.2.2 A quantum system whose state $|\phi\rangle$ is known exactly is said to be in a pure state; the density operator in this case is given simply by $\rho = |\phi\rangle\langle\phi|$. Otherwise, we say that ρ is in a mixed state.

At this point, we are able to describe the postulates of quantum mechanics based on the formalism of density operator.

Postulate 2.2.1 *Associated with any isolated physical system is a complex Hilbert space known as the state space of the system. The system is completely described by its density operator, which is a positive operator with trace one, acting on the state space of the system. If the system is in the state ρ_i with probability p_i, then the density operator for the system is given by $\sum_i p_i\rho_i$.*

Postulate 2.2.2 *The evolution of a closed quantum system is described by means of a unitary transformation: the state ρ at time t_1 is related to σ at time t_2 by a unitary operator U,*

$$\sigma = U\rho U^\dagger.$$

The operator U depends only on the times t_1 and t_2.

Postulate 2.2.3 *The quantum measurements are described by a collection $\{M_m\}$, where the M_m's are called measurements operators. The operators M_m's act on the state space of the system being measured. The index m refers to the measurement outcomes possible to occur. If the state of the system is in ρ, then the probability of occurrence of m is equal to*

$$p(m) = \mathrm{Tr}(M_m^\dagger M_m \rho),$$

and the state after the measurement is

$$\sigma = \frac{M_m \rho M_m^\dagger}{\mathrm{Tr}(M_m^\dagger M_m \rho)}.$$

The operators M_m's satisfy the completeness equation

$$\sum_m M_m^\dagger M_m = I.$$

Postulate 2.2.4 *Given any composite physical system, its state of space is the tensor product of the state spaces of the component physical systems. Additionally, in the case in which the systems are numbered 1 to n, and the system number i is prepared in the state ρ_i, then the join state is described by*

$$\rho_1 \otimes \rho_2 \otimes \ldots \otimes \rho_n.$$

2.3 Universal Quantum Gates

In this subsection we recall some important quantum gates, essential for quantum mechanics. We assume the reader is familiar with theory of quantum gates (single and multiple qubit quantum gates). For more details with respect to this topic, we suggest the textbook [121].

2.3.1 Single Qubit Gates

The single qubit gates, as the own name says, are gates acting on one qubit state. The *quantum NOT gate* X can be represented in terms of matrices as

$$X \equiv \begin{bmatrix} 0 & 1 \\ 1 & 0 \end{bmatrix}.$$

Hence, if we have a generic quantum state $|v\rangle = \alpha|0\rangle + \beta|1\rangle$ represented as a column matrix

$$|v\rangle = \begin{bmatrix} \alpha \\ \beta \end{bmatrix},$$

then the action of X in $|v\rangle$ is

$$X \begin{bmatrix} \alpha \\ \beta \end{bmatrix} = \begin{bmatrix} \beta \\ \alpha \end{bmatrix}.$$

The quantum Z gate is described as

$$Z \equiv \begin{bmatrix} 1 & 0 \\ 0 & -1 \end{bmatrix}.$$

The action of Z on $|v\rangle$ is given by

$$Z \begin{bmatrix} \alpha \\ \beta \end{bmatrix} = \begin{bmatrix} \alpha \\ -\beta \end{bmatrix}.$$

Remark 2.3.1 As it was mentioned previously (see Definition 1.7.5), the X and the Z gates are examples of Pauli matrices, which play a fundamental role in quantum mechanics.

Another important single qubit gate is the *Hadamard* gate

$$H \equiv \frac{1}{\sqrt{2}} \begin{bmatrix} 1 & 1 \\ 1 & -1 \end{bmatrix}.$$

The action of H in $|v\rangle$ is

$$H \begin{bmatrix} \alpha \\ \beta \end{bmatrix} = \frac{1}{\sqrt{2}} \begin{bmatrix} \alpha + \beta \\ \alpha - \beta \end{bmatrix},$$

which can be also described as $H|v\rangle = \alpha \frac{|0\rangle + |1\rangle}{\sqrt{2}} + \beta \frac{|0\rangle - |1\rangle}{\sqrt{2}}$, where

$$|0\rangle = \begin{bmatrix} 1 \\ 0 \end{bmatrix} \quad \text{and} \quad |1\rangle = \begin{bmatrix} 0 \\ 1 \end{bmatrix}.$$

Note that all these gates are unitary matrices (or operators), i.e., $X^\dagger X = I, Z^\dagger Z = I$ and $H^\dagger H = I$ (see Definition 1.7.10). In fact, any unitary matrix specifies a valid quantum gate, since unitary operators preserve unitary vectors, and unitary vectors (in complex Hilbert spaces) are the quantum states of quantum mechanics. Since there exist infinity many unitary matrices of order two, by the previous remark, there also exist infinity many single qubit gates. However, it is possible to build an arbitrary single qubit gate by utilizing only a finite set of quantum gates. This means that the quantum computation is *universal*: an arbitrary quantum computation on qubits can be performed by a finite set of quantum gates.

The following result states that an arbitrary single quantum gate can be described in terms of suitable two by two matrices.

Theorem 2.3.1 *Let U be an* 2×2 *unitary matrix with complex entries. Then U can be decomposed as*

$$U = e^{ia} \begin{bmatrix} e^{-ib/2} & 0 \\ 0 & e^{ib/2} \end{bmatrix} \begin{bmatrix} \cos(\frac{c}{2}) & -\sin(\frac{c}{2}) \\ \sin(\frac{c}{2}) & \cos(\frac{c}{2}) \end{bmatrix} \begin{bmatrix} e^{-id/2} & 0 \\ 0 & e^{id/2} \end{bmatrix},$$

where a, b, c, d are real numbers.

Exercise 2.3.1 Show Theorem 2.3.1.

Remark 2.3.2 Notice that the matrices in the decomposition of U in Theorem 2.3.1 are (or can be understood as) rotations.

2.3.2   Multiple Qubit Gates

The first gate that we recall is the well-known *Controlled-NOT* or CNOT gate that acts on two qubits.

$$U_{\text{CN}} = \begin{bmatrix} 1 & 0 & 0 & 0 \\ 0 & 1 & 0 & 0 \\ 0 & 0 & 0 & 1 \\ 0 & 0 & 1 & 0 \end{bmatrix}.$$

The action of the CNOT gate in the basis vectors is shown in the sequence:

$$U_{\text{CN}}(|00\rangle) = |00\rangle;$$
$$U_{\text{CN}}(|01\rangle) = |01\rangle;$$
$$U_{\text{CN}}(|10\rangle) = |11\rangle;$$
$$U_{\text{CN}}(|11\rangle) = |10\rangle.$$

As in the case of single qubit, the CNOT gate U_{CN} is, of course, unitary. The *Toffoli gate* is an example of a three qubit quantum gate.

$$U_{\text{CN}} = \begin{bmatrix} 1 & 0 & 0 & 0 & 0 & 0 & 0 & 0 \\ 0 & 1 & 0 & 0 & 0 & 0 & 0 & 0 \\ 0 & 0 & 1 & 0 & 0 & 0 & 0 & 0 \\ 0 & 0 & 0 & 1 & 0 & 0 & 0 & 0 \\ 0 & 0 & 0 & 0 & 1 & 0 & 0 & 0 \\ 0 & 0 & 0 & 0 & 0 & 1 & 0 & 0 \\ 0 & 0 & 0 & 0 & 0 & 0 & 0 & 1 \\ 0 & 0 & 0 & 0 & 0 & 0 & 1 & 0 \end{bmatrix}.$$

A remarkable result valid in quantum computation is the universality.

Theorem 2.3.2 *An arbitrary multiple qubit gate can be composed into CNOT and single qubit gates.*

Exercise 2.3.2 Show Theorem 2.3.2.

2.3.3 Quantum Channels

There exist several types of quantum noises. The more common are the *bit flip channel*, which will be studied in Sect. 3.2.1 (see Definition 3.2.1), the *phase shift channel*, which will be presented in Sect. 3.2.2 (see Definition 3.2.2) and the *depolarizing channel*, presented here.

We show how the depolarizing channel corrupts the quantum message. We start with a single qubit which has probability p to suffer depolarization through the channel and probability $1 - p$ to left unchanged. After passing through the channel, the state is corrupted by the noise and becomes

$$\varepsilon(\rho) = \frac{pI}{2} + (1 - p)\rho. \tag{2.1}$$

Since

$$\frac{I}{2} = \frac{\rho + X\rho X + Y\rho Y + Z\rho Z}{4},$$

replacing such equation in Eq. (2.1) we have

$$\varepsilon(\rho) = \left(1 - \frac{3p}{4}\right)\rho + \frac{p}{4}(X\rho X + Y\rho Y + Z\rho Z).$$

If we want to make explicit the state ρ with its corresponding probability to remain unchanged $1 - p$, we write

$$\varepsilon(\rho) = (1 - p)\rho + \frac{p}{3}(X\rho X + Y\rho Y + Z\rho Z). \tag{2.2}$$

Notice that in Eq. (2.2), the Pauli matrices X, Y and Z act with probability $p/3$.

Chapter 3
Quantum Error-Correcting Codes

3.1 Introduction

In this chapter, we introduce the concept of quantum error-correcting codes and show several examples chronologically. This chapter is fundamental for the development of this work, since the main aim of this book is to present several types of different constructions of quantum block and convolutional codes, as well as constructions of asymmetric quantum codes.

In order to proceed further, we need to recall the concept of *pure quantum state*.

Definition 3.1.1 If a state $|v\rangle$ in a quantum system is known exactly, we say that the system is in a pure state.

There exist some apparent difficulties to construct quantum error-correcting codes. The first situation is the impossibility of copying qubits, as we see in the well-known No-cloning theorem.

Theorem 3.1.1 (No-cloning theorem) *Assume that $|v\rangle$ is a pure (unknown) state. Then there is no unitary operator taking $|v\rangle \otimes |s\rangle$ to $|v\rangle \otimes |v\rangle$ (quantum copy of $|v\rangle$) for all quantum states $|v\rangle$, where $|s\rangle$ is a standard pure state. More precisely, given two particular pure quantum states $|v\rangle$ and $|w\rangle$ to be copied, it follows that the quantum copy process is possible only if $|v\rangle = |w\rangle$ or if $|v\rangle$ and $|w\rangle$ are orthogonal.*

Therefore, based on Theorem 3.1.1, it is not possible to construct a repetition quantum code due to the impossibility of copying arbitrary qubits.

The second apparent difficulty is that the set of errors is continuous. Thus, at a first glance, it seems that the quantum code must correct an infinity of different types of errors.

The third difficulty is that the measurements of qubits destroy the quantum information.

Fortunately, (to guarantee our jobs) all these situations can be solved. In the sequence, we present the first quantum code displayed in the literature which is capable of correcting one arbitrary quantum error in any single qubit, the Shor code.

© Springer Nature Switzerland AG 2020

G. G. La Guardia, *Quantum Error Correction*, Quantum Science and Technology,

https://doi.org/10.1007/978-3-030-48551-1_3

3.2 The Shor Code

In this subsection we describe the construction of the Shor code. In order to proceed further, we must construct first the *three qubit bit flip code* and after, the *three qubit phase flip code*. For more details about the encoding–decoding process, we refer the reader to [121].

3.2.1 Three Qubit Bit Flip Code

The *bit flip channel* is defined below. Roughly speaking, this channel represents the action of the Pauli operator X

$$X \equiv \begin{bmatrix} 0 & 1 \\ 1 & 0 \end{bmatrix}.$$

Let us define formally such quantum noise.

Definition 3.2.1 Let $|v\rangle = a|0\rangle + b|1\rangle$ be a qubit state (qubit for short). The bit flip channel acts on $|v\rangle$ as follows:

- $|v\rangle \xrightarrow{\text{channel}} X|v\rangle = a|1\rangle + b|0\rangle$ with probability p;
- $|v\rangle \xrightarrow{\text{channel}} |v\rangle$ with probability $1 - p$.

In order to protect qubits against the effects of the bit flip channel, one utilizes the three qubit bit flip code. We begin by recalling that we write $|v_1 v_2 v_3\rangle$ to denote $|v_1\rangle \otimes |v_2\rangle \otimes |v_3\rangle$, as previously specified.

Let us consider a single qubit given by $|v\rangle = a|0\rangle + |1\rangle$. Assume that $|v\rangle$ was encoded as $|v\rangle_{enc} = a|000\rangle + b|111\rangle$, where, as usual, we define $|000\rangle$ as being the *logical zero* $|0_L\rangle$ and $|111\rangle$ as the *logical one* $|1_L\rangle$, i.e., $|0_L\rangle := |000\rangle$ and $|1_L\rangle := |111\rangle$. To this end, we have encoded $|0\rangle \xrightarrow{encode} |000\rangle$ and $|1\rangle \xrightarrow{encode} |111\rangle$.

The channel is assumed to be independent, that is, each qubit passes through an independent copy of it. Assume that one error (or none) has occurred to the encoded state $|v\rangle_{enc}$. For this channel one has four error syndromes corresponding to the four projection operators.

3.2.1.1 First Procedure of Measurement

To detect the error (if there exists), we perform a measurement in order to know which qubit was corrupted. The result of the measurement is said to be *error syndrome*. There exist four error syndromes corresponding to the following four projection operators:

- $P_0 := |000\rangle\langle000| + |111\rangle\langle111|$ (associated with no occurrence of error);
- $P_1 := |100\rangle\langle100| + |011\rangle\langle011|$ (error in the first qubit);
- $P_2 := |010\rangle\langle010| + |101\rangle\langle101|$ (error in the second qubit);
- $P_3 := |001\rangle\langle001| + |110\rangle\langle110|$ (error in the third qubit);

Let us see how the detection process works. Assume without loss of generality (w.l.o.g.) that an error corrupted the second qubit; then the state

$$|v\rangle_{enc} = a|000\rangle + b|111\rangle$$

becomes

$$|w\rangle = a|010\rangle + b|101\rangle.$$

Hence (according to Postulate 2.1), applying P_2 to $|w\rangle$ we have

$$
\begin{aligned}
p(2) &= \langle w|P_2|w\rangle \\
&= (a^*\langle010| + b^*\langle101|)|010\rangle\langle010|(a|010\rangle + b|101\rangle)) \\
&\quad + (a^*\langle010| + b^*\langle101|)|101\rangle\langle101|(a|010\rangle + b|101\rangle)) \\
&= |a|^2 + |b|^2 = 1.
\end{aligned}
$$

Therefore, we know that the error occurred in the second qubit.

Remark 3.2.1 In this detection process, it is interesting to observe that the corrupted state $a|010\rangle + b|101\rangle$ is not affected by the syndrome measurement. In fact, the syndrome contains only information about the corrupted qubit, but no information about the state being measured (a and b are not known). This is excellent, since none of the measurements applied for the decoding operation destroys the superpositions of quantum states that must be preserved by applying the encoding process.

To recover the original encoded state $|v\rangle_{enc}$, note that since the error has occurred in the second qubit and since the channel flips the qubit, then it suffices to flip to second qubit again. Thus, the encoded state $|v\rangle_{enc}$ is recovered. It is clear that this procedure holds in general, independently in which qubit the error has occurred. If no error occurs in this process, by applying the operator P_0 we have $p(0) = \langle w|P_0|w\rangle = 1$, that is, we know that (probability one) no error has occurred. Proceeding similarly, we can recover the original encoded state in all cases.

3.2.1.2 Second Procedure of Measurement

We next present an alternative way to proceed with the measurement process. Assume that we replace the four measurements operators P_0, P_1, P_2, P_3 by the observables $Z_1Z_2 := Z \otimes Z \otimes I$ and $Z_2Z_3 := I \otimes Z \otimes Z$, both with eigenvalues -1 and $+1$. To perform the measurement, we first apply Z_1Z_2 and, in the sequence, the observable Z_2Z_3.

The first operator has the following spectral decomposition:

$$Z_1 Z_2 = [(|00\rangle\langle 00| + |11\rangle\langle 11|) \otimes I] - [(|01\rangle\langle 01| + |10\rangle\langle 10|) \otimes I]$$

There are two possibilities for the result of the measurement of $Z_1 Z_2$: the eigenvalue is $+1$ or -1. Let us analyze all the situations.

Recall that the original encoded vector is $|v\rangle_{enc} = a|000\rangle + b|111\rangle$. If the channel corrupted the first qubit, then the corresponding qubit state is $|w\rangle = a|100\rangle + b|011\rangle$, so the result of the measurement of $Z_1 Z_2$ is -1 because

$$p(-1)$$
$$= \langle w|Z_1 Z_2|w\rangle$$
$$= (a^*\langle 100| + b^*\langle 011|) - [(|01\rangle\langle 01| + |10\rangle\langle 10|) \otimes I](a|100\rangle + b|011\rangle))$$
$$= (-[b^*\langle 1| + a^*\langle 0|], -[a|0\rangle + b|1\rangle)])$$
$$= |a|^2 + |b|^2 = 1.$$

Analogously, if the channel corrupted the second qubit, the result of the measurement of $Z_1 Z_2$ is also -1. Thus, if the eigenvalue equals -1, the first and the second qubit are distinct. If the eigenvalue is $+1$ then such qubits are equal. Analogously, when performing the measurement of the observable $Z_2 Z_3$, if the eigenvalue is $+1$, then the second and the third qubit are equal; if it is -1, they are distinct.

Next, we deduce in which (if any) qubit the error has occurred. Assume that the result of the measurements of $Z_1 Z_2$ and $Z_2 Z_3$ are both $+1$. Then all the three qubits are equal and no error has occurred. If the eigenvalues are $+1$ and -1, respectively, then the error corrupted with high probability the third qubit; if the eigenvalues are -1 and $+1$, respectively, the error occurred with high probability in the first qubit. Finally, if the results are -1 and -1 then (with high probability) the second qubit was corrupted. Note that none of the measurements give information about the states being measured like the first procedure. To recover the quantum state it suffices to proceed as in the first case, i.e., the corrupted qubit can be flipped again.

It is interesting to note that, in the latter procedure, we only need to use two observables to detect the error, whereas in the first case we need to have four operators for measurement. This is an advantage offered in the second procedure.

3.2.2 Three Qubit Phase Flip Code

The *phase flip channel* represents the action of the Pauli operator Z

$$Z \equiv \begin{bmatrix} 1 & 0 \\ 0 & -1 \end{bmatrix}.$$

This quantum noise shown is defined in the sequence.

Definition 3.2.2 Let $|v\rangle = a|0\rangle + b|1\rangle$ be a qubit. The phase flip channel acts on $|v\rangle$ as follows:

- $|v\rangle \xrightarrow{\text{channel}} Z|v\rangle = a|0\rangle - b|1\rangle$ with probability p;
- $|v\rangle \xrightarrow{\text{channel}} |v\rangle$ with probability $1 - p$.

The procedure to recover the encoded state is to turn the phase flip channel into a bit flip channel. In order to do this, let us consider the qubit basis $|+\rangle = (|0\rangle + |1\rangle)/2$ and $|-\rangle = (|0\rangle - |1\rangle)/2$. Since $Z|+\rangle = |-\rangle$ and $Z|-\rangle = |+\rangle$, Z acts as a bit flip in such vectors. We then perform the encoding:

$$|0\rangle \xrightarrow{\text{encode}} |0_L\rangle := |+++\rangle$$

and

$$|1\rangle \xrightarrow{\text{encode}} |1_L\rangle := |---\rangle.$$

In this manner we can protect at least one qubit against phase flip errors. From this moment, the encoding, detection and the recovery process is the same as the bit flip channel with respect to the basis $|+\rangle$ and $|-\rangle$.

3.2.3 The Shor Code

Here we present the Shor code, the first quantum error-correcting code to protect an arbitrary single qubit against an arbitrary quantum error. The code is constructed by means of concatenation of qubits as we can see in the following.

The construction of this code is based on the three qubit bit flip and the three qubit phase flip codes presented in Sects. 3.2.1 and 3.2.2, respectively.

The stages of construction of the Shor code is given in the sequence.

(1) The first stage is to utilize the three qubit phase flip code to encode the qubit, that is, $|0\rangle \xrightarrow{\text{encode}} |+++\rangle$ and $|1\rangle \xrightarrow{\text{encode}} |---\rangle$.

(2) Each of these qubits (namely, $|+\rangle$ and $|-\rangle$) are encoded by applying the three qubit phase flip code, i.e., $|+\rangle \xrightarrow{\text{encode}} (|000\rangle + |111\rangle)/\sqrt{2}$ and $|-\rangle \xrightarrow{\text{encode}} (|000\rangle - |111\rangle)/\sqrt{2}$.

Thus, the resulting code is the Shor code given by

$$|0\rangle \xrightarrow{\text{encode}} |0_L\rangle := \frac{(|000\rangle + |111\rangle)(|000\rangle + |111\rangle)(|000\rangle + |111\rangle)}{2\sqrt{2}}$$

and

$$|1\rangle \xrightarrow{\ encode\ } |1_L\rangle := \frac{(|000\rangle - |111\rangle)(|000\rangle - |111\rangle)(|000\rangle - |111\rangle)}{2\sqrt{2}}.$$

We explain now how the Shor code can correct phase flip and bit flip errors on any single qubit. In fact, the analysis we perform here is in the same spirit with as the second procedure of measurement shown in Sect. 3.2. We present a scheme in order to clarify the understanding of how the code can recover the initial state.

First Case—Correcting bit flip errors. Assume w.l.o.g. that an error has occurred in the seventh qubit. We then perform the measurement of the observable $Z_7 Z_8$, finding the eigenvalue -1. After this, we follow by measuring $Z_8 Z_9$ obtaining therefore the eigenvalue $+1$; so the seventh qubit is the corrupted one. Applying bit flip again in the seventh qubit one has the initial state. Proceeding similarly, we can detect and recover any (single) state which was corrupted by bit flip errors, by means of the measurement of the observables $Z_1 Z_2$, $Z_2 Z_3$, $Z_4 Z_5$, $Z_5 Z_6$, $Z_7 Z_8$ and $Z_8 Z_9$.

Second Case—Correcting phase flip errors. Assume that an error occurs in the second qubit for example. Due to the properties of tensor product, the first block of three qubits $|000\rangle + |111\rangle$ becomes $|000\rangle - |111\rangle$ and $|000\rangle - |111\rangle$ becomes $|000\rangle + |111\rangle$. In other words, the two basis states now read as

$$|0_L\rangle \xrightarrow{\ channel\ } \frac{(|000\rangle - |111\rangle)(|000\rangle + |111\rangle)(|000\rangle + |111\rangle)}{2\sqrt{2}}$$

and

$$|1_L\rangle \xrightarrow{\ channel\ } \frac{(|000\rangle + |111\rangle)(|000\rangle - |111\rangle)(|000\rangle - |111\rangle)}{2\sqrt{2}}.$$

After this, we compare the sign of the first and the second blocks of qubits, i.e., $|000\rangle - |111\rangle$ is compared with $|000\rangle + |111\rangle$ (has distinct sign) and $|000\rangle + |111\rangle$ is compared with $|000\rangle - |111\rangle$ (has distinct sign). Here, we consider that the block has the same sign in the cases $(|000\rangle + |111\rangle)(|000\rangle + |111\rangle)$ and $(|000\rangle - |111\rangle)(|000\rangle - |111\rangle)$ and they have different signs in the cases $(|000\rangle + |111\rangle)(|000\rangle - |111\rangle)$ and $(|000\rangle - |111\rangle)(|000\rangle + |111\rangle)$. Next, we perform a comparison between the sign of the second and the third blocks of qubits, i.e., $(|000\rangle + |111\rangle)(|000\rangle + |111\rangle)$ (has the same sign) and $(|000\rangle - |111\rangle)(|000\rangle - |111\rangle)$ (has the same sign). Thus, we know that the phase flip has corrupted one of the three first qubits. To recover the initial encoded state it suffices to flip the sign of the first block of three qubits.

Such procedure to detect phase flip errors presented above is similar to perform the measurement of the observables $X_1 X_2 X_3 X_4 X_5 X_6$ and $X_4 X_5 X_6 \, X_7 X_8 X_9$.

Third Case—Phase flip and bit flip on the same qubit. This case is a direct application of the previous ones. More precisely, it suffices to apply the procedure shown in the **First Case** to recover the qubit affected by the bit flip action of the channel after applying the **Second Case** to correct the phase flip error occurred. These facts are true because both error-correction process are independent.

Therefore, the stabilizer for the Shor nine qubit code is

- $Z \otimes Z \otimes I \otimes I \otimes I \otimes I \otimes I \otimes I \otimes I$
- $I \otimes Z \otimes Z \otimes I \otimes I \otimes I \otimes I \otimes I \otimes I$
- $I \otimes I \otimes I \otimes Z \otimes Z \otimes I \otimes I \otimes I \otimes I$
- $I \otimes I \otimes I \otimes I \otimes Z \otimes Z \otimes I \otimes I \otimes I$
- $I \otimes I \otimes I \otimes I \otimes I \otimes I \otimes Z \otimes Z \otimes I$
- $I \otimes I \otimes I \otimes I \otimes I \otimes I \otimes I \otimes Z \otimes Z$
- $X \otimes X \otimes X \otimes X \otimes X \otimes X \otimes I \otimes I \otimes I$
- $I \otimes I \otimes I \otimes X \otimes X \otimes X \otimes X \otimes X \otimes X$

The Shor code has parameters $[[9, 1, 3]]$, that is, the code utilizes nine qubits to encode a qubit, and it is capable of correcting one arbitrary quantum error.

Until now we have seen only errors of the types phase and bit flip. But the question is: Is the Shor code capable of correcting an arbitrary error? The answer for this question is yes!

To see this, note that the Pauli matrices I, X, Y, Z span $M_2(\mathbb{C})$, the vector space of the matrices of order 2 with complex entries. Since $XZ = -iY$ then the matrices I, X, Z, XZ also span $M_2(\mathbb{C})$. Thus, given an error matrix E in one qubit we can write $E = a_1 I + a_2 X + a_3 Z + a_4 XZ$, where $a_i \in \mathbb{C}$ for all $i = 1, 2, 3, 4$. Therefore, if $|v\rangle$ is a qubit, we have the quantum state $E|v\rangle = a_1|v\rangle + a_2 X|v\rangle + a_3 Z|v\rangle + a_4 XZ|v\rangle$.

By the measurement of the error syndrome the state $E|v\rangle$ collapses to one of the states $|v\rangle, X|v\rangle, Z|v\rangle$ or $XZ|v\rangle$. Since these operators are invertible, we then apply the inverse operator to recover the initial quantum state. In other words, if the code is capable of correcting errors of the type bit flip, phase flip and bit-phase flip combined in a given qubit, then the code is capable of correcting all arbitrary errors of such qubit. This is an interesting feature of quantum codes: if the code C corrects a suitable discrete subset of errors then C corrects all types of (*continuum*) errors. This fact is essential in quantum error correction; based on this property, it is possible to construct efficient quantum codes against arbitrary quantum errors.

Theorem 3.2.1 ([121, Theorem 10.1]) (Quantum error-correction conditions) *Let C be a quantum code, and let P be the projector onto C. Assume that E is a quantum operation with operation elements $\{E_i\}$. Then there exists an error-correction operation R correcting E on C if and only if $PE_i^\dagger E_j P = a_{ij} P$ for some Hermitian matrix $A = [a_{ij}]$ with complex entries. The operation elements $\{Ei\}$ are called errors for the noise E. If R exists then $\{Ei\}$ is called* a correctable set of errors.

This previous discussion can be summarized in the next result.

Theorem 3.2.2 ([121, Theorem 10.2]) *Let C be a quantum code and let R be the error-correction operation constructed in the proof of Theorem 3.2.1 to recover* *from a noise process E with operation elements $\{E_i\}$. Assume that F is a quantum operation with operation elements F_j which are linear combinations of the E_i. Then R also corrects for the effects of the noise process F on C.*

3.3 The Steane Code

The Steane $[[7, 1, 3]]$ seven qubit code is an example of the application of the Calderbank–Shor–Steane (CSS) quantum code construction that will be presented in Sect. 3.6. For more details on how to encode an arbitrary state or to learn syndrome measurement strategies for the Steane code, see [24, 158, 159].

The basis states for this code are given in the sequence:

$$|0_L\rangle := \frac{1}{\sqrt{8}}[|0000000\rangle + |1010101\rangle + |0110011\rangle + |1100110\rangle$$
$$+|0001111\rangle + |1011010\rangle + |0111100\rangle + |1101001\rangle]$$

and

$$|1_L\rangle := \frac{1}{\sqrt{8}}[|1111111\rangle + |0101010\rangle + |1001100\rangle + |0011001\rangle$$
$$+|1110000\rangle + |0100101\rangle + |1000011\rangle + |0010110\rangle]$$

The stabilizer for the seven qubit code due to Steane is

- $I \otimes I \otimes I \otimes X \otimes X \otimes X \otimes X$
- $I \otimes X \otimes X \otimes I \otimes I \otimes X \otimes X$
- $X \otimes I \otimes X \otimes I \otimes X \otimes I \otimes X$
- $I \otimes I \otimes I \otimes Z \otimes Z \otimes Z \otimes Z$
- $I \otimes Z \otimes Z \otimes I \otimes I \otimes Z \otimes Z$
- $Z \otimes I \otimes Z \otimes I \otimes Z \otimes I \otimes Z$

The code has parameters $[[7, 1, 3]]$, that is, it can correct an arbitrary error in a single qubit and utilizes seven qubits in the encoding process. The classical self-orthogonal code utilized in the encoding process is the $[7, 4, 3]$ Hamming code with parity check matrix

$$H = \begin{bmatrix} 0 & 0 & 0 & 1 & 1 & 1 & 1 \\ 0 & 1 & 1 & 0 & 0 & 1 & 1 \\ 1 & 0 & 1 & 0 & 1 & 0 & 1 \end{bmatrix}.$$

3.4 Five-Qubit Code

The five-qubit code [88] was the first optimal quantum code exhibited in the literature. Such code was constructed by Laflamme et al.. The stabilizer operators for the five-qubit code are

- $X \otimes Z \otimes Z \otimes X \otimes I$
- $I \otimes X \otimes Z \otimes Z \otimes X$
- $X \otimes I \otimes X \otimes Z \otimes Z$
- $Z \otimes X \otimes I \otimes X \otimes Z$
- $X \otimes X \otimes X \otimes X \otimes X$
- $Z \otimes Z \otimes Z \otimes Z \otimes Z$

The five-qubit code has parameters [[5, 1, 3]] and it is a maximum distance separable (MDS) code, because the parameters attain the quantum Singleton bound with equality.

3.5 Stabilizer Codes

Let us consider the Pauli matrices $\{I, X, Y, Z\}$. Then it is easy to see that the set $G_1 = \{\pm I, \pm i I, \pm X, \pm i X, \pm Y, \pm i Y, \pm Z, \pm i Z\}$ endowed with the operation of matrix multiplication is a group. Based on this fact, we can state the following definition.

Definition 3.5.1 The Pauli group on 1 qubit is defined by the group $(G_1, \cdot_m)$, where $\cdot_m$ is the product of matrices.

More generally, we can define the general Pauli group on n qubits.

Definition 3.5.2 The general Pauli group on n qubits, denoted by G_n, is the group consisting of all n-fold tensor products of Pauli matrices with coefficients ± 1 or $\pm i$.

Exercise 3.5.1 Show that G_n given in Definition 3.5.2 is a group.

It is interesting to note that until now we have dealt with quantum bits or tensor product of quantum bits. With this theory in mind one can construct good quantum error-correcting codes but only in the binary alphabet. Because of this limitation, it is necessary to have a theory to construct quantum codes in nonbinary alphabets. A giant step toward this was the work by Calderbank, Rains, Shor and Sloane [25].

In the past two decades, many authors [9–11, 53, 83–86, 132, 133] tried to obtain different types of models in order to incorporate quantum code over nonbinary alphabets. Finally, in 2006, a brilliant work by Ketkar, Klappenecker, Kumar and Sarvepalli [80] did this task. In fact, the authors generalized in several ways the formalism of stabilizers for binary and nonbinary alphabets by applying Galois theory. In fact, this paper contains a unified theory of stabilizers and, because of this, we present it here. We adopt the notation given in [80] to maintain the coherence of the text.

Notation. We first fix some notation. As always, p denotes a prime number, q denotes a prime power, $\mathbb{F}_q$ is the finite field with q elements, $\mathbb{C}^q$ is the complex vector space of dimension q (quantum mechanical system scenario), $|x_i\rangle$ are the vectors of an orthonormal basis of $\mathbb{C}^q$, where x_i range over all elements of $\mathbb{F}_q$, and $\mathbb{C}^{q^n}$ denotes the n-tensor product $\mathbb{C}^{q^n} = \mathbb{C}^q \otimes \mathbb{C}^q \otimes \cdots \mathbb{C}^q$.

Let us recall the concept of trace map.

Definition 3.5.3 The trace map $\mathrm{tr}_{q^m/q} : \mathbb{F}_{q^m} \longrightarrow \mathbb{F}_q$ is defined as

$$\mathrm{tr}_{q^m/q}(a) := \sum_{i=0}^{m-1} a^{q^i}.$$

Keeping this notation throughout this section we can start with the stabilizer theory. Quantum codes are important to protect quantum digits against noise produced by the channel. A quantum error-correcting code is a K-dimensional vector space of $\mathbb{C}^{q^n}$.

We next present the error model utilized in quantum mechanics. To do this, we need to define *error bases*. The error model is a natural generalization of Pauli matrices to nonbinary alphabets, as we will see in the sequence.

Let $q = p^m$ be a prime power and assume that a and b are elements of $\mathbb{F}_q$. We then define two unitary operators:

$$X(a) : \mathbb{C}^q \longrightarrow \mathbb{C}^q$$

$$|x_i\rangle \longrightarrow X(a)|x_i\rangle = |x_i + a\rangle$$

and

$$Z(b) : \mathbb{C}^q \longrightarrow \mathbb{C}^q$$

$$|x_i\rangle \longrightarrow Z(b)|x_i\rangle = \omega^{\mathrm{tr}(bx_i)}|x_i\rangle,$$

where tr : $\mathbb{F}_{p^m} \longrightarrow \mathbb{F}_p$ is the trace map and $\omega = \exp(2\pi i/p)$ is a primitive pth root of unity.

Remark 3.5.1 Note that the definitions of operators $X(a)$ and $Z(b)$ are natural generalizations of the Pauli matrices X and Z, respectively, to q-ary alphabets. In fact, $X(a)$ acts by changing the vectors of the orthonormal basis, and $Z(b)$ changes the phase of the vectors of the basis.

We can now define the set of error operators.

Definition 3.5.4 Let $X(a)$ and $Z(b)$ be the operators defined above. The set

$$\varepsilon = \{X(a)Z(b)|a, b \in \mathbb{F}_q\}$$

is called the set of error operators.

In the sequence we define the concept of nice error basis.

Definition 3.5.5 Let β be a set of q^2 unitary matrices. We say that β is a *nice error basis* if β satisfies the following conditions:

(1) $I_q \in \beta$, where I_q is the identity matrix of order q;
(2) if $A, B \in \beta$ then AB is a scalar multiple of another element of β;
(3) if $A, B \in \beta$, with $A \neq B$, then $\mathrm{Tr}(A^\dagger B) = 0$, where Tr denotes the trace of the matrix.

The set ε given in Definition 3.5.4 is a nice error basis.

Proposition 3.5.1 *The set $\varepsilon = \{X(a)Z(b)|a, b \in \mathbb{F}_q\}$ satisfies the three conditions of Definition 3.5.5, i.e., ε is a nice error basis on $\mathbb{C}^q$.*

Exercise 3.5.2 Show Proposition 3.5.1.

In order to improve the understanding of the text we give here an example of a nice error basis for $q = 4$. This is, in fact, the Example 2 of the paper by Ketkar et al. [80].

Example 3.5.1 Let us consider the finite field with four elements $\mathbb{F}_4 = \{0, 1, \alpha, \overline{\alpha}\}$. According to the notation adopted above, a basis for $\mathbb{C}^4$ can be written as $|0\rangle$, $|1\rangle$, $|\alpha\rangle$ and $|\overline{\alpha}\rangle$. Let

$$\mathbb{I}_2 = \begin{bmatrix} 1 & 0 \\ 0 & 1 \end{bmatrix}, \quad \sigma_X = \begin{bmatrix} 0 & 1 \\ 1 & 0 \end{bmatrix}, \quad \text{and} \quad \sigma_Z = \begin{bmatrix} 1 & 0 \\ 0 & -1 \end{bmatrix}.$$

By a simple computation we have
$X(0) = \mathbb{I}_2 \otimes \mathbb{I}_2, X(1) = \mathbb{I}_2 \otimes \sigma_X, X(\alpha) = \mathbb{I}_2 \otimes \mathbb{I}_2, X(\overline{\alpha}) = \sigma_X \otimes \sigma_X,$
$Z(0) = \mathbb{I}_2 \otimes \mathbb{I}_2, Z(1) = \sigma_Z \otimes \mathbb{I}_2, Z(\alpha) = \sigma_Z \otimes \sigma_Z, X(\overline{\alpha}) = \mathbb{I}_2 \otimes \sigma_Z.$

We must know how the errors act on multiple qudits. In other words, it is necessary to know if the tensor products of a finite number of nice error basis is also a nice error basis.

Proposition 3.5.2 *Let β_1 and β_2 be two sets of nice error bases on $\mathbb{C}^q$. Then the set*

$$\beta = \{E_1 \otimes E_2 | E_1 \in \beta_1, \ E_2 \in \beta_2\}$$

is also a nice error basis.

By applying induction, we know that Proposition 3.5.2 also holds for a finite number of tensor products. Assuming that $\mathbf{a} = (a_1, a_2, \ldots, a_n)$ is a vector in $\mathbb{F}_q^n$, we denote $\mathcal{X}(\mathbf{a}) = X(a_1) \otimes X(a_2) \otimes \cdots \otimes X(a_n)$ and $\mathcal{Z}(\mathbf{a}) = Z(a_1) \otimes Z(a_2) \otimes \cdots \otimes Z(a_n)$ for tensor products of n error operators.

Corollary 3.5.1 *Assume the notation above. Then the set*

$$\varepsilon_n = \{\mathcal{X}(\boldsymbol{a})\mathcal{Z}(\boldsymbol{b}) | \boldsymbol{a}, \boldsymbol{b} \in \mathbb{F}_q^n\}$$

is a nice error basis on the complex vector space $\mathbb{C}^{q^n}$.

Hence, we have a complete characterization (the model) of the errors that can corrupt the quantum digits.

In the sequence we define the concept of stabilizer code. We start with the group G_n generated by the matrices of ε_n:

$$G_n = \{\omega \mathcal{X}(\mathbf{a})\mathcal{Z}(\mathbf{b}) | \mathbf{a}, \mathbf{b} \in \mathbb{F}_q^n, c \in \mathbb{F}_p\},$$

which is called *error group* associated with ε_n. A stabilizer code is the joint eigenspace with eigenvalue 1 of some subgroup of G_n, as we see in the following.

Definition 3.5.6 Let S be a subgroup of the error group G_n. A stabilizer code $\mathcal{Q} \neq \{0\}$ is a subspace of $\mathbb{C}^{q^n}$ satisfying the equality

$$\mathcal{Q} = \bigcap_{E \in S} \{|v\rangle \in \mathbb{C}^{q^n} : E|v\rangle = |v\rangle\}.$$

We need to define the weight of an element in the error group G_n. To this end, let $\mathbf{a}, \mathbf{b}$ be two vectors in $\mathbb{F}_q^n$ and consider the vector $(\mathbf{a}|\mathbf{b}) \in \mathbb{F}_q^{2n}$.

Definition 3.5.7 The symplectic weight swt$((\mathbf{a}|\mathbf{b}))$ of $(\mathbf{a}|\mathbf{b})$ is the number of nonzero ordered pairs of the form (a_i, b_i), where $i = 1, 2, \ldots, n$, i.e.,

$$\text{swt}((\mathbf{a}|\mathbf{b})) = |\{i \, | (a_i, b_i) \neq (0, 0)\}|.$$

Definition 3.5.8 Let $E = \omega^c \mathcal{X}(\mathbf{a})\mathcal{Z}(\mathbf{b})$ be an element in the error group G_n. Then the weight wt(E) of E is defined as wt$(E) = \text{swt}((\mathbf{a}|\mathbf{b}))$.

It is interesting to note that $(a_i, b_i) \neq (0, 0)$ if and only if $(X(a_i), Z(b_i)) \neq (I_q, I_q)$. Thus, from Definition 3.5.8, the weight $\text{wt}(E)$ of E can be interpreted as the number of nonidentity tensor components, as expected.

Definition 3.5.9 A quantum error-correcting code (QC) Q is a K-dimensional subspace of $\mathbb{C}^{q^n}$. If Q has minimum distance d, then we say that Q is an $((n, K, d))_q$ code. If $K = q^k$ we write $[[n, k, d]]_q$. The length n, the dimension K and minimum distance d are the parameters of Q. The code Q is said to be *pure* to l if and only if its stabilizer group does not contain non-scalar matrices of weight less than l; Q is pure if and only if it is pure to its minimum distance.

Definition 3.5.10 We say that a quantum code Q has minimum weight d if it can detect all errors in G_n of weight less than d, but it cannot detect some error of weight d.

A QC with minimum distance d corrects all errors of weight $\lfloor (d-1)/2 \rfloor$ or less.

We next recall two important definitions in group theory. The center $Z(G_n)$ of the group G_n is the subgroup given by

$$Z(G_n) = \{E \in G_n | EF = FE, \forall F \in G_n\}.$$

In words, $Z(G_n)$ consists of the elements in G_n that commute with all elements of G_n.

Let S be a subgroup of G_n. The centralizer $C_{G_n}(S)$ of S in G_n is defined as

$$C_{G_n}(S) = \{E \in G_n | EF = FE, \forall F \in S\}.$$

Analogously, the elements of $C_{G_n}(S)$ are the elements of G_n that commute with all elements of S. Further, let us consider $SZ(G_n)$ as the group generated by S and $Z(G_n)$. The following result gives necessary and sufficient conditions for error-detection.

Theorem 3.5.1 *Let S be a subgroup of G_n such that S is the stabilizer group of a stabilizer code Q of dimension greater than 1. A necessary and sufficient condition in order to Q detects an error $E \in G_n$ is either $E \in SZ(G_n)$ or $E \notin C_{G_n}(S)$.*

Exercise 3.5.3 Show Theorem 3.5.1.

There exist some well-known bounds with respect to the parameters of a quantum code. The quantum Singleton bound will be utilized several times throughout this book.

Lemma 3.5.1 *Let C be an $[[n, k, d]]_q$ quantum code. Then the quantum Singleton bound asserts that the parameters of C satisfy $k + 2d \leq n + 2$. If C attains the quantum Singleton bound, i.e., $k + 2d = n + 2$, then it is called a quantum maximum distance separable (MDS) code.*

More generally, we have the following result to stabilizer codes.

Lemma 3.5.2 (Quantum Singleton bound) ([80, Corollary 28]) *The parameters of an $((n, K, d))_q$ stabilizer code with $K > 1$ satisfy the inequality*

$$K \leq q^{n-2d+2}.$$

Another bound much utilized for quantum codes is the quantum Hamming bound. Although we do not utilize such bound in the quantum code construction presented in this book, we present it here for completeness.

Lemma 3.5.3 (Quantum Hamming Bound) (see [38, Lemma 12]; see also [80, p. 9]) *Let C be an $((n, K, d))_q$ pure stabilizer code. Then the parameters of C satisfy the inequality:*

$$\frac{q^n}{K} \geq \sum_{i=0}^{\lfloor (d-1)/2 \rfloor} (q^2 - 1)^i \binom{n}{i}.$$

3.6 Calderbank–Shor–Steane Construction

The Calderbank–Shor–Steane (CSS for short) code construction is one of the most interesting code constructions shown in the literature [121]. In fact, it is the first construction method exhibited in the literature in the sense that one can derive families of quantum codes by applying such construction, and not only few codes with specific parameters. Such method utilizes two classical linear nested codes (or an Euclidean self-orthogonal linear code) to address the problem of correcting phase and qubit flip errors. The CSS codes form a subclass of the class of the stabilizer codes (see for instance [80]). We next present a detailed construction of these codes.

The process starts with two binary linear codes C_1 and C_2 with parameters $[n, k_1, d_1]$ and $[n, k_2, d_2]$, respectively, such that $C_2 \subset C_1$ and both C_1 and $C_2^{\perp}$ correct t errors. Based on these classical codes we define a quantum code $CSS(C_1, C_2)$ (this notation is not usual in the literature but we prefer to adopt it here to maintain the notation of the textbook [121]) as follows.

Assume that $c, c' \in C_1$ are two codewords of C_1. Define the following relation on $C_1 : c \approx c' \iff c - c' \in C_2$. It is easy to see that $\approx$ is an equivalence relation on C_1. Moreover, it is not difficult to see that the equivalence class $\bar{c}$ determined by a codeword $c \in C_1$ is equal to $\bar{c} = c + C_2 := \{c + x | x \in C_2\}$. Thus, the cosets $c + C_2$ form a partition of C_1. Since we are working with quantum states (qubits or tensor product of qubits) then we must adapt the notation. We then define the quantum state $|c + C_2\rangle$ (already normalized) as

$$|c + C_2\rangle := \frac{1}{\sqrt{|C_2|}} \sum_{x \in C_2} |c + x\rangle,$$

where $+$ denotes the componentwise addition modulo 2.

Exercise 3.6.1 Show that the relation defined above is an equivalence relation whose equivalence class of a codeword $c \in C_1$ is the coset $c + C_2$

Suppose c and c' belong to the disjoint cosets of C_2; this implies that there is no $x, x' \in C_2$ such that $c + x = c' + x'$, otherwise,

$$c - c' = x' - x \in C_2 \Longrightarrow c + C_2 = c' + C_2,$$

which is a contradiction. Hence, for distinct $c \neq c'$, the corresponding quantum states $|c + C_2\rangle$ and $|c' + C_2\rangle$ are orthonormal. Thus, we define the quantum code CSS(C_1, C_2) to be the vector space spanned by the states $|c + C_2\rangle$ for all $c \in C_1$. Since the number of cosets is $\frac{|C_1|}{|C_2|}$, the dimension of CSS(C_1, C_2) equals $\frac{|C_1|}{|C_2|} = 2^{k_1 - k_2}$. Therefore, the quantum code CSS(C_1, C_2) has parameters $[n, k_1 - k_2]$, capable to correct errors on t qubits.

Let us see how the code works in the error correction. Assume that the initial state of a quantum system is $|c + C_2\rangle$. After passing through the channel, the original state can suffer some kind of error (bit flip and/or phase flip). In the error model it is assumed that bit flip errors e_b are binary vectors of length n (the code length) such that the component is 1 where the bit flip occurs and 0 otherwise. The phase flip errors are also binary vectors e_p of length n with 1 in the coordinate that a phase flip occurs and 0 otherwise. Note that both binary vectors cannot have more than t ones. Adopting this model, we know that the corrupted state is

$$|c + C_2\rangle \xrightarrow{channel} \frac{1}{\sqrt{|C_2|}} \sum_{x \in C_2} (-1)^{(c+x) \cdot e_p} |c + x + e_b\rangle.$$

Bit flip Detection. We introduce a sufficient large ancilla system capable of storing the syndrome for C_1, which is initially in the all zero state $|0\rangle$. Applying the parity check matrix H_1 of C_1 to all state $|c + x + e_b\rangle$, since $H_1(c + x) = 0$ we have

$$|c + x + e_b\rangle |0\rangle \longrightarrow |c + x + e_b\rangle |H_1(c + x + e_b)\rangle = |c + x + e_b\rangle |H_1 e_b\rangle,$$

that is, the error was isolated. Thus, we obtain the state

$$\frac{1}{\sqrt{|C_2|}} \sum_{x \in C_2} (-1)^{(c+x) \cdot e_p} |c + x + e_b\rangle |H_1 e_b\rangle.$$

Performing the measurement of the ancilla, we obtain $H_1 e_b$; discarding the ancilla we return to the quantum state

$$\frac{1}{\sqrt{|C_2|}} \sum_{x \in C_2} (-1)^{(c+x) \cdot e_p} |c + x + e_b\rangle.$$

Since the classical syndrome $H_1 e_1$ is known, then C_1 tells us the vector error e_b. To recover the state, we apply the NOT gate in all the qubits corrupted by the error, which leads to the state

$$\frac{1}{\sqrt{|C_2|}} \sum_{x \in C_2} (-1)^{(c+x) \cdot e_p} |c + x\rangle.$$

Our next task is to detect the qubits corrupted by the phase flip error.

Phase flip Detection. Starting from the state

$$\frac{1}{\sqrt{|C_2|}} \sum_{x \in C_2} (-1)^{(c+x) \cdot e_p} |c + x\rangle,$$

the Hadamard gate

$$H = \frac{1}{\sqrt{2}} \begin{bmatrix} 1 & 1 \\ 1 & -1 \end{bmatrix}$$

is applied to each qubit producing the state

$$\frac{1}{\sqrt{|C_2| 2^n}} \sum_{v \in \mathbb{F}_2^n} \sum_{x \in C_2} (-1)^{(c+x) \cdot (e_p+v)} |v\rangle,$$

which can be rewritten as

$$\frac{1}{\sqrt{|C_2| 2^n}} \sum_{w \in \mathbb{F}_2^n} \sum_{x \in C_2} (-1)^{(c+x) \cdot w} |w + e_p\rangle,$$

where $w = v + e_p$ (notice that $e_p + e_p$ is the zero vector). If $w \in C_2^{\perp}$ then $w \cdot x = 0$ for all $x \in C_2$; so

$$\sum_{x \in C_2} (-1)^{xw} = \underbrace{1 + 1 + \cdots + 1}_{|C_2| \text{ times}} = |C_2|.$$

On the other hand, if $w \notin C_2^{\perp}$ then it is easy to see that

$$\sum_{x \in C_2} (-1)^{xw} = 0.$$

Hence, the state can be rewritten as

$$\frac{1}{\sqrt{2^n / |C_2|}} \sum_{w \in C_2^{\perp}} (-1)^{c \cdot w} |w + e_p\rangle,$$

which is similar to a bit flip type of error. To correct it, we proceed analogously to the correction of e_b, i.e., we introduce an ancilla and apply the parity check matrix G_2 for $C_2^\perp$ (which is a generator matrix for the code C_2) as done previously, to obtain $G_2 e_p$ and to correct the error e_p, obtaining the quantum state

$$\frac{1}{\sqrt{2^n/|C_2|}} \sum_{w \in C_2^\perp} (-1)^{c \cdot w} |w\rangle.$$

Applying again the Hadamard gate to each qubit results in the original state

$$|c + C_2\rangle := \frac{1}{\sqrt{|C_2|}} \sum_{x \in C_2} |c + x\rangle.$$

Therefore, the resulting quantum code $\text{CSS}(C_1, C_2)$ has parameters $[[n, k_1 - k_2]]$ and can correct arbitrary errors up to t qubits.

After some years of research, the CSS construction was generalized to nonbinary alphabets, as we can see in the following lemma.

Lemma 3.6.1 ([25, 80]) *Let q be a prime power. Let C_1 and C_2 denote two classical linear codes both over the field $\mathbb{F}_q$, with parameters $[n, k_1, d_1]_q$ and $[n, k_2, d_2]_q$, respectively, such that $C_2 \subset C_1$. Then there exists an $[[n, K = k_1 - k_2, D]]_q$ CSS quantum code, where $D = \min\{wt(c) : c \in (C_1 \backslash C_2) \cup (C_2^\perp \backslash C_1^\perp)\}$.*

Remark 3.6.1 As it was said previously, the CSS codes shown in Lemma 3.6.1 are constructed over nonbinary alphabets. In fact, the original version of the CSS code construction (as we have presented in this subsection) was presented for binary alphabets. For nonbinary alphabets, a quantum state is called quantum digits or qudits (for short).

Remark 3.6.2 In the following section (Sect. 4) we present carefully the background concerning linear codes mentioned in Lemma 3.6.1. We prefer not to present Theory of Linear Codes before in order to maintain the quantum theory without interruption.

Chapter 4
Linear Block Codes

4.1 Introduction

As usual, in this chapter we consider that p denotes a prime number, q is a prime power and $\mathbb{F}_q$ is the finite field with q elements. We begin by defining the general concept of a code over $\mathbb{F}_q$; after this, we present the concept of a linear code that is the more important class of codes in coding and information theory due to its vector space structure. In this subsection, the vectors are written in bold.

Definition 4.1.1 Assume that $\mathbb{F}_q^n$ is the vector space (over $\mathbb{F}_q$) of all n-tuples in $\mathbb{F}_q$. Then an (n, M) code C over $\mathbb{F}_q$ is a subset of $\mathbb{F}_q^n$ of size M. If $\mathbf{c} = (a_1, a_2, \ldots, a_n) \in \mathbb{F}_q^n$ is such that $\mathbf{c} \in C$, then the vector $\mathbf{c}$ is called *codeword* of C.

To work with codes without structures is really hard. In this book, we only deal with *linear codes*, i.e., codes with a vector space structure.

Definition 4.1.2 A linear code C over $\mathbb{F}_q$ of length n and dimension k is a k-dimensional $\mathbb{F}_q$-subspace of $\mathbb{F}_q^n$. Such codes are denoted by $[n, k]$. We say that n and k are the parameters of the code.

Since the dimension of C is equal to k and because the field $\mathbb{F}_q$ has q elements, it follows that the number of codewords of (a linear code) C is equal to q^k.

There exist two usual ways of defining a linear code: by means of generator matrices or based on parity check matrices.

Definition 4.1.3 Let C be a linear code over $\mathbb{F}_q$ with parameters $[n, k]$. A generator matrix G for C is a $k \times n$ matrix with entries in $\mathbb{F}_q$ such that the rows of G form a basis for C.

In general, a generator matrix of a linear code is not unique. However, in the case below one has the uniqueness.

Definition 4.1.4 Let C be a linear code over $\mathbb{F}_q$ with parameters $[n, k]$. If the first k coordinates form an information set, then C has a unique generator matrix of the

© Springer Nature Switzerland AG 2020
G. G. La Guardia, *Quantum Error Correction*, Quantum Science and Technology,
https://doi.org/10.1007/978-3-030-48551-1_4

form $[I_k|A]$ with entries in $\mathbb{F}_q$, where I_k is the identity matrix of order k. This matrix is said to be in standard form.

Another way of defining a linear code is by means of parity check matrices.

Definition 4.1.5 Let C be a linear code over $\mathbb{F}_q$ with parameters $[n, k]$. A parity check matrix for C is an $(n - k) \times n$ matrix H with entries in $\mathbb{F}_q$ defined by

$$C = \{\mathbf{c} \in \mathbb{F}_q^n | H\mathbf{c}^T = \mathbf{0}\},$$

where $\mathbf{c}^T$ denotes the transpose of vector $\mathbf{c}$.

The rows of H are also linearly independent (they form a basis for the (Euclidean) dual $C^\perp$ of C). Evidently, in general, a parity check matrix of a given code is not unique.

Theorem 4.1.1 *Let C be a linear code over $\mathbb{F}_q$ with parameters $[n, k]$. If $G = [I_k|A]$ is a generator matrix for C, then $H = [-A^T|I_{n-k}]$ is a parity check matrix for C.*

Exercise 4.1.1 Show Theorem 4.1.1.

Example 4.1.1 An example of a linear code is the well-known binary Hamming code with parameters $[7, 4]$ with generator matrix (in standard form)

$$G = \begin{bmatrix} 1\,0\,0\,0\,0\,1\,1 \\ 0\,1\,0\,0\,1\,0\,1 \\ 0\,0\,1\,0\,1\,1\,0 \\ 0\,0\,0\,1\,1\,1\,1 \end{bmatrix}$$

and parity check matrix (in standard form)

$$H = \begin{bmatrix} 0\,1\,1\,1\,1\,0\,0 \\ 1\,0\,1\,1\,0\,1\,0 \\ 1\,0\,1\,1\,0\,1\,0. \end{bmatrix}.$$

As we will see later, the Hamming code has minimum distance three (see Proposition 4.1.2).

Another important parameter of a linear code is the minimum distance. To define this concept we need first to define Hamming distance.

Definition 4.1.6 The Hamming distance $d(\mathbf{v}, \mathbf{w})$ between two vectors $\mathbf{v}, \mathbf{w}$ in $\mathbb{F}_q^n$ is the number of coordinates in which $\mathbf{v}$ and $\mathbf{w}$ differ.

Exercise 4.1.2 Show that the Hamming distance is, in fact, a metric (cf. Definition 1.6.1) on $\mathbb{F}_q^n$.

We are now ready to define the minimum distance of a code, which is totally correlated with the power of error-correction of the code.

Definition 4.1.7 The minimum distance of a code C (not necessarily linear) is the smallest Hamming distance between distinct codewords of C.

Definition 4.1.8 Let $\mathbf{v} \in \mathbb{F}_q^n$. The Hamming weight wt($\mathbf{v}$) of $\mathbf{v}$ is defined as the number of nonzero coordinates in $\mathbf{v}$.

It is easy to see that, for all vectors $\mathbf{v}, \mathbf{w} \in \mathbb{F}_q^n$, it follows that

$$d(\mathbf{v}, \mathbf{w}) = \text{wt}(\mathbf{v} - \mathbf{w}).$$

Since in the linear case, $\forall\, \mathbf{v}, \mathbf{w} \in C$ implies $\mathbf{v} - \mathbf{w} \in C$, we have the following result.

Proposition 4.1.1 *Let C be a linear code over $\mathbb{F}_q$. Then the minimum distance of C is equal to the minimum weight of all nonzero codewords of C.*

Thus, if the code is linear, its minimum distance is also called the minimum weight of the code.

We have now the complete set of parameters of a linear code, i.e., length, dimension and minimum distance: a linear code C over $\mathbb{F}_q$, of length n, dimension k and minimum distance d is denoted by $[n, k, d]_q$.

There exists a well-known way to find the minimum distance of a linear code.

Proposition 4.1.2 *Let C be a linear code over $\mathbb{F}_q$ with parity check matrix H. Then C has minimum weight d if and only if H has a set of d linearly dependent columns but no set of $d - 1$ linearly dependent columns.*

The minimum distance of a code is totally correlated with the error-correcting capacity of the code.

Theorem 4.1.2 *A code C having minimum distance d can correct $\left\lfloor \frac{(d-1)}{2} \right\rfloor$ errors. If d is even, the code can simultaneously correct $\frac{(d-2)}{2}$ errors and detect $d/2$ errors.*

By applying Proposition 4.1.2 in the parity check matrix

$$H = \begin{bmatrix} 0 & 1 & 1 & 1 & 1 & 0 & 0 \\ 1 & 0 & 1 & 1 & 0 & 1 & 0 \\ 1 & 0 & 1 & 1 & 0 & 1 & 0 \end{bmatrix}$$

of the binary $[7, 4]$ Hamming code of Example 4.1.1, we see that the code has minimum distance 3, i.e., the Hamming code is a single-error-correcting code due to Theorem 4.1.2.

Exercise 4.1.3 Show Theorem 4.1.2.

4.2 Dual Codes

Let C be an $[n, k, d]_q$ linear code over $\mathbb{F}_q$ with parity check matrix H. Since the rows of H are linearly independent, H is a generator matrix of some code, called *Euclidean dual code* of C, denoted by $C^\perp$. The dual code $C^\perp$ has length n and dimension $n - k$.

The dual code can be also defined by an alternative way by means of the usual (Euclidean) inner product on $\mathbb{F}_q^n$ in the following way. Recall that if $\mathbf{v} = (v_1, v_2, \ldots, v_n)$ and $\mathbf{w} = (w_1, w_2, \ldots, w_n)$ are two vectors in $\mathbb{F}_q^n$, then the Euclidean inner product $\mathbf{v} \cdot \mathbf{w}$ of $\mathbf{v}$ and $\mathbf{w}$ is defined as

$$\mathbf{v} \cdot \mathbf{w} = \sum_{i=1}^n v_i w_i.$$

Based on Definition 4.1.5, we have the following.

Definition 4.2.1 The Euclidean dual code $C^\perp$ of o linear code C is defined as

$$C^\perp = \{\mathbf{v} \in \mathbb{F}_q^n | \mathbf{v} \cdot \mathbf{c} = 0 \ \forall \mathbf{c} \in C\}.$$

It is easy to see that if G and H are generator and parity check matrices, respectively, for a given code C, then it follows that H and G are generator and parity check matrices, respectively, for the dual $C^\perp$.

Let $C \subseteq \mathbb{F}_{q^2}^n$ be a linear code defined over $\mathbb{F}_{q^2}$. In this case, we can also define the dual code $C^{\perp_H}$ of C with respect to the *Hermitian inner product*. To do this, let $\mathbf{v}, \mathbf{w} \in \mathbb{F}_{q^2}^n$ be two vectors.

Definition 4.2.2 The Hermitian inner product $\langle \mathbf{v} | \mathbf{w} \rangle_H$ of $\mathbf{v}, \mathbf{w} \in \mathbb{F}_{q^2}^n$ is defined as

$$\langle \mathbf{v} | \mathbf{w} \rangle_H = \mathbf{v}^q \cdot \mathbf{w} = \sum_{i=1}^n v_i^q w_i,$$

where $\mathbf{v}^q = (v_1^q, v_2^q, \ldots, v_n^q)$.

Based on the Hermitian inner product, one has the *Hermitian dual code* $C^{\perp_H}$ of C.

Definition 4.2.3 Let $C \subseteq \mathbb{F}_{q^2}^n$ be a linear code. The Hermitian dual code $C^{\perp_H}$ of C is defined by

$$C^{\perp_H} = \{\mathbf{v} \in \mathbb{F}_{q^2}^n | \mathbf{v}^q \cdot \mathbf{c} = 0 \ \forall \mathbf{c} \in C\}.$$

4.3 New Codes from Old

In this subsection, we describe some well-known techniques to obtain new codes from old ones, namely, puncturing, extending, code expansion, direct sum, the $(\mathbf{u}|\mathbf{u} + \mathbf{v})$ construction and the direct product code construction. The reader who is interested to investigate more details of such techniques can consult the textbooks [67, 114].

4.3.1 Puncturing Codes

We begin by the technique of puncturing codes.

Definition 4.3.1 Assume that C is an $[n, k, d]_q$ linear code over $\mathbb{F}_q$. Then if the same coordinate i is deleted in each codewords of C we say that C was punctured. The punctured code derived from C by deleting the ith coordinate will be denoted by C_i^P.

Let C be as in Definition 4.3.1. If G is a generator matrix for C, then by deleting the ith column of G and omitting a possible zero or duplicate row we obtain a generator matrix G^P for C_i^P.

The following theorem gives us information about the parameters of the punctured code.

Theorem 4.3.1 Let C be an $[n, k, d]_q$ linear code over $\mathbb{F}_q$. Assume that C_i^P is the punctured code on the ith coordinate. Then the following hold:

(1) If $d > 1$, then C_i^P is an $[n - 1, k, d_i]_q$ code, where $d_i = d - 1$ if C has a minimum weight codeword with a nonzero ith coordinate, and $d_i = d$ otherwise.
(2) If $d = 1$, then C_i^P is an $[n - 1, k, 1]_q$ code if C has no codeword of weight 1 whose nonzero entry is in the coordinate i; otherwise, if $k > 1$, then C_i^P is an $[n - 1, k - 1, d_i]_q$ code with $d_i \geq 1$.

4.3.2 Code Extension

A linear code C over $\mathbb{F}_q$ can be extended in several ways. However, the most common method is by adding an extra symbol such that the sum (over $\mathbb{F}_q$) of all coordinates of the new code is equal to zero.

Definition 4.3.2 Assume that C is an $[n, k, d]_q$ linear code over $\mathbb{F}_q$. Then the *extended* code C^e derived from C is defined as

$$C^e = \{(x_1, \ldots, x_n, x_{n+1}) \in \mathbb{F}_q^{n+1} | (x_1, \ldots, x_n) \in C, x_1 + \cdots + x_n + x_{n+1} = 0\}.$$

Exercise 4.3.1 Show that the code C^e given in Definition 4.3.2 is linear and has parameters $[n + 1, k, d^e]_q$, where $d^e = d$ or $d^e = d + 1$.

Let C be a linear code over $\mathbb{F}_q$ with generator matrix G and parity check matrix H. Then a generator G^e and a parity check matrix H^e for C^e can be obtained from G and H, respectively, as

$$G^e = [G | \mathbf{v}^T],$$

where $\mathbf{v}^T$ is a column vector such that the sum (over $\mathbb{F}_q$, of course) of the elements of all rows of G^e is equal to zero, and

$$H^e = \begin{bmatrix} 1 \cdots 1 & 1 \\ & 0 \\ H & \vdots \\ & 0 \end{bmatrix}.$$

In the sequence, we present two concepts that will be utilized in our quantum code construction shown in Sect. 6.7.

Definition 4.3.3 Let $\mathbf{v} = (v_1, \ldots, v_n)$ be a vector in $\mathbb{F}_q^n$. We say that $\mathbf{v}$ is even-like if it satisfies the equality $\sum_{i=1}^{n} v_i = 0$, and odd-like otherwise.

Definition 4.3.4 Let C be an $[n, k, d]_q$ linear code. Then the minimum weight of the even-like codewords of C is called minimum even-like weight of the code and it is denoted by d_{even} (or $(d)_{even}$). Similarly, the minimum weight of the odd-like codewords of C is called minimum odd-like weight of the code, denoted by d_{odd} (or $(d)_{odd}$).

4.3.3   Code Expansion

We begin by recalling the concept of *dual basis* [110].

Definition 4.3.5 Let $\beta = \{b_1, b_2, \ldots, b_m\}$ be a basis of $\mathbb{F}_{q^m}$ over $\mathbb{F}_q$. A dual basis of β is defined as $\beta^\perp = \{b_1{}^*, b_2{}^*, \ldots, b_m{}^*\}$, where $\operatorname{tr}_{q^m/q}(b_i b_j{}^*) = \delta_{ij}$, for all $i, j \in \{1, \ldots, m\}$. A self-dual basis β is a basis satisfying $\beta = \beta^\perp$.

If C is an $[n, k, d_1]_{q^m}$ linear code and $\beta = \{b_1, b_2, \ldots, b_m\}$ is a basis of $\mathbb{F}_{q^m}$ over $\mathbb{F}_q$, then the q-ary expansion $\beta(C)$ of C with respect to β is an $[mn, mk, d_2 \geq d_1]_q$ linear code given by $\beta(C) := \{(c_{ij})_{i,j} \in \mathbb{F}_q^{mn} \mid \mathbf{c} = (\sum_j c_{ij} b_j)_i \in C\}$.

The next lemma was presented in different works [12, 56, 94]. It is important in order to compute the dual code of the q-ary expansion of some linear codes.

Lemma 4.3.1 *Let $C = [n, k, d]_{q^m}$ be a linear code over $\mathbb{F}_{q^m}$, where q is a prime power. Let $C^\perp$ be the dual of the code C. Then the dual code of the q-ary expansion $\beta(C)$ of the code C with respect to the basis β is the q-ary expansion $\beta^\perp(C^\perp)$ of the dual code $C^\perp$ with respect to $\beta^\perp$.*

4.3.4 Direct Sum and Direct Product

To perform the direct sum of linear codes is a well-known method to obtain more linear codes. Let us recall this technique.

Definition 4.3.6 Let $C_1 = [n_1, k_1, d_1]_q$ and $C_2 = [n_2, k_2, d_2]_q$ be two linear codes both over $\mathbb{F}_q$. Then the direct sum code $C_1 \oplus C_2$ is the linear code given by

$$C_1 \oplus C_2 = \{(\mathbf{c}_1, \mathbf{c}_2) | \mathbf{c}_1 \in C_1, \mathbf{c}_2 \in C_2\}.$$

From construction, it is easy to see that the code $C_1 \oplus C_2$ has parameters $[n_1 + n_2, k_1 + k_2, \min\{d_1, d_2\}]_q$.

Assume that C_i has generator matrix G_i and parity check H_i, for $i = 1, 2$. Then $C_1 \oplus C_2$ has a generator matrix of the form

$$G_1 \oplus G_2 = \begin{bmatrix} G_1 & 0 \\ 0 & G_2 \end{bmatrix},$$

and a parity check matrix given by

$$H_1 \oplus H_2 = \begin{bmatrix} H_1 & 0 \\ 0 & H_2 \end{bmatrix}.$$

Exercise 4.3.2 Show that the direct sum code is linear. Show also that $G_1 \oplus G_2$ and $H_1 \oplus H_2$ are, in fact, a generator and a parity check matrix for $C_1 \oplus C_2$.

We next define the product code obtained by means of tensor product of matrices.

Definition 4.3.7 Let us assume that $C_1 = [n_1, k_1, d_1]_q$ and $C_2 = [n_2, k_2, d_2]_q$ are two linear codes over $\mathbb{F}_q$. Then the direct product $C_1 \otimes C_2$ is a linear code over $\mathbb{F}_q$ with parameters $[n_1 n_2, k_1 k_2, d_1 d_2]_q$. The codewords of $C_1 \otimes C_2$ consist of all $n_1 \times n_2$ arrays such that the columns belong to C_1 and the rows to C_2.

Let G_i be a generator matrix for the code C_i, for $i = 1, 2$. Then the Kronecker product $G_1 \otimes G_2$ (cf. Definition 1.7.14) is a generator matrix for the code $C_1 \otimes C_2$.

Exercise 4.3.3 Prove that $C_1 \otimes C_2$ is linear and has parameters $[n_1 n_2, k_1 k_2, d_1 d_2]_q$. Show also that $G_1 \otimes G_2$ is a generator matrix for $C_1 \otimes C_2$.

4.3.5 The (u|u + v) Construction

The (u|u + v) construction combines two linear codes of same length and defined over the same field to produce a new linear code whose length is twice as the previous one.

Definition 4.3.8 Let C_1 and C_2 be two linear codes both over $\mathbb{F}_q$ with parameters $[n, k_1, d_1]_q$ and $[n, k_2, d_2]_q$, respectively. Define a code C as follows:

$$C = \{(\mathbf{u}, \mathbf{u} + \mathbf{v}) | \mathbf{u} \in C_1, \mathbf{v} \in C_2\}.$$

The new linear code C has parameters $[2n, k_1 + k_2, \min\{2d_1, d_2\}]_q$.

In order to simplify the notation, the code produced by applying the (u|u + v) construction to the codes C_1 and C_2 is denoted by $(C_1|C_1 + C_2)$.

Suppose that C_i has generator matrix G_i and parity check matrix H_i, for $i = 1, 2$. Then $(C_1|C_1 + C_2)$ has a generator matrix $G_1|G_2$ of the form

$$G_1|G_2 = \begin{bmatrix} G_1 & G_1 \\ 0 & G_2 \end{bmatrix}$$

and a parity check matrix $H_1|H_2$ given by

$$H_1|H_2 = \begin{bmatrix} H_1 & 0 \\ -H_2 & H_2 \end{bmatrix}.$$

We observe that the notations utilized here for a generator $G_1|G_2$ and for a parity check matrix $H_1|H_2$ for C are not usual in the literature, but we decide to adopt them to simplify the understanding of the text.

Exercise 4.3.4 Show that $G_1|G_2$ and $H_1|H_2$ are a generator and a parity check matrix for C, respectively.

4.4 Cyclic Codes

In this part, we review the concept of cyclic and BCH codes. For more details, the reader can consult the textbooks [67, 114].

We assume that q is a prime power and $\mathbb{F}_q$ is the finite field with q elements. Cyclic codes form an important class of linear codes. Throughout this book, we always assume that $\gcd(q, n) = 1$, where n is the code length. Recall that the multiplicative order of q modulo n $\mathrm{ord}_n(q)$ is the smallest positive integer m such that $n|(q^m - 1)$.

Definition 4.4.1 The minimal polynomial over $\mathbb{F}_q$ of $\beta \in \mathbb{F}_{q^m}$ is the monic polynomial of smallest degree, $M(x)$, with coefficients in $\mathbb{F}_q$ such that $M(\beta) = 0$. If

$\beta = \alpha^i$ for some primitive element $\alpha \in \mathbb{F}_{q^m}$, then the minimal polynomial of $\beta = \alpha^i$ is denoted by $M^{(i)}(x)$.

Irreducible polynomials are generated in the following way.

Theorem 4.4.1 $x^{q^m} - x = $ *product of all monic, irreducible polynomials over* $\mathbb{F}_q$, *whose degree divides m.*

The following result is well known.

Theorem 4.4.2 $x^n - 1 = \prod_j M^{(j)}(x)$, *where* $M^{(j)}(x)$ *denotes the minimal polynomial of* $\alpha^j \in \mathbb{F}_{q^m}$ *and j runs through the coset representatives mod n.*

In order to proceed further, the reader can recall some known concepts of algebra such as ideals in commutative rings and quotient rings. These concepts can be found in the Appendix of this book.

Let $\mathbb{F}_q[x]$ denote the ring of polynomials in $\mathbb{F}_q$ and consider the quotient ring $R_n = \mathbb{F}_q[x]/(x^n - 1)$. From this context we define the concept of cyclic code.

Definition 4.4.2 A cyclic code C of length n over $\mathbb{F}_q$ is a nonzero ideal in R_n.

It is well-known that there exists only one polynomial $g(x)$ with minimal degree in C; $g(x)$ is a generator polynomial of C. Moreover, $g(x)$ is a factor of $x^n - 1$. The dimension of a cyclic code C is equal to $n - \deg(g(x))$, where $\deg(g(x))$ is the degree of the polynomial $g(x)$.

The dual code $C^{\perp}$ of a cyclic code C is also cyclic and has generator polynomial given by

$$g(x)^{\perp} = x^{\deg h(x)} h(x^{-1}), \tag{4.1}$$

where $h(x) = (x^n - 1)/g(x)$.

Definition 4.4.3 Two codes C and C^* are called equivalent if they differ only in the arrangement of symbols. More precisely, if C is the row space of a matrix G, then C^* is a code equivalent to C if and only if C^* is the row space of a matrix G^* that is obtained from G by rearranging columns.

Based on Definition 4.4.3 and from Eq. (4.1), it follows that the code with generator polynomial $h(x)$ is equivalent to the (Euclidean) dual code $C^{\perp}$.

Let us recall the well-known BCH bound theorem.

Theorem 4.4.3 (The BCH bound) *Let q be a prime power and α a primitive nth root of unity. Let C be a cyclic code with generator polynomial $g(x)$ such that, for some integers $b \geq 0$ and $\delta \geq 1$, and for $\alpha \in \mathbb{F}_q$, we have*

$$g(\alpha^b) = g(\alpha^{b+1}) = \ldots = g(\alpha^{b+\delta-2}) = 0,$$

that is, the code has a sequence of $\delta - 1$ consecutive powers of α as zeros. Then the minimum distance of C is greater than or equal to δ.

In the sequence, we present the definition of a BCH code.

Definition 4.4.4 Let q be a prime power and let n be a positive integer such that $\gcd(q, n) = 1$. Assume that α is a primitive nth root of unity. A cyclic code C of length n over $\mathbb{F}_q$ is a BCH code with designed distance δ if, for some integer $b \geq 0$, we have

$$g(x) = \text{l. c. m.}\{M^{(b)}(x), M^{(b+1)}(x), \ldots, M^{(b+\delta-2)}(x)\},$$

that is, $g(x)$ is the monic polynomial of smallest degree over $\mathbb{F}_q$ having α^b, α^{b+1}, $\ldots, \alpha^{b+\delta-2}$ as zeros.

Therefore, $c \in C$ if and only if $c(\alpha^b) = c(\alpha^{b+1}) = \ldots = c(\alpha^{b+\delta-2}) = 0$. Thus the code has a string of $\delta - 1$ consecutive powers of α as zeros, Hence, from the BCH bound, its minimum distance is at least δ. If $n = q^l - 1$, then the BCH code is called primitive and if $b = 1$ it is called narrow-sense.

Definition 4.4.5 The q-ary cyclotomic coset (or q-ary coset or q-coset) modulo n containing an element s is defined by $\{s, sq, sq^2, sq^3, \ldots, sq^{m_s-1}\}$, where m_s is the smallest positive integer such that $sq^{m_s} \equiv s \mod n$. If s is the smallest number in coset, this coset is denoted by $\mathbb{C}_s$.

In terms of cyclotomic cosets, the generator polynomial of a BCH code is of the form

$$g(x) = \prod_{z \in Z}(x - \alpha^z),$$

where $Z = \mathbb{C}_b \cup \mathbb{C}_{b+1} \cup \cdots \cup \mathbb{C}_{b+\delta-2}$ is the defining set of the code.

A parity check matrix for C is given by

$$H_{\delta,b} = \begin{bmatrix} 1 & \alpha^b & \alpha^{2b} & \cdots & \alpha^{(n-1)b} \\ 1 & \alpha^{(b+1)} & \alpha^{2(b+1)} & \cdots & \alpha^{(n-1)(b+1)} \\ \vdots & \vdots & \vdots & \vdots & \vdots \\ 1 & \alpha^{(b+\delta-2)} & \cdots & \cdots & \alpha^{(n-1)(b+\delta-2)} \end{bmatrix},$$

where each entry is replaced by the corresrponding column of $l = \text{ord}_n(q)$ elements from $\mathbb{F}_q$, then removing any linearly dependent rows. The rows of the resulting matrix over $\mathbb{F}_q$ are the parity checks satisfied by C.

4.5 Algebraic Geometry Codes

In this part, we recall necessary concepts and results on algebraic geometry codes that will be utilized in our constructions. More detailed results concerning such codes can be found in [120, 152]. We follow the notation of [152].

The class of algebraic geometry (AG) codes was introduced by Goppa in his seminal work [52]. These codes have nice properties; among them is the fact that such codes are asymptotically good. There exist several works dealing with investigations concerning AG codes (see for instance [71, 106, 119, 125]). We next present this well-known class of codes. For this subsection, we assume that the reader is familiar with the concepts of field extension and algebraic extension of fields. For details, we refer to [125].

Definition 4.5.1 Let F and K be fields. An algebraic function field (function field for short) F/K of one variable over K is a field extension $K \subseteq F$ such that F is a finite algebraic extension of $K(x)$ for some element $x \in F$ which is transcendental over K.

Definition 4.5.2 A valuation ring of the algebraic function field F/K is a ring $\mathcal{O} \subseteq F$ satisfying the following conditions:

(i) $K \subsetneq \mathcal{O} \subsetneq F$;
(ii) $\forall z \in F, z \in \mathcal{O}$ or $z^{-1} \in \mathcal{O}$.

It is well-known that a valuation ring is a local ring, i.e., it has a unique maximal ideal (see [152, Proposition 1.1.5.]). In our context we always consider that $K = \mathbb{F}_q$ is the finite field with q elements.

Definition 4.5.3 Let F/K be an algebraic function field. A place P of F/K is the maximal ideal of some valuation ring $\mathcal{O}_P$ of F/K.

We write $\mathbb{P}_F$ to denote the set of places, i.e., $\mathbb{P}_F := \{P \mid P \text{ is a place of } F/K\}$. As usual, we denote by $\mathbb{Z}$ the ring of integers.

Definition 4.5.4 Let F/K be a function field. A discrete valuation of F/K is a function $v : F \longrightarrow \mathbb{Z} \cup \{\infty\}$ satisfying the following conditions:

(1) $v(x) = \infty \Longleftrightarrow x = 0$;
(2) $\forall x, y \in F, v(xy) = v(x) + v(y)$;
(3) $\forall x, y \in F, v(x + y) \geq \min\{v(x), v(y)\}$;
(4) there exists $z \in F$ such that $v(z) = 1$;
(5) $\forall a \in K, a \neq 0, v(a) = 0$.

In the following, we recall the concept of *divisor*.

Definition 4.5.5 A divisor of F/K is a formal sum of places given by $D := \sum_{P \in \mathbb{P}_F} n_P P$, where n_P is an integer number and almost all $n_P = 0$.

We denote by $\mathcal{D}_F$ the free group of divisors of F/K. The *support* and the *degree* of a divisor D are defined, respectively, by $\operatorname{supp} D := \{P \in \mathbb{P}_F | n_P \neq 0\}$ and $\deg(D) := \sum_{P \in \mathbb{P}_F} n_P \deg(P)$, where $\deg(P)$ is the degree of the place P.

For each $x \in F/K$, the principal divisor (x) of x is defined by

$$(x) := \sum_{P \in \mathbb{P}_F} v_P(x)P,$$

where v_P is the discrete valuation corresponding to the place P.

For every $x \in \mathcal{O}_P$, we define $x(P) \in \mathcal{O}_P/P$ as the residue class of x modulo P; if $x \in F - \mathcal{O}_P$, we put $x(P) := \infty$.

Definition 4.5.6 Let A be a divisor of F/K. The Riemann–Roch space associated to A is defined as

$$\mathcal{L}(A) := \{x \in F | (x) \geq -A\} \cup \{0\}.$$

The integer $l(A) := \dim \mathcal{L}(A)$ is called the dimension of the divisor A.

We here define the *genus* of a function field, which is the most important invariant of a function field.

Definition 4.5.7 Let F/K be a function field. The genus g of F/K is defined as

$$g := \max\{\deg(A) - l(A) + 1 \mid A \in \mathcal{D}_F\}.$$

Definition 4.5.8 An adele of F/K is a mapping $\alpha : \mathbb{P}_F \longrightarrow F$, defined by $\alpha(P) = \alpha_P$, such that $\alpha_P \in \mathcal{O}_P$ for almost all $P \in \mathbb{P}_F$.

An adele can be considered as an element of the direct product $\prod_{P \in \mathbb{P}_F} F$; hence, we utilize the notation $\alpha = (\alpha_P)_{P \in \mathbb{P}_F}$ or $\alpha = (\alpha_P)$.

The set $\mathcal{A}_F := \{\alpha | \alpha \text{ is an adele of } F/K\}$ is said to be the *adele space* of F/K.

Definition 4.5.9 Let $A \in \mathbb{P}_F$. Then we define $\mathcal{A}_F(A) := \{\alpha \in \mathcal{A}_F | v_P(\alpha) \geq -v_P (A) \; \forall \; P \in \mathbb{P}_F\}$, where $v_P(A) = n_P$ and $v_P(\alpha) := v_P(\alpha_P)$.

It is easy to see that $\mathcal{A}_F(A)$ is a K-subspace of $\mathcal{A}_F$. We next present the concept of *Weil differential*.

Definition 4.5.10 Let F/K be a function field. A Weil differential of F/K is a K-linear map $\omega : \mathcal{A}_F \longrightarrow K$ which vanishes on $\mathcal{A}_F(A) + F$ for some divisor $A \in \mathcal{D}_F$.

Let Ω_F be the differential space of F/K, i.e.,

$$\Omega_F := \{w \mid w \text{ is a Weil differential of } F/K\}.$$

A divisor W is called *canonical* if $W = (w)$ for some $w \in \Omega_F$. For every nonzero differential w, its canonical divisor is denoted by $(w) := \sum_{P \in \mathbb{P}_F} v_P(w)P$, where $v_P(w) := v_P((w))$. All canonical divisors are equivalent and have degree $2g - 2$. Given a divisor G, we define

$$\Omega_F(G) := \{w \in \Omega_F \mid w = 0 \text{ or} (w) \geq G\}.$$

The dimension of $\Omega_F(G)$ as $\mathbb{F}_q$-vector space is denoted by $i(G)$.

A fundamental result concerning AG codes is the well-known Riemann–Roch theorem:

Theorem 4.5.1 (Riemann–Roch) (Thm. 1.5.15 of [152]) *Let W be a canonical divisor of $F/\mathbb{F}_q$. Then for each divisor $A \in \mathcal{D}_F$, the dimension of $\mathcal{L}(A)$ is given by $l(A) = \deg(A) + 1 - g + l(W - A)$.*

The definition of an algebraic geometry (AG) code is given in the sequence.

Definition 4.5.11 Let $P_1, \ldots, P_n$ be pairwise distinct places of $F/\mathbb{F}_q$ of degree 1, and let $D = P_1 + \ldots + P_n$ be a divisor. Let G be a divisor of $F/\mathbb{F}_q$ such that $\text{supp}\, G \cap \text{supp}\, D = \emptyset$. The algebraic geometry code $C_{\mathcal{L}}(D, G)$ associated with D and G is defined by

$$C_{\mathcal{L}}(D, G) := \{(x(P_1), \ldots, x(P_n)) \mid x \in \mathcal{L}(G)\} \subseteq \mathbb{F}_q^n.$$

If $G = mQ$, where Q is a rational place (place of degree 1) not belonging in the support of D, then the code $C_{\mathcal{L}}(D, G)$ is said to be a one-point AG code.

The following result establishes the parameters of an AG code.

Proposition 4.5.1 ([152, Thm. 2.2.2./Cor.2.2.3]) *Let $F/\mathbb{F}_q$ be an algebraic function field of genus g. Then $C_{\mathcal{L}}(D, G)$ is an $[n, k, d]_q$ code with $k = l(G) - l(G - D)$ and $d \geq n - \deg(G)$. In addition, if $2g - 2 < \deg(G) < n$, then one has $k = \deg(G) + 1 - g$. Moreover, if $\{x_1, \ldots, x_k\}$ is a basis of $\mathcal{L}(G)$, then the matrix*

$$H_{\delta, b} = \begin{bmatrix} x_1(P_1) & x_1(P_2) & \ldots & x_1(P_n) \\ \vdots & \vdots & & \vdots \\ x_k(P_1) & x_k(P_2) & \ldots & x_k(P_n) \end{bmatrix}$$

is a generator matrix for $C_{\mathcal{L}}(D, G)$.

The Weierstrass semigroup of a divisor Q plays an important role in our constructions.

Definition 4.5.12 Assume that Q is a divisor of degree 1 of $F/\mathbb{F}_q$. Let us consider $\mathcal{L}(\infty Q) = \bigcup_{r \geq 0} \mathcal{L}(rQ)$ be the space of rational functions having poles only at Q. Then the Weierstrass semigroup of Q is defined as

$$(Q) := \{-v_Q(f)|f \in \mathcal{L}(\infty Q)\} = \{0 = \rho_1 < \rho_2 < \cdots\},$$

where v_Q denotes the valuation at Q.

Another interesting type of AG code is the code $C_\Omega(D, G)$, defined in the sequence.

Definition 4.5.13 Let $F/\mathbb{F}_q$ be a function field of genus g. Let G and D be divisors as in Definition 4.5.11. The code $C_\Omega(D, G) \subseteq \mathbb{F}_q^n$ is defined by

$$C_\Omega(D, G) := \{(resp_{P_1}(w), \ldots, resp_{P_n}(w))|w \in \Omega_F(G - D)\},$$

where $resp_{P_i}(w)$ is the residue of w at P_i.

Theorem 4.5.2 computes the parameters of the code $C_\Omega(D, G)$.

Theorem 4.5.2 ([152, Thm. 2.2.7.]) *Assume the notation of Definition 4.5.11. The code $C_\Omega(D, G)$ is an $[n, k^*, d^*]_q$ code, where $k^* = i(G - D) - i(G)$ and $d^* \geq \deg(G) - (2g - 2)$. Additionally, if $2g - 2 < \deg(G) < n$ one has $k^* = n + g - 1 - \deg(G)$.*

$C_\mathcal{L}(D, G)$ and $C_\Omega(D, G)$ have an important correlation.

Theorem 4.5.3 ([152, Thm. 2.2.8.]) *The codes $C_\mathcal{L}(D, G)$ and $C_\Omega(D, G)$ are (Euclidean) dual of each other, i.e., $C_\Omega(D, G) = C_\mathcal{L}(D, G)^\perp$.*

The following result characterizes when an AG code is self-orthogonal.

Proposition 4.5.2 ([152, Cor. 2.2.11.]) *Suppose there exists a Weil differential η such that $2G - D \leq (\eta)$ and $\eta_{P_i}(1) = 1$ for $i = 1, 2, \ldots, n$. Then $C_\mathcal{L}(D, G)$ is Euclidean self-orthogonal, that is, $C_\mathcal{L}(D, G) \subseteq C_\mathcal{L}^\perp(D, G)$.*

Remark 4.5.1 Note that the Hermitian self-orthogonal condition to AG code can be easily obtained from the Euclidean condition due to the fact that $C_\mathcal{L}(D, G) \subseteq C_\mathcal{L}^{\perp_H}(D, G)$ if and only if $C_\mathcal{L}^q(D, G) \subseteq C_\mathcal{L}^\perp(D, G)$.

Chapter 5
Quantum Code Constructions

Quantum codes are fundamental to the error protection in quantum computers. Several families of good or optimal quantum codes were constructed by several researches over time [4, 17, 25, 27, 28, 30, 39, 45, 46, 54, 55, 62, 72–74, 77–81, 89, 91, 105, 106, 111, 138, 147, 148, 162, 165]. The Calderbank–Shor–Steane (CSS) construction is a remarkable technique to construct quantum codes derived from classical ones. As was said previously, such technique has been applied by a great number of quantum coding researchers in order to derive efficient quantum codes. With the possible advent of efficient quantum computers, it is extremely important to investigate how to obtain families of efficient quantum codes. Based on these facts, we present here some constructions of quantum codes derived from several classes of classical codes by means of the CSS construction. The classical codes utilized here are the well-known Bose–Chaudhuri–Hocquenghem (BCH) (Sects. 5.1, 5.2, 5.3), algebraic geometry codes (Sect. 5.4) and quantum synchronizable codes (Sect. 5.5).

We invite the reader to start our journey through the quantum code constructions. The first families of quantum codes exhibited in Sect. 5.1 are obtained by applying the CSS construction to suitable families of (classical) Bose–Chaudhuri–Hocquenghem codes constructed carefully in order to attain codes which have, at the same time, large dimension and minimum distance.

5.1 BCH Codes—Part I

In this subsection, we present five quantum code constructions generating several families of nonbinary quantum BCH (see [22, 23, 63] for the first papers which originated such class of cyclic codes) with good parameters. The first two ones are based on the CSS construction derived from two nonprimitive BCH codes. The third construction is based on Steane's enlargement of nonbinary CSS codes applied to suitable sub-families of nonprimitive non-narrow-sense BCH codes. The fourth construction is obtained from suitable sub-families of Hermitian dual-containing nonprimitive

© Springer Nature Switzerland AG 2020
G. G. La Guardia, *Quantum Error Correction*, Quantum Science and Technology,
https://doi.org/10.1007/978-3-030-48551-1_5

non-narrow-sense BCH codes constructed here. The fifth and last construction is based on finding cyclic codes whose defining set consists of only one coset with at least two consecutive integers. The content presented in Sect. 5.1.1–5.1.4 can be found in our paper [100] and the material of Sect. 5.1.5 is based on our paper [103].

The key of most of these constructions is the investigation of suitable properties of cyclotomic cosets. More precisely, we need to know the cardinality of each of them and also (when it is possible) to compute a large quantity of consecutive elements in the union of all cosets (the defining set) of a given cyclic code C in order to obtain the exact dimension of the code as well as a lower bound for the minimum distance of it. In other words, knowing the cardinality of the defining set Z of a given BCH code C and by computing the maximum sequence of consecutive integers belonging to Z we know the dimension and the maximum lower bound for the minimum distance of C, respectively.

Since until now the true minimum distance of BCH code is not known and since the computation of its dimension in all cases is also not known, this is an interesting area of research. In other words, there is much challenge to be surpassed.

As we will see in the sequence, in a particular case, we apply the concept of linear congruence, to prove (for codes of prime length) the existence of at least one q-ary coset containing two consecutive integers. This is interesting because by means of this result we can construct families of nonbinary quantum codes with good parameters. To be more precise, our families of quantum BCH codes have parameters described in the following sequence:

- $[[n, n - 4(c - 2) - 2, d \geq c]]_q$, where $q \geq 4$ is a prime power, n is an integer with $\gcd(q, n) = 1$, $(q - 1) \mid n$, $m = \mathrm{ord}_n(q) = 2$ and $2 \leq c \leq r$, where r is such that $n = r(q - 1)$;
- $[[n, n - 2mr, d \geq r + 2]]_q$, where $m = \mathrm{ord}_n(q) \geq 2$, n is a prime number and r is the number of cosets satisfying suitable conditions (see Theorem 5.1.4);
- $[[n, n - m(2r - 1), d \geq r + 2]]_q$, where $m = \mathrm{ord}_n(q) \geq 2$, n is a prime number and $q \geq 3$;
- $[[n, n - 4c, d \geq c + 2]]_q$, where $n > q$ is an integer with $\gcd(q, n) = 1$, $(q - 1) \mid n$, $m = \mathrm{ord}_n(q) = 2$, $1 \leq c \leq r - 3$ and $r > 3$ satisfies $n = r(q - 1)$;
- $[[n, n - 4c - 2, d \geq c + 2]]_q$, where $2 \leq c \leq r - 2$, $q > 3$, $n = r(q^2 - 1)$, $r > 1$ and $m = \mathrm{ord}_n(q^2) = 2$;
- $[[n, n - 2mr, d \geq r + 2]]_q$, where $q \geq 3$ is a prime power, $n > q^2$ is a prime number such that $\gcd(q, n) = 1$, $m = \mathrm{ord}_n(q^2) \geq 2$ and r is the number of cosets satisfying suitable conditions (see Theorem 5.1.9).
- $[[n, n - 2m^*, d \geq r + 2]]_q$, where $q \geq 3$ is a prime power and $n > m$ ($m = \mathrm{ord}_n(q) \geq r + 2$) is a positive integer such that $\gcd(q, n) = 1$, $\gcd(q^{a_i} - 1, n) = 1$ for every $i = 1, 2, \ldots, r$, where $1 \leq r, a_1, a_2, \ldots, a_r < m$ are integers, and $n \mid \gcd(t_2, \ldots, t_r)$, where $t_j = [(j - (j - 1)q^{a_j})(q^{a_j} - 1)^{-1} - (q^{a_1} - 1)^{-1}]$ for every $j = 2, \ldots, r$ (the operations are performed modulo n).

Before proceeding further, we communicate to the reader that we will utilize freely the expressions such as q-ary coset, q-ary coset, q-coset, or even coset when the context is clear, meaning, of course, a cyclotomic coset. Another important remark is that we always assume that the code length and the cardinality of the alphabet are relatively prime, i.e., $\gcd(q, n) = 1$, because this condition ensures that C has simple roots. Additionally, throughout this book, we utilize the notation $\mathbb{C}_{[a]}$ to denote the cyclotomic coset containing a, where a is not necessarily the smallest number in $\mathbb{C}_{[a]}$.

5.1.1 Construction I

In this subsection we construct new families of nonbinary CSS codes derived from two distinct classical BCH codes, not necessarily dual-containing. To proceed further, let us recall the so-called CSS construction given in Lemma 3.6.1.

Let q be a prime power. Let C_1 and C_2 denote two classical linear codes both over the field $\mathbb{F}_q$ with parameters $[n, k_1, d_1]_q$ and $[n, k_2, d_2]_q$, respectively, such that $C_2 \subset C_1$. Then there exists an $[[n, K = k_1 - k_2, D]]_q$ CSS quantum code where $D = \min\{wt(c) : c \in (C_1 \backslash C_2) \cup (C_2^\perp \backslash C_1^\perp)\}$.

We start by showing Lemma 5.1.1.

Lemma 5.1.1 *Let $q \geq 3$ be a prime power and let $n > q$ be an integer such that $\gcd(q, n) = 1$. Assume also that $(q - 1) \mid n$ and $m = \operatorname{ord}_n(q) \geq 2$ hold. Then each of the q-ary cosets $\mathbb{C}_{[lr]}$ has only one element, where r is given by $n = r(q - 1)$, and $1 \leq l \leq q - 2$ is an integer.*

Proof Since $rq = n + r$ holds, one has

$$(lr)q = l(n + r) \equiv lr \mod n;$$

hence

$$(lr)q^t \equiv lr \mod n$$

for each $1 \leq t \leq m - 1$, proving the lemma. $\qquad\qquad\qquad\square$

Lemma 5.1.1 is applied in the proof of Theorem 5.1.1.

Theorem 5.1.1 *Assume that $q > 3$ is a prime power and $n > q$ is an integer relatively prime with q. Assume also that $(q - 1) \mid n$ and $m = \operatorname{ord}_n(q) = 2$ are true. Then there exists a quantum code with parameters $[[n, n - 4(r - 2) - 2, d \geq r]]_q$, where r is such that $n = r(q - 1)$.*

Proof Since $n \mid (q^2 - 1)$ and because we consider only nonprimitive BCH codes, it follows that $r \leq q$. As $\gcd(q, n) = 1$, one has $r < q$, so the inequalities $(r - 2)q < n$ and $r + (r - 2)q < n$ hold. We next show that all the q-ary cosets (modulo n of course) given by $\mathbb{C}_{[0]} = \{0\}, \mathbb{C}_{[1]} = \{1, \quad q\}, \mathbb{C}_{[2]} = \{2, \quad 2q\}, \mathbb{C}_{[3]} = \{3, \quad 3q\}, \ldots, \mathbb{C}_{[r-2]} = \{r - 2, \quad (r - 2)q\}, \mathbb{C}_{[r]} = \{r\}, \mathbb{C}_{[r+1]} = \{r + 1, \quad r + q\},$

$\mathbb{C}_{[r+2]} = \{r + 2, \ r + 2q\}, \ldots, \mathbb{C}_{[2r-2]} = \{2r - 2, \ r + (r - 2)q\}$, are mutually disjoint and, with exception of the cosets $\mathbb{C}_{[0]} = \{0\}$ and $\mathbb{C}_{[r]} = \{r\}$, each of them has two elements.

The cosets $\mathbb{C}_{[0]}$ and $\mathbb{C}_{[r]}$ have only one element. Let us show that each of the other cosets has two elements. Since $(r - 2)q < n$, then the congruence $l \equiv lq \mod n$ implies that $l = lq$, where $1 \leq l \leq r - 2$, which is a contradiction. If $r + s \equiv (r + s)q \mod n$, where $1 \leq s \leq r - 2$, then $r + s = r + sq$, which is a contradiction.

From now on, we show that all these cosets given above and $\mathbb{C}_{[0]}$ and $\mathbb{C}_{[r]}$ are mutually disjoint. We only consider the case $\mathbb{C}_{[r+l]} = \mathbb{C}_{[r-s]}$, where $1 \leq l, s \leq r - 2$, since the other cases are similar to this one. Seeking a contradiction, we assume that $\mathbb{C}_{[r+l]} = \mathbb{C}_{[r-s]}$, where $1 \leq l, s \leq r - 2$. If the congruence $(r + l) \equiv (r - s) \mod n$ holds, we obtain

$$(r + l) \equiv (r - s) \mod n \Longrightarrow n \mid (l + s).$$

If $l + s \neq 0$, one has $n \leq l + s$, which is a contradiction. If $l + s = 0$, this implies $l = -s$, which is a contradiction.

On the other hand, if $(r + l)q \equiv r - s \mod n$ holds, we have

$$(r + l)q \equiv r - s \Longrightarrow lq \equiv -s \mod n$$
$$\Longrightarrow n \mid (lq + s).$$

Since $l, s \leq r - 2$ and $r < q$ are true, if $lq + s \neq 0$ holds, it follows that $lq + s < n$, which is a contradiction. If $lq + s = 0$ then $lq = -s$, which is a contradiction. Thus all the q-ary cosets $\mathbb{C}_{[0]}, \mathbb{C}_{[1]}, \ldots, \mathbb{C}_{[r-2]}$, are disjoint from each of the q-ary cosets $\mathbb{C}_{[r]}, \mathbb{C}_{[r+1]}, \ldots, \mathbb{C}_{[2r-2]}$. Additionally, all the q-ary cosets $\mathbb{C}_{[0]}, \mathbb{C}_{[1]}, \ldots, \mathbb{C}_{[r-2]}$, are mutually disjoint and all the q-ary cosets $\mathbb{C}_{[r]}, \mathbb{C}_{[r+1]}, \ldots, \mathbb{C}_{[2r-2]}$ are also mutually disjoint.

Let C_1 be the cyclic code generated by the product of the minimal polynomials

$$M^{(0)}(x)M^{(1)}(x) \cdot \ldots \cdot M^{(r-2)}(x),$$

and C_2 be the cyclic code generated by $g_2(x)$, that is the product of the minimal polynomials

$$g_2(x) = \prod_i M^{(i)}(x),$$

where $i \notin \{r, r + 1, \ldots, 2r - 2\}$ and i runs through the coset representatives mod n. From construction it follows that $C_2 \subsetneq C_1$. From the BCH bound, the minimum distance of C_1 is greater than or equal to r, because its defining set contains the sequence $0, 1, \ldots, r - 2$, of $r - 1$ consecutive integers. Similarly, the defining set of the code C generated by the polynomial $h(x) = \frac{x^n - 1}{g_2(x)}$ contains the sequence $r, r + 1, \ldots, 2r - 2$, of $r - 1$ consecutive integers and so, from the BCH bound, C

also has minimum distance greater than or equal to r. Since the code $C_2^{\perp}$ is equivalent to C, $C_2^{\perp}$ also has minimum distance greater than or equal to r. Therefore, the resulting CSS code has minimum distance greater than or equal to r.

Next we compute the dimension of the corresponding CSS code. We know that the degree of the generator polynomial of a cyclic code is equal to the cardinality of its defining set. Furthermore, the defining set Z_1 of C_1 has $r - 1$ disjoint cyclotomic cosets. Moreover, all of them (except coset $\mathbb{C}_0$) have two elements; hence Z_1 has $2(r - 2) + 1$ elements. Therefore, C_1 has dimension $k_1 = n - 2(r - 2) - 1$. Similarly, C_2 has dimension $k_2 = 2(r - 2) + 1$. Thus the dimension of the corresponding CSS code is $n - 4(r - 2) - 2$. Applying the CSS construction to the codes C_1 and C_2, one can get a quantum code with parameters $[[n, n - 4(r - 2) - 2, d \geq r]]_q$. The proof is complete.   □

We illustrate Theorem 5.1.1 by means of a graphical scheme:

$$
\overbrace{\underbrace{\mathbb{C}_{[0]}\mathbb{C}_{[1]}}_{} \mathbb{C}_{[2]} \; \cdots \; \mathbb{C}_{[r-2]}}^{C_1}
$$

$$
\overbrace{\mathbb{C}_{[r]} \; \mathbb{C}_{[r+1]} \cdots \; \mathbb{C}_{[2r-2]}}^{C} \underbrace{\mathbb{C}_{[a_1]} \cdots \mathbb{C}_{[a_n]}}_{C_2} \cdot
$$

Observe that the union of the q-cosets $\mathbb{C}_{[0]}, \mathbb{C}_{[1]}, \ldots, \mathbb{C}_{[r-2]}$ is the defining set of code C_1; the union of the cosets $\mathbb{C}_{[0]}, \mathbb{C}_{[1]}, \ldots, \mathbb{C}_{[r-2]}, \mathbb{C}_{[a_1]}, \ldots, \mathbb{C}_{[a_n]}$ is the defining set of C_2, where $\mathbb{C}_{[a_1]}, \ldots, \mathbb{C}_{[a_n]}$ are the remaining cosets in order to complete the set of all q-cosets. Finally, the union of the cosets $\mathbb{C}_{[r]}, \mathbb{C}_{[r+1]}, \ldots, \mathbb{C}_{[2r-2]}$ is the defining set of C.

As an immediate result we have

Corollary 5.1.1 *Assume that all the hypotheses of Theorem 5.1.1 are valid. Then there exists a quantum code with parameters $[[n, n - 4(c - 2) - 2, d \geq c]]_q$, where $2 \leq c < r$.*

Proof Let C_1 be the cyclic code generated by the product of the minimal polynomials

$$
M^{(0)}(x)M^{(1)}(x) \cdot \ldots \cdot M^{(c-3)}(x)M^{(c-2)}(x),
$$

and C_2 be the cyclic code generated by the product of the minimal polynomials

$$
\prod_i M^{(i)}(x),
$$

where $i \notin \{r, r + 1, \ldots, r + c - 2\}$ and i runs through the coset representatives mod n. Proceeding similarly as in the proof of Theorem 5.1.1, the result follows.   □

Example 5.1.1 As an example, let us consider that $q = 9$ and $n = 40$; then $\gcd(9, 40) = 1, 8 \mid 40$ and $ord_{40}(9) = 2$. In this case we have $r = 5$. Theorem 5.1.1 asserts the existence of a quantum code with parameters $[[40, 26, d \geq 5]]_9$. Consider next $q = 11$ and $n = 30$. Let C_1 be the cyclic code generated by the product of the minimal polynomials $M^{(0)}(x)M^{(1)}(x)\ldots M^{(6)}(x)$ and C_2 be the cyclic code generated by the product of the minimal polynomials $\prod_i M^{(i)}(x)$, where $i \notin \{7, 10, 15, 16, 18, 19, 21\}$ and i runs through the coset representatives mod 30. Proceeding similarly as in the proof of Theorem 5.1.1, an $[[30, 8, d \geq 8]]_{11}$ quantum code can be constructed.

5.1.2 Construction II

Here the attention is focused on cyclic codes of prime length. Among the contributions exhibited in this subsection, we prove there exists at least one q-ary coset containing two consecutive integers (see Lemma 5.1.2). In order to proceed further, let us recall a well-known result from number theory.

Theorem 5.1.2 *A linear congruence $ax \equiv b \ (mod \ m)$, where $a \neq 0$, admits an integer solution if and only if $d = \gcd(a, m)$ divides b.*

Applying Theorem 5.1.2 we prove Lemma 5.1.2.

Lemma 5.1.2 *Assume that $q \geq 3$ is a prime power, $n > q$ is a prime number and consider that $m = \operatorname{ord}_n(q) \geq 2$. Then there exists at least one q-ary coset containing two consecutive integers.*

Proof Note first that $\gcd(q, n) = 1$. In order to prove this lemma, it suffices to show that the congruence $xq \equiv x + 1 (\ mod\ n)$ has at least one solution for some $0 \leq x \leq n - 1$ or, equivalently, the congruence $(q - 1)x \equiv 1 \ (mod\ n)$ has at least one solution. We know that $\gcd(q - 1, n) = 1$, because $n > q$ and n is prime. Since $q - 1 \neq 0$, it follows from Theorem 5.1.2 that $(q - 1)x \equiv 1 \ (mod\ n)$ has an integer solution x_0. Applying the division algorithm for x_0 and n we have $x_0 = ns_0 + r_0$, where r_0 and s_0 are integers and $0 \leq r_0 \leq n - 1$. Since $(q - 1)x_0 \equiv 1 \ (mod\ n)$ holds then the congruence $(q - 1)r_0 \equiv 1 \ (mod\ n)$ also holds. Therefore, the result follows. $\square$

Remark 5.1.1 Note that in Lemma 5.1.2 it is not necessary to assume that n is a prime number. In fact, we need only to suppose that $\gcd(q - 1, n) = 1$ and $\gcd(q, n) = 1$ hold. However, since the corresponding q-ary cosets of BCH codes of prime length have nice properties, we have assumed that n is prime. Nevertheless, if one assumes that $\gcd(q - 1, n) = 1$ and $\gcd(q, n) = 1$ hold, more good quantum codes can be constructed.

Theorem 5.1.3 *Let $q \geq 3$ be a prime power, $n > q$ be a prime number and consider $m = \mathrm{ord}_n(q) \geq 2$. Suppose also that the q-cosets $\mathbb{C}_{[s]}$ and $\mathbb{C}_{[-s]}$ are disjoint, where $\mathbb{C}_{[s]}$ is a q-coset containing two consecutive integers. Then there exists an $[[n, n - 2m, d \geq 3]]_q$ quantum code.*

Proof Note that $\gcd(q, n) = 1$. Let C_1 be the code generated by $M^{(s)}(x)$ and C_2 generated by $\prod_i M^{(i)}(x)$, where $i \neq -s$ and i runs through the coset representatives mod n. It is easy to see that the cosets $\mathbb{C}_{[s]}$ and $\mathbb{C}_{[-s]}$ contain m elements. Proceeding similarly as in the proof of Theorem 5.1.1, the result follows. $\qquad\square$

Theorem 5.1.4 *Let $q \geq 3$ be a prime power, $n > q$ be a prime and consider that $m = \mathrm{ord}_n(q) \geq 2$. Let $\mathbb{C}_{[s]}$ be the q-coset containing s and $s + 1$. Suppose also that all the q-ary cosets $\mathbb{C}_{[s]}, \mathbb{C}_{[s+2]}, \ldots, \mathbb{C}_{[s+r]}, \mathbb{C}_{[-s]}, \mathbb{C}_{[-s-2]}, \ldots, \mathbb{C}_{[-s-r]}$, are mutually disjoint. Then there exists a quantum code with parameters $[[n, n - 2mr, d \geq r + 2]]_q$.*

Proof We know that $\gcd(q, n) = 1$ and the coset $\mathbb{C}_{[-s]}$ also contains two consecutive integers, namely, $-s - 1$ and $-s$. Let C_1 be the cyclic code generated by

$$M^{(s)}(x) M^{(s+2)}(x) \cdot \ldots \cdot M^{(s+r)}(x),$$

and let C_2 be the cyclic code generated by the polynomial $g_2(x)$, that is the product of the minimal polynomials

$$g_2(x) = \prod_j M^{(j)}(x),$$

where $j \notin \{-s - r, \ldots, -s - 2, -s\}$ and j runs through the coset representatives mod n.

From the BCH bound, the minimum distance of C_1 is greater than or equal to $r + 2$ because its defining set contains the sequence of $r + 1$ consecutive integers given by $s, s + 1, s + 2, \ldots, s + r$. Similarly, the defining set of the code C generated by the polynomial $h_2(x) = (x^n - 1)/g_2(x)$, contains a sequence of $r + 1$ consecutive integers given by $-s - r, \ldots, -s - 2, -s - 1, -s$. Again, from the BCH bound, C has minimum distance greater than or equal to $r + 2$. Since C is equivalent to $C_2^\perp$, it follows that $C_2^\perp$ also has minimum distance greater than or equal to $r + 2$. Therefore, the resulting CSS code have minimum distance greater than or equal to $r + 2$. If $s \in [1, n - 1]$ satisfies $\gcd(s, n) = 1$ then the coset $\mathbb{C}_s$ has cardinality m. In fact, if $|\mathbb{C}_s| = c < m$ it follows that $n | s(q^c - 1)$, so $n | (q^c - 1)$, a contradiction. Thus, since n is prime, each of the cosets $\mathbb{C}_s$, where $s \in [1, n - 1]$, has cardinality m. Additionally, from the hypotheses, all the q-ary cosets $\mathbb{C}_{[s]}, \mathbb{C}_{[s+2]}, \ldots, \mathbb{C}_{[s+r]}$, are mutually disjoint. Thus C_1 has dimension $k_1 = n - mr$ and C_2 has dimension $k_2 = mr$, since there exist r disjoint q-cosets not contained in the defining set of C_2, where each of them has cardinality m. Therefore, the corresponding CSS code has dimension $K = n - 2mr$. Since the cosets $\mathbb{C}_{[s]}, \mathbb{C}_{[s+2]}, \ldots, \mathbb{C}_{[s+r]}, \mathbb{C}_{[-s]}, \mathbb{C}_{[-s-2]}, \ldots, \mathbb{C}_{[-s-r]},$

are mutually disjoint, it follows that $C_2 \subsetneq C_1$. Applying the CSS construction to C_1 and C_2, one obtains an $[[n, n - 2mr, d \geq r + 2]]_q$ quantum code, and we are done.   □

Example 5.1.2 Theorem 5.1.4 has variants as follows: to construct an $[19, 13, d \geq 3]]_7$ quantum code, let us consider that $q = 7$, $n = 19$ and $m = 3$. The cosets are given by $\mathbb{C}_2 = \{2, 14, 3\}$ and $\mathbb{C}_{16} = \{5, 16, 17\}$. Let C_1 be generated by $M^{(2)}(x)$ and C_2 generated by $g_2(x) = \prod_i M^{(i)}(x)$, where $i \notin \{16\}$ and i runs through the coset representatives mod 19. Then an $[[19, 13, d \geq 3]]_7$ quantum code can be constructed. Proceeding similarly, one can get quantum codes with parameters $[[31, 25, d \geq 3]]_5$, $[[71, 61, d \geq 3]]_5$, $[[11, 1, d \geq 4]]_3$, $[[31, 19, d \geq 4]]_5$, $[[31, 13, d \geq 5]]_5$, $[[71, 51, d \geq 4]]_5$, $[[71, 41, d \geq 6]]_5$.

5.1.3 Construction III

In this subsection, we construct families of quantum BCH codes of prime length by applying Steane's enlargement of nonbinary CSS construction [62, Corollary 4]. These new families have parameters better than the parameters of the quantum BCH codes available in the literature. Let us recall the Steane enlargement code construction applied to nonbinary alphabets.

Corollary 5.1.2 ([62, Corollary 4]) *Assume that we have an $[N_0, K_0]$ linear code L which contains its Euclidean dual, $L^{\perp} \leq L$, and which can be enlarged to an $[N_0, K_0']$ linear code L', where $K_0' \geq K_0 + 2$. Then there exists a quantum code with parameters $[[N_0, K_0 + K_0' - N_0, d \geq min\{d, \lceil \frac{q+1}{q}d' \rceil\}]]$, where $d = w(L\backslash L'^{\perp})$ and $d' = w(L'\backslash L'^{\perp})$.*

Euclidean dual-containing cyclic codes can be derived from Lemma 5.1.3.

Lemma 5.1.3 *[4, Lemma 1] Assume that $\gcd(q, n) = 1$. A cyclic code of length n over $\mathbb{F}_q$ with defining set Z contains its Euclidean dual code if and only if $Z \cap Z^{-1} = \emptyset$, where $Z^{-1} = \{-z \mod n \mid z \in Z\}$.*

In Lemma 5.1.2 of Sect. 5.1.2 we have shown the existence of, at least, one q-ary cyclotomic coset containing two consecutive integers provided the code length is prime. In what follows, we show how to construct good quantum codes of prime length by applying Steane's code construction. We begin by presenting an illustrative example.

Example 5.1.3 Assume that $n = 31$ and $q = 5$. From Lemma 5.1.2, there exists a coset containing at least two consecutive integers; here it is the coset $\mathbb{C}_8 = \{8, 9, 14\}$. Let C be the cyclic code generated by the product of the minimal polynomials $C = \langle g(x) \rangle = \langle M^{(4)}(x)M^{(8)}(x) \rangle$. C has defining set $Z = \mathbb{C}_4 \cup \mathbb{C}_8 = \{4, 7, 8, 9, 14, 20\}$

and has parameters $[31, 25, d \geq 4]_5$. From Lemma 5.1.3, it is easy to check that C is Euclidean dual-containing. Furthermore, C can be enlarged to a code C' with parameters $[31, 28, d \geq 3]_5$, whose generator polynomial is $M^{(8)}(x)$. Applying Corollary 5.1.2 to C and C', we obtain an $[[31, 22, d \geq 4]]_5$ quantum code.

Theorem 5.1.5 *Let $q \geq 3$ be a prime power, $n > q$ be a prime and consider that $m = \mathrm{ord}_n(q) \geq 2$. Let $\mathbb{C}_{[s]}$ be the q-ary coset containing s and $s + 1$ and let $Z = \mathbb{C}_{[s]} \cup \mathbb{C}_{[s+2]}$, where $\mathbb{C}_{[s]} \neq \mathbb{C}_{[s+2]}$. Assume also that $Z \cap Z^{-1} = \emptyset$ holds. Then there exists an $[[n, n - 3m, d \geq 4]]_q$ code.*

Proof We know that $\gcd(q, n) = 1$. Let C be the cyclic code generated by $\langle M^{(s)}(x) M^{(s+2)}(x) \rangle$. By hypothesis and from Lemma 5.1.3, we know that C is Euclidean dual-containing; C has parameters $[n, n - 2m, d \geq 4]_q$. Let C' be the cyclic code generated by $M^{(s)}(x)$. We know that C' is an enlargement of C and it has parameters $[n, n - m, d \geq 3]_q$. Since $m \geq 2$, it follows that $k' - k = m \geq 2$, where k' is the dimension of C' and k is the dimension of C. Applying the Steane code construction to C and C', since $\frac{q+1}{q} > 1$, we get an $[[n, n - 3m, d \geq 4]]_q$ quantum code. $\qquad\square$

Theorem 5.1.5 can be generalized in the following way.

Theorem 5.1.6 *Assume that $q \geq 3$ is a prime power, $n > q$ is a prime number and consider that $m = \mathrm{ord}_n(q) \geq 2$. Let $\mathbb{C}_{[s]}$ be the coset containing s and $s + 1$. Assume that $Z = \mathbb{C}_{[s]} \cup \mathbb{C}_{[s+2]} \cup \ldots \cup \mathbb{C}_{[s+r]}$, where all the q-cosets $\mathbb{C}_{[s+i]}$, $i = 0, 2, 3, \ldots, r$, are mutually disjoint. Assume also that $Z \cap Z^{-1} = \emptyset$. Then there exists an $[[n, n - m(2r - 1), d \geq r + 2]]_q$ quantum code.*

Proof We know that $\gcd(q, n) = 1$. Let C be the cyclic code generated by

$$M^{(s)}(x) M^{(s+2)}(x) \cdot \ldots \cdot M^{(s+r)}(x).$$

Since $Z \cap Z^{-1} = \emptyset$, it follows from Lemma 5.1.3 that C is Euclidean dual-containing. From the hypotheses, all the q-ary cosets $\mathbb{C}_{[s]}, \mathbb{C}_{[s+2]}, \ldots, \mathbb{C}_{[s+r]}$ are mutually disjoint; hence, C has dimension $k = n - mr$ and its minimum distance is lower bounded by $d \geq r + 2$, i.e., C is an $[n, n - mr, d \geq r + 2]_q$ code. Let C' be the cyclic code generated by

$$M^{(s)}(x) M^{(s+2)}(x) \cdot \ldots \cdot M^{(s+r-1)}(x).$$

We know that C' is an enlargement of C and it has parameters $[n, n - m(r - 1), d \geq r + 1]_q$. Since $m \geq 2$, we have $k' - k = m \geq 2$, where k' is the dimension of C' and k is the dimension of C. Applying the Steane's construction to C and C' we obtain an $[[n, n - m(2r - 1), d \geq r + 2]]_q$ code, as required. $\qquad\square$

Example 5.1.4 In this example we construct an $[[31, 16, d \geq 5]]_5$ quantum code. For this purpose we take $n = 31$ and $q = 5$; then $m = \mathrm{ord}_n(q) = 3$. Let C be the cyclic code generated by $M^{(4)}(x)M^{(6)}(x)M^{(8)}(x)$. It is easy to see that C is Euclidean dual-containing and has parameters $[31, 22, d \geq 5]_5$. Let C' be the cyclic code generated by $M^{(4)}(x)M^{(8)}(x)$. The code C' has parameters $[31, 25, d \geq 4]_5$. Thus there exists an $[[31, 16, d \geq 5]]_5$ quantum code.

We next establish Theorem 5.1.7, an analogous to Theorem 5.1.1.

Theorem 5.1.7 *Suppose that $q \geq 5$ is a prime power and $n > q$ is an integer such that $\gcd(q, n) = 1$. Assume also that $(q - 1) \mid n$ and $m = \operatorname{ord}_n(q) = 2$ hold. Then there exists a quantum code with parameters $[[n, n - 4c, d \geq c + 2]]_q$, where $1 \leq c \leq r - 3$ and $r > 3$ is such that $n = r(q - 1)$.*

Proof We only prove the existence of an $[[n, n - 4(r - 3), d \geq r - 1]]_q$ code, since the constructions of the other codes are quite similar.

Let C be the cyclic code generated by

$$M^{(r)}(x)M^{(r+1)}(x) \cdot \ldots \cdot M^{(2r-3)}(x).$$

From Lemma 5.1.1 and from the proof of Theorem 5.1.1, we know that the q-cosets given by $\mathbb{C}_{[r]} = \{r\}$, $\mathbb{C}_{[r+1]} = \{r + 1, \quad r + q\}$, $\mathbb{C}_{[r+2]} = \{r + 2, \quad r + 2q\}, \ldots,$ $\mathbb{C}_{[2r-3]} = \{2r - 3, \quad r + (r - 3)q\}$ are mutually disjoint and each of them has two elements. Therefore, C has dimension $k = n - 2(r - 3) - 1$ and minimum distance $d \geq r - 1$.

Let us prove that C is Euclidean dual-containing. In fact, if $(r + i) \equiv -(r + j)$ mod n, where $0 \leq i, j \leq r - 3$, it follows that $2r + i + j \equiv 0 \mod n$. Since the inequality $2r + i + j < n$ holds because $q \geq 5$, one has a contradiction. On the other hand, if $(r + i)q \equiv -(r + j) \mod n$ holds then

$$(iq + j)(q - 1) \equiv 0 \mod n \Longrightarrow$$
$$i(q^2 - q) + j(q - 1) \equiv 0 \mod n \Longrightarrow$$
$$j(q - 1) \equiv i(q - 1) \mod n,$$

where the latter congruence holds because $\operatorname{ord}_n(q) = 2$. Then the unique solution is when $i = j$. Let us investigate this case. Seeking a contradiction, we assume that the congruence $(r + i)q \equiv -(r + i) \mod n$ is true. Then we obtain

$$(r + i)q \equiv -(r + i) \mod n \Longrightarrow$$
$$2r + i(q + 1) \equiv 0 \mod n \Longrightarrow$$
$$r(q - 3) \equiv i(q + 1) \mod n.$$

If $0 \leq i \leq r - 4$, then

$$r(q - 3) - i(q + 1) \geq r(q - 3) - (r - 4)(q + 1) = 4q - 4r + 4 > 0,$$

where the latter inequality holds because $r < q$ since we only consider nonprimitive BCH codes. Moreover, the inequality $r(q - 3) - i(q + 1) < n$ also holds, which is a contradiction. If $i = r - 3$ then the congruence $r(q - 3) \equiv (r - 3)(q + 1) \mod n$ holds, that is, $4r \equiv 3(q + 1) \mod n$ holds. Since $r \mid (q + 1)$ and $q + 1 > r$ hold, it implies that $q + 1 \geq 2r$ so, $3(q + 1) - 4r \geq 2r > 0$. Moreover, the inequality $3(q + 1) - 4r < n$ holds, which is a contradiction. Therefore, C is Euclidean dual-containing.

Let C' be the cyclic code generated by

$$M^{(r)}(x)M^{(r+1)}(x) \cdot \ldots \cdot M^{(2r-4)}(x).$$

C' is an enlargement of C; C' has dimension $k' = n - 2(r - 4) - 1$ and minimum distance $d' \geq r - 2$. Since $m = 2$ then $k' - k = 2$, where k' denotes the dimension of C' and k is the dimension of C. We know that $\lceil \frac{q+1}{q} d' \rceil \geq r - 1$. Thus, applying the Steane's construction one has an $[[n, n - 4(r - 3), d \geq r - 1]]_q$ quantum code, as required. $\qquad\square$

Recall that an $[[n, k, d]]_q$ code C satisfies the quantum Singleton bound given by $k + 2d \leq n + 2$. If C attains the quantum Singleton bound, i.e., $k + 2d = n + 2$, then it is called a quantum maximum distance separable (MDS) code. In the following two examples we construct quantum MDS-BCH codes:

Example 5.1.5 Applying Theorem 5.1.7 for $q = 9$ and $n = 40$ one has $r = 5$. Thus there exists an $[[40, 36, 3]]_9$ quantum MDS-BCH code. Analogously, applying Theorem 5.1.7 for $q = 11$ and $n = 60$ one obtains an $[[60, 56, 3]]_{11}$ quantum MDS-BCH code. Additionally, an $[[60, 48, d \geq 5]]_{11}$ and an $[[60, 52, d \geq 4]]_{11}$ quantum codes can be constructed.

5.1.4 Construction IV

In this subsection we present the fourth proposed construction, which is based on finding good Hermitian dual-containing BCH codes. Let us recall some useful concepts.

Lemma 5.1.4 ([4, Lemma 13]) *Assume that* $\gcd(q, n) = 1$. *A cyclic code of length n over $\mathbb{F}_{q^2}$ with defining set Z contains its Hermitian dual code if and only if $Z \cap Z^{-q} = \emptyset$, where $Z^{-q} = \{-qz \mod n \mid z \in Z\}$.*

Lemma 5.1.5 ([4, Lemma 17c]) *(Hermitian Construction) If there exists a classical linear $[n, k, d]_{q^2}$ code D such that $D^{\perp_H} \subset D$, then there exists an $[[n, 2k - n, \geq d]]_q$ stabilizer code that is pure to d. If the minimum distance $d^{\perp_H}$ of $D^{\perp_H}$ exceeds d, then the stabilizer code is pure and has minimum distance d.*

Let us start with an example of how Lemma 5.1.1 can be applied together the Hermitian construction in order to construct good codes. Assume that $q = 7, n = 144$, $m = 3$ and $r = 3$; the q^2-ary cosets $\mathbb{C}_3, \mathbb{C}_6, \mathbb{C}_9$ and $\mathbb{C}_{12}$ contain only one element. The other q-cosets are $\mathbb{C}_4 = \{4, 52, 100\}$, $\mathbb{C}_5 = \{5, 101, 53\}$, $\mathbb{C}_7 = \{7, 55, 103\}$, $\mathbb{C}_8 = \{8, 104, 56\}$, $\mathbb{C}_{10} = \{10, 58, 106\}$, $\mathbb{C}_{11} = \{11, 107, 59\}$. Let C be the cyclic code generated by $M^{(3)}(x)M^{(4)}(x)M^{(5)}(x)M^{(6)}(x)M^{(7)}(x) \quad M^{(8)}(x)M^{(9)}(x) \cdot M^{(10)}(x)$ $M^{(11)}(x)M^{(12)}(x)$. It is straightforward to show that C is Hermitian dual-containing and has parameters $[144, 122, d \geq 11]_{7^2}$. Thus, applying the Hermitian construction, we obtain an $[[144, 100, d \geq 11]]_7$ quantum code. Similarly one can construct quantum codes with parameters $[[144, 102, d \geq 10]]_7$, $[[144, 108, d \geq 9]]_7$, $[[144, 114, d \geq 8]]_7$, $[[144, 116, d \geq 7]]_7$, $[[144, 122, d \geq 6]]_7$, $[[144, 128, d \geq 5]]_7$, $[[144, 130, d \geq 4]]_7$ and $[[144, 136, d \geq 3]]_7$.

Theorem 5.1.8 *Suppose that $q > 3$ is a prime power and $n > q^2$ is an integer such that $\gcd(q^2, n) = 1$. Assume also that $(q^2 - 1) \mid n$ and $m = \text{ord}_n(q^2) = 2$ hold. Then there exists a quantum code with parameters $[[n, n - 4(r - 2) - 2, d \geq r]]_q$, where r satisfies $n = r(q^2 - 1)$.*

Proof Let C be the cyclic code generated by

$$M^{(r)}(x)M^{(r+1)}(x) \cdot \ldots \cdot M^{(2r-2)}(x).$$

We first show that C is Hermitian dual-containing. For this, let us consider the defining set Z of C consisting of the q^2-ary cosets given by $\mathbb{C}_{[r]} = \{r\}$, $\mathbb{C}_{[r+1]} = \{r + 1, \ r + q^2\}$, $\mathbb{C}_{[r+2]} = \{r + 2, \ r + 2q^2\}$, $\ldots$, $\mathbb{C}_{[2r-2]} = \{2r - 2, \ r + (r - 2)q^2\}$.

We know that $\gcd(q, n) = 1$ holds. From Lemma 5.1.4, it suffices to show that $Z \cap Z^{-q} = \emptyset$. Seeking a contradiction, we assume that $Z \cap Z^{-q} \neq \emptyset$. Then there exist i, j, where $0 \leq i, j \leq r - 2$, such that $(r + j)q^l \equiv -q(r + i) \bmod n$, where $l = 0$ or $l = 2$. If $l = 0$, one has $r + j \equiv -q(r + i) \bmod n$, so $q(r + i) + r + j \equiv 0 \bmod n$. Since both $q(r + i) + r + j < n$ and $q(r + i) + r + j \neq 0$ are true, one has a contradiction. If $l = 2$, it implies that $(r + j)q^2 \equiv -q(r + i) \bmod n$ and since $\gcd(q^2, n) = 1$ and $rq^2 \equiv r \bmod n$ one obtains

$$(r + j)q^2 \equiv -q(r + i) \quad \bmod n$$
$$\implies r + jq^2 \equiv -q(r + i) \quad \bmod n$$
$$\implies (q + 1)r \equiv -q(i + jq) \quad \bmod n$$
$$\implies -q(i + jq)(q - 1) \equiv 0 \quad \bmod n$$
$$\implies n \mid q(i + jq)(q - 1)$$
$$\implies r(q + 1) \mid q(i + jq).$$

Since $\gcd(r, q) = 1$ and $\gcd(q + 1, q) = 1$ hold, it follows that $r(q + 1) \mid (i + jq)$, which is a contradiction because $i + jq < r(q + 1)$. Thus C is Hermitian dual-containing.

It is easy to see that these cosets are mutually disjoint. With exception of $\mathbb{C}_{[r]}$, the other q-cosets have two elements. Thus, C has dimension $k = n - 2(r - 2) - 1$. By construction, the defining set of C contains the sequence $r, r + 1, \ldots, 2r - 2$, of $r - 1$ consecutive integers and, so the minimum distance of C is greater than or equal to r, that is, C is an $[n, n - 2(r - 2) - 1, d \geq r]_{q^2}$ code. Applying the Hermitian construction to the code C, one can get an $[[n, n - 4(r - 2) - 2, d \geq r]]_q$ code, as desired.   □

Corollary 5.1.3 *Suppose $q > 3$ is a prime power and $n > q^2$ is an integer such that* $\gcd(q^2, n) = 1$. *Assume also $(q^2 - 1) \mid n$ and $m = \mathrm{ord}_n(q^2) = 2$. Then there exist quantum codes with parameters $[[n, n - 4c - 2, d \geq c + 2]]_q$, where $2 \leq c < r - 2$ and $n = r(q^2 - 1)$.*

Proof Let C be the BCH code generated by

$$M^{(r)}(x)M^{(r+1)}(x) \cdot \ldots \cdot M^{(r+c)}(x).$$

Proceeding similarly as in the proof of Theorem 5.1.8, the result follows.   □

Theorem 5.1.9 *Let $q \geq 3$ be a prime power, $n > q^2$ be a prime number and consider that $m = \mathrm{ord}_n(q^2) \geq 2$. Let $\mathbb{C}_{[s]}$ be the q-coset containing s and $s + 1$. Assume that $Z = \mathbb{C}_{[s]} \cup \mathbb{C}_{[s+2]} \cup \ldots \cup \mathbb{C}_{[s+r]}$, where all the q-ary cosets $\mathbb{C}_{[s+i]}$, $i = 0, 2, 3, \ldots, r$, are mutually disjoint, and suppose that $Z \cap Z^{-q} = \emptyset$. Then there exists an $[[n, n - 2mr, d \geq r + 2]]_q$ quantum code.*

Proof We know that $\gcd(q, n) = 1$ holds. Let C be the cyclic code generated by

$$M^{(s)}(x)M^{(s+2)}(x) \cdot \ldots \cdot M^{(s \mid r)}(x).$$

Since $Z \cap Z^{-q} = \emptyset$ holds, it follows from Lemma 5.1.4 that C is Hermitian dual-containing. From the BCH bound, the minimum distance of C is greater than or equal to $r + 2$. It is easy to see that all the cosets $\mathbb{C}_{[s+i]}$, where $i = 0, 2, 3, \ldots, r$, have m elements and they are mutually disjoint. Thus C has parameters $[n, n - mr, d \geq r + 2]_{q^2}$. Applying the Hermitian construction one can get an $[[n, n - 2mr, d \geq r + 2]]_q$ quantum code.   □

We finish this subsection by showing how Lemma 5.1.2 works for constructing quantum MDS-BCH codes.

Example 5.1.6 Let us consider $q = 5$ and $n = 13$. Since $\gcd(13, 24) = 1$, the linear congruence $(q^2 - 1)x \equiv 1 \bmod n$ has a solution, so there exists at least one q^2-ary coset containing two consecutive integers, namely, the coset $\mathbb{C}_{[6]} = \{6, 7\}$. Let $C = \langle M^{(6)}(x) \rangle$. Since $\mathbb{C}_{[4]}$ and $\mathbb{C}_{[6]}$ are disjoint, C is Hermitian dual-containing and has parameters $[13, 11, d \geq 3]_5$. Applying the Hermitian construction, an $[[13, 9, 3]]_5$ quantum MDS-BCH code is constructed. Similarly, we can also construct an $[[17, 13, 3]]_4$ and an $[[17, 9, 5]]_4$ quantum MDS-BCH code.

5.1.5 Construction V

In this subsection, we show the existence of (classical) cyclic codes whose defining set consists of only one cyclotomic coset containing at least two consecutive integers. This fact allows us to construct quantum codes with good parameters.

Lemma 5.1.6 in the following is a particular case of the CSS construction.

Lemma 5.1.6 ([4, Lemma 17]) *If there exists a classical linear $[n, k, d]_q$ code C such that $C^\perp \subset C$, then there exists an $[[n, 2k - n, \geq d]]_q$ stabilizer code that is pure to d.*

Theorem 5.1.10 establishes conditions for the existence of a cyclic code whose defining set consists of only one q-coset containing at least two consecutive integers. This fact produces conditions to construct quantum codes with good parameters (in the sense of the QSB).

Theorem 5.1.10 *Let $q \geq 3$ be a prime power and $n > m$ be a positive integer such that $\gcd(q, n) = 1$ and $\gcd(q^{a_i} - 1, n) = 1$ for every $i = 1, 2, \ldots, r$, where $m = \mathrm{ord}_n(q) \geq r + 2$ and $1 \leq r, a_1, a_2, \ldots, a_r < m$ are integers. If $n \mid \gcd(t_2, \ldots, t_r)$, where $t_j = [(j - (j - 1)q^{a_j})(q^{a_j} - 1)^{-1} - (q^{a_1} - 1)^{-1}]$ for every $j = 2, \ldots, r$ (the operations are performed modulo n), then there exists an $[n, n - m^*, d \geq r + 2]_q$ cyclic code, where m^* is the cardinality of the q-coset containing $r + 1$ consecutive integers.*

Proof We will investigate the following system of congruences:

$$xq^{a_1} \equiv (x + 1) \quad \mathrm{mod}\ n$$
$$(x + 1)q^{a_2} \equiv (x + 2) \quad \mathrm{mod}\ n$$
$$(x + 2)q^{a_3} \equiv (x + 3) \quad \mathrm{mod}\ n$$
$$\vdots$$
$$(x + r - 1)q^{a_r} \equiv (x + r) \quad \mathrm{mod}\ n,$$

where $1 \leq r, a_1, a_2, \ldots, a_r < m$. Since $\gcd(q^{a_i} - 1, n) = 1$ for every $i = 1, 2, \ldots, r$, it follows that such system is equivalent to

$$x \equiv (q^{a_1} - 1)^{-1} \quad \mathrm{mod}\ n$$
$$x \equiv (2 - q^{a_2})(q^{a_2} - 1)^{-1} \quad \mathrm{mod}\ n$$
$$x \equiv (3 - 2q^{a_3})(q^{a_3} - 1)^{-1} \quad \mathrm{mod}\ n$$
$$\vdots$$
$$x \equiv [r - (r - 1)q^{a_r}](q^{a_r} - 1)^{-1} \quad \mathrm{mod}\ n,$$

where $(q^{a_i} - 1)^{-1}$ denotes the multiplicative inverse of $(q^{a_i} - 1)$ modulo n.

We know that the last system has a solution if and only if

$$[j - (j - 1)q^{a_j}](q^{a_j} - 1)^{-1} \equiv [i - (i - 1)q^{a_i}](q^{a_i} - 1)^{-1} \mod n$$

for all $i, j = 2, \ldots, r$ and

$$(q^{a_1} - 1)^{-1} \equiv [i - (i - 1)q^{a_i}](q^{a_i} - 1)^{-1} \mod n$$

for all $i = 2, \ldots, r$. This fact means that

$$n | [((j - (j - 1)q^{a_j})(q^{a_j} - 1)^{-1} - (q^{a_1} - 1)^{-1}]$$

for every $j = 2, \ldots, r$, i.e., $n | \gcd(t_2, \ldots, t_r)$, where

$$t_j = [(j - (j - 1)q^{a_j})(q^{a_j} - 1)^{-1} - (q^{a_1} - 1)^{-1}]$$

for all $j = 2, \ldots, r$.

Let C be the cyclic code whose defining is the q-coset $\mathbb{C}_x$. From construction, the defining set of C, i.e., the coset $\mathbb{C}_x$, contains the sequence $x, x + 1, \ldots, x + r$ of $r + 1$ consecutive integers. From the BCH bound, the minimum distance d of C satisfies $d \geq r + 2$. Since $|\mathbb{C}_x| = m^*$, C has dimension $n - m^*$. We then obtain an $[n, n - m^*, d \geq r + 2]_q$ code, as required. $\qquad\square$

If the code length is prime we have the following particular case of Theorem 5.1.10.

Corollary 5.1.4 *Let $q \geq 3$ be a prime power and $n > m$ be a prime number such that $\gcd(q, n) = 1$, where $m = \mathrm{ord}_n(q) \geq r + 2$ and $1 \leq r, a_1, a_2, \ldots, a_r < m$ are integers. Assume that $n | \gcd(t_2, \ldots, t_r)$, where $t_j = [(j - (j - 1)q^{a_j})(q^{a_j} - 1)^{-1} - (q^{a_1} - 1)^{-1}]$ for every $j = 2, \ldots, r$ and $a_1, a_2, \ldots, a_r$ are integers such that $1 \leq a_1 + a_2 + \cdots + a_r < m$ (the operations are performed modulo n). Then there exists an $[n, n - m^*, d \geq r + 2]_q$ cyclic code.*

Proof Notice that since n is prime, it follows that $\gcd(q^{a_i} - 1, n) = 1$ for every $i = 1, 2, \ldots, r$, because $a_1, a_2, \ldots, a_r < m$. We next apply Theorem 5.1.10 to obtain the desired result. $\qquad\square$

In order to proceed further, we will denote by $\mathbb{C}_{-x}$ the coset of $-x$, where $-x$ is taken modulo n. With this notation we have the following result.

Theorem 5.1.11 *Assume all hypotheses of Theorem 5.1.10 hold. Let C be the cyclic code with defining set $\mathbb{C}_x$, where $\mathbb{C}_x$ is a coset containing $r + 1$ consecutive integers. If $\mathbb{C}_x \neq \mathbb{C}_{-x}$ then there exists an $[[n, n - 2m^*, d \geq r + 2]]_q$ quantum code.*

Proof From [4, Lemma 1], C contains its Euclidean dual code $C^\perp$. The dimension and the minimum distance of the corresponding quantum code follow directly from Theorem 5.1.10 and from Lemma 5.1.6. $\qquad\square$

Let us present some examples of how our construction works.

Example 5.1.7 Consider that $q = 5$ and $n = 11$; hence $m = \text{ord}_{11}(5) = 5$. The 5-cosets are given by $\mathbb{C}_0 = \{0\}$, $\mathbb{C}_1 = \{1, 5, 3, 4, 9\}$ and $\mathbb{C}_2 = \{2, 10, 6, 8, 7\}$. If C_1 is the cyclic code with defining set $\mathbb{C}_1$, then C_1 is a dual-containing code with parameters $[11, 6, d \geq 4]_5$. From Lemma 5.1.6, one can get an $[[11, 1, d \geq 4]]_5$ code.

Let us now take $q = 17$ and $n = 19$; so $m = \text{ord}_{19}(17) = 9$. If C_1 is the code with defining set $\mathbb{C}_1 = \{1, 17, 4, 11, 16, 6, 7, 5, 9\}$ we obtain an $[[19, 1, d \geq 5]]_{17}$ code.

Similarly, we can construct an $[61, 56, d \geq 3]_9$ code C_2 with defining set $\mathbb{C}_8 = \{8, 11, 38, 37, 28\}$. We know that C_2 is a dual-containing code, so an $[[61, 51, d \geq 3]]_9$ quantum code exists.

We can also construct an $[67, 64, d \geq 3]_{29}$ dual-containing code with defining set $\mathbb{C}_{12} = \{12, 13, 42\}$. Hence, there exists an $[[67, 61, d \geq 3]]_{29}$ quantum code.

The existence of an $[35, 31, d \geq 3]_{13}$ dual-containing code generates an $[[35, 27, d \geq 3]]_{13}$ quantum code. An $[35, 31, d \geq 3]_{27}$ dual-containing code with defining set $\mathbb{C}_3 = \{3, 11, 17, 4\}$ guarantees the existence of an $[[35, 27, d \geq 3]]_{27}$ quantum code. An $[73, 70, d \geq 3]_{64}$ dual-containing code with defining set $\mathbb{C}_{21} = \{22, 21, 30\}$ exists, so there exists an $[[73, 67, d \geq 3]]_{64}$ quantum code.

Example 5.1.8 In this example, we construct cyclic codes whose defining set consists of two q-cosets (the idea is the same as that presented in Theorem 5.1.10). An $[35, 27, d \geq 4]_{27}$ dual-containing code C with defining set consisting of $\mathbb{C}_2$ and $\mathbb{C}_3$ ensures the existence of an $[[35, 19, d \geq 4]]_{27}$ quantum code. Taking the cosets $C_{14} = \{14, 20, 30\}$ and C_{21} one has an $[[73, 61, d \geq 4]]_{64}$ quantum code. Similarly, an $[[63, 51, d \geq 3]]_{11}$ quantum code (coset C_{43}) and an $[[63, 39, d \geq 4]]_{11}$ code (cosets C_{43} and C_{20}) can be constructed. Analogously, an $[[63, 51, d \geq 3]]_{23}$ and an $[[63, 45, d \geq 4]]_{23}$ quantum code (cosets C_4 and C_{27}) can be constructed.

5.1.6   Code Comparison

In this section, we compare the parameters of our quantum BCH codes with the ones available in the literature. The codes available in the literature derived from Steane's code construction are generated by the same method presented in [148, Table I] by considering the criterion for classical Euclidean dual-containing BCH codes given in [4, Theorems 3 and 5].

Let us fix the notation:

- $[[n, k, d]]_q$ are the parameters of the new quantum codes;
- $[[n', k', d']]_q = [[n', n' - 2m(\lceil (\delta - 1)(1 - 1/q) \rceil), d' \geq \delta]]_q$ are the parameters of quantum codes available in [4];
- $[[n'', k'', d'']]_q$ are the parameters of quantum BCH codes derived from Steane's code construction shown in [62, Corollary 4].

Table 5.1 displays a comparison of the parameters of some CSS codes constructed here with the parameters of the CSS codes shown in [4], and Table 5.2 shows a comparison between our CSS codes with the quantum codes derived from the nonbinary Steane's construction (see Corollary 5.1.2).

Tables 5.3 and 5.4 show our codes obtained from Construction I and from Theorem 5.1.4 in Construction II, respectively. Table 5.5 presents some codes generated from Construction III, and Table 5.6 displays some codes generated from Construction IV. Finally, Table 5.7 exhibited some codes derived from Construction V.

Checking the parameters of our quantum BCH codes tabulated, one can see that our codes have parameters better than the ones available in the literature. In other words, fixing the code length n and the minimum distance d (or the lower bound for the minimum distance d, since the true minimum distance of BCH are not known in general), the quantum BCH codes constructed here achieve greater values of the number of qudits than the quantum BCH codes available in the literature.

Remark 5.1.2 The procedure of code comparison exhibited above will be adopted throughout the entire book in order to perform the comparison among the parameters of the quantum codes constructed here with the parameters of the quantum codes available in the literature. In other words: to compare the parameters of an $[n, k_1, d]_q$ quantum code $\mathbb{Q}_1$ constructed here, we perform searching for a quantum code of length n and minimum distance d. If such code $\mathbb{Q}_2$ is an $[n, k_2, d]_q$ code with $k_1 > k_2$, then $\mathbb{Q}_1$ is better than $\mathbb{Q}_2$; if $k_2 > k_1$ it implies that $\mathbb{Q}_2$ is better than $\mathbb{Q}_1$. In many cases we fix the code length and the lower bound for the minimum distance as it was said above (see Tables 5.1, 5.2, 5.3, 5.4, 5.5, 5.6 and 5.7 to see this), after comparing the code dimension, as was done earlier. This criterion of code comparison is usual in the literature.

Note that our $[[1093, 1079, d \geq 3]]_3$ code has the same parameters of the corresponding Hamming code; our $[[71, 61, d \geq 3]]_5$ code can be compared with distance three codes obtained by shortening Hamming codes.

The codes $[[67, 61, d \geq 3]]_{29}$ and $[[73, 67, d \geq 3]]_{64}$ shown in Table 5.7 have parameters satisfying $n + 2 - k - 2d \leq 2$; the parameters of the codes $[[11, 1, d \geq 4]]_5$, $[[35, 27, d \geq 3]]_{13}$ and $[[35, 27, d \geq 3]]_{27}$, satisfy $n + 2 - k - 2d \leq 4$. The $[[11, 1, d \geq 4]]_5$ code is comparable to the $[[17, 9, 4]]_5$ code shown in [17], and the $[[61, 51, d \geq 3]]_9$ code is comparable to the $[[65, 51, 4]]_9$ code shown in [17].

Summarizing the results: in this subsection, we have presented five quantum code constructions generating families of quantum BCH codes with good parameters.

Table 5.1 Code comparison

Our CSS codes	CSS codes in [4]
$[[n, k, d]]_q$	$[[n', k', d']]_q$
$[[40, 30, d \geq 4]]_9$	$[[40, 28, d' \geq 4]]_9$
$[[40, 20, d \geq 7]]_9$	—
$[[30, 7, d \geq 8]]_{11}$	—
$[[61, 55, d \geq 3]]_{13}$	$[[61, 49, d' \geq 3]]_{13}$
$[[84, 74, d \geq 4]]_{13}$	$[[84, 72, d' \geq 4]]_{13}$
$[[84, 70, d \geq 5]]_{13}$	$[[84, 68, d' \geq 5]]_{13}$
$[[84, 66, d \geq 6]]_{13}$	$[[84, 64, d' \geq 6]]_{13}$
$[[91, 85, d \geq 3]]_{16}$	$[[91, 79, d' \geq 3]]_{16}$
$[[144, 126, d \geq 6]]_{17}$	$[[144, 124, d' \geq 6]]_{17}$
$[[144, 122, d \geq 7]]_{17}$	$[[144, 120, d' \geq 7]]_{17}$
$[[144, 118, d \geq 8]]_{17}$	$[[144, 116, d' \geq 8]]_{17}$
$[[127, 121, d \geq 3]]_{19}$	$[[127, 115, d' \geq 3]]_{19}$

Table 5.2 Code comparison

Our CSS codes	q-ary Steane's construction
$[[n, k, d]]_q$	$[[n'', k'', d'']]_q$
$[[19, 13, d \geq 3]]_7$	—
$[[13, 7, d \geq 3]]_9$	—
$[[19, 13, d \geq 3]]_{11}$	—
$[[61, 55, d \geq 3]]_{13}$	$[[61, 52, d'' \geq 3]]_{13}$
$[[91, 85, d \geq 3]]_{16}$	$[[91, 82, d'' \geq 3]]_{16}$
$[[127, 121, d \geq 3]]_{19}$	$[[127, 118, d'' \geq 3]]_{19}$
$[[13, 5, d \geq 3]]_5$	—
$[[13, 5, d \geq 3]]_8$	—
$[[13, 7, d \geq 3]]_3$	—
$[[43, 31, d \geq 3]]_7$	—
$[[73, 61, d \geq 3]]_9$	—
$[[1093, 1079, d \geq 3]]_3$	$[[1093, 1072, d'' \geq 3]]_3$

5.2 BCH Codes—Part II

The material of this subsection is based on the results shown in our paper [104]. We investigate here several properties of q-ary cyclotomic cosets modulo $n = q^m - 1$, where $q \neq 2$ is a prime power. As applications, several families of nonbinary Calderbank–Shor–Steane (CSS) quantum codes derived from two distinct Bose–Chaudhuri–Hocquenghem (BCH) codes are constructed.

Table 5.3 Code comparison

Our CSS codes	CSS codes in [4]
$[[n, k, d]]_q$	$[[n', k', d']]_q$
$[[11, 1, d \geq 4]]_3$	—
$[[13, 1, d \geq 4]]_3$	—
$[[1093, 1079, d \geq 3]]_3$	$[[1093, 1065, d' \geq 3]]_3$
$[[31, 19, d \geq 4]]_5$	$[[31, 13, d' \geq 4]]_5$
$[[31, 13, d \geq 5]]_5$	$[[31, 7, d' \geq 5]]_5$
$[[71, 61, d \geq 3]]_5$	$[[71, 51, d' \geq 3]]_5$
$[[71, 51, d \geq 4]]_5$	$[[71, 41, d' \geq 4]]_5$
$[[73, 61, d \geq 4]]_8$	$[[73, 55, d' \geq 4]]_8$
$[[73, 55, d \geq 5]]_8$	$[[73, 49, d' \geq 5]]_8$
$[[73, 49, d \geq 6]]_8$	$[[73, 43, d' \geq 6]]_8$
$[[73, 43, d \geq 7]]_8$	$[[73, 37, d' \geq 7]]_8$

Table 5.4 Code comparison

Our CSS codes	Steane's code construction
$[[n, k, d]]_q$	$[[n'', k'', d'']]_q : L, L'$
$[[31, 19, d \geq 4]]_5$	$[[31, 16, d'' \geq 4]]_5 : [31, 22, 4]_5, [31, 25, 3]_5$
$[[31, 13, d \geq 5]]_5$	$[[31, 10, d'' \geq 5]]_5 : [31, 19, 5]_5, [31, 22, 4]_5$
$[[73, 61, d \geq 4]]_8$	$[[73, 58, d'' \geq 4]]_8 : [73, 64, 4]_8, [73, 67, 3]_8$
$[[73, 55, d \geq 5]]_8$	$[[73, 52, d'' \geq 5]]_8 : [73, 61, 5]_8, [73, 64, 4]_8$
$[[73, 49, d \geq 6]]_8$	$[[73, 46, d'' \geq 6]]_8 : [73, 58, 6]_8, [73, 61, 5]_8$
$[[73, 43, d \geq 7]]_8$	$[[73, 40, d'' \geq 7]]_8 : [73, 55, 7]_8, [73, 58, 6]_8$

Table 5.5 Code comparison

Our codes (Construction III)	Steane's code construction
$[[n, k, d]]_q$	$[[n'', k'', d'']]_q$
$[[31, 22, d \geq 4]]_5$	$[[31, 16, d'' \geq 4]]_5$
$[[31, 16, d \geq 5]]_5$	$[[31, 10, d'' \geq 5]]_5$
$[[71, 56, d \geq 4]]_5$	$[[71, 46, d'' \geq 4]]_5$
$[[73, 64, d \geq 4]]_8$	$[[73, 58, d'' \geq 4]]_8$
$[[73, 58, d \geq 5]]_8$	$[[73, 52, d'' \geq 5]]_8$
$[[40, 36, 3]]_9$ (MDS)	
$[[60, 56, 3]]_{11}$ (MDS)	

Table 5.6 Code comparison

Our Hermitian Codes (Construction IV)	Hermitian Codes in [4]
$[[n, k, d]]_q$	$[[n', k', d']]_q$
$[[17, 13, 3]]_4$ (MDS)	
$[[17, 9, 5]]_4$ (MDS)	
$[[13, 9, 3]]_5$ (MDS)	
$[[312, 298, d \geq 5]]_5$	$[[312, 296, d' \geq 5]]_5$
$[[312, 294, d \geq 6]]_5$	$[[312, 292, d' \geq 6]]_5$
$[[312, 290, d \geq 7]]_5$	$[[312, 288, d' \geq 7]]_5$
$[[312, 286, d \geq 8]]_5$	$[[312, 284, d' \geq 8]]_5$
$[[312, 282, d \geq 9]]_5$	$[[312, 280, d' \geq 9]]_5$
$[[312, 278, d \geq 10]]_5$	$[[312, 276, d' \geq 10]]_5$
$[[312, 274, d \geq 11]]_5$	$[[312, 272, d' \geq 11]]_5$
$[[312, 270, d \geq 12]]_5$	$[[312, 268, d' \geq 12]]_5$
$[[144, 128, d \geq 5]]_7$	$[[144, 120, d \geq 5]]_7$
$[[144, 122, d \geq 6]]_7$	$[[144, 114, d \geq 6]]_7$
$[[144, 116, d \geq 7]]_7$	$[[144, 108, d \geq 7]]_7$
$[[144, 114, d \geq 8]]_7$	$[[144, 102, d \geq 8]]_7$
$[[144, 108, d \geq 9]]_7$	$[[144, 96, d \geq 9]]_7$
$[[144, 102, d \geq 10]]_7$	$[[144, 90, d \geq 10]]_7$
$[[144, 100, d \geq 11]]_7$	$[[144, 84, d \geq 11]]_7$

Table 5.7 Our quantum codes

Parameters of the new codes
$[[11, 1, d \geq 4]]_5$
$[[19, 1, d \geq 5]]_{17}$
$[[35, 27, d \geq 3]]_{13}$
$[[35, 27, d \geq 3]]_{27}$
$[[35, 19, d \geq 4]]_{27}$
$[[51, 35, d \geq 3]]_{32}$
$[[61, 51, d \geq 3]]_9$
$[[63, 51, d \geq 3]]_{11}$
$[[63, 39, d \geq 4]]_{11}$
$[[63, 51, d \geq 3]]_{23}$
$[[63, 45, d \geq 4]]_{23}$
$[[67, 61, d \geq 3]]_{29}$
$[[73, 67, d \geq 3]]_{64}$
$[[73, 61, d \geq 4]]_{64}$

Properties of cyclotomic cosets have been studied in many areas of research, especially in theory of classical and quantum error-correcting codes. For instance, the q-cosets were investigated by several researchers in order to obtain efficient classical cyclic codes [15, 110, 114, 115, 129, 141–143, 165, 166], as well as to construct efficient quantum codes [4, 17, 25, 30, 56, 80, 81, 89, 108, 113, 148, 153, 155, 156, 162, 163].

In this subsection we show properties of q-ary cosets modulo $n = q^m - 1$, where $q \neq 2$ in order to construct families of good quantum BCH codes by applying the CSS construction. Our codes have parameters

- $[[q^2 - 1, q^2 - 4c + 5, d \geq c]]_q$,
  where $2 \leq c \leq q$ and $q \geq 4$ is a prime power;
- $[[n, n - 2m(c - 2) - m/2 - 1, d \geq c]]_q$,
  where $n = q^m - 1$, $q \geq 4$, $2 \leq c \leq q$ and $m \geq 2$ is an even integer;
- $[[q^m - 1, q^m - 6m - 3, d \geq 5]]_q$,
- $[[q^m - 1, q^m - 2m - 3, d \geq 3]]_q$,
- $[[q^m - 1, q^m - 4m - 3, d \geq 4]]_q$,
  where $q \geq 5$ is an odd prime power;
- $[[n, n - m(2c - 3) - 1, d \geq c]]_q$,
  where $n = q^m - 1$, $q \geq 3$ is a prime power, $m \geq 3$ and $2 \leq c \leq q$.

To the reader's convenience, we present a brief organization of the topics. Section 5.2.1 establishes properties concerning q-cosets modulo $q^m - 1$. These results will be applied for constructing q-ary CSS quantum codes with good parameters. In Sect. 5.2.2, we explain how to obtain families of CSS codes by utilizing such properties. In Sect. 5.2.3, the parameters of our CSS codes are compared with the ones available in the literature.

5.2.1 Properties of q-cosets

We here explore the structure of q-ary cosets in order to show nice properties of them. As already said, we are interested in the study of q-ary cosets modulo $n = q^m - 1$.

Proposition 5.2.1 *Let q be an odd prime power and $\mathbb{C}_s$ be a q-ary coset with representative s. Then s is even if and only if $\forall\ t \in \mathbb{C}_s$, t is even.*

Proof Suppose first that $s = 2k$, where k is an integer and let t be an element of the coset $\mathbb{C}_s = \{s, qs, q^2 s, q^3 s, \ldots, q^{m_s - 1} s\}$ without considering the modulo operation. Then it follows that $t = sq^l = 2kq^l$, where $0 \leq l \leq m_s - 1$. Applying the division algorithm for t and $q^m - 1$, one has $2kq^l = (q^m - 1)a + r$, where r is an integer such that $0 \leq r < q^m - 1$. Hence, it follows that $r = 2kq^l - (q^m - 1)a$. Since $q^m - 1$ is even, r is also even, as required.

Conversely, suppose that each t, where $t \in \mathbb{C}_s$, (considering the modulo operation) is of the form $t = 2k$, with k integer. Applying again the division algorithm for sq^l

and $q^m - 1$, it follows that $sq^l = (q^m - 1)a + t$, where $0 \leq t < q^m - 1$ is even. Since t and $q^m - 1$ are even, then also is sq^l; because q^l is odd, it follows that s is even. The proof is complete.   □

From Proposition 5.2.1, we obtain two useful results.

Corollary 5.2.1 *If q is an odd prime power, then there is no consecutive integers belonging to the same q-ary coset modulo $n = q^m - 1$.*

Proof Immediate.   □

Remark 5.2.1 Note that, in the binary case, there exists at least one coset containing two consecutive elements, namely, $\mathbb{C}_1$.

Corollary 5.2.2 *Suppose that $\mathbb{C}_x$ and $\mathbb{C}_y$ are two q-ary cosets, where q is an odd prime power. Assume also that $a \in \mathbb{C}_x$ and $b \in \mathbb{C}_y$. If $a \not\equiv b \bmod 2$, then $\mathbb{C}_x \neq \mathbb{C}_y$.*

Proof It follows directly from Proposition 5.2.1.   □

The next result shows that the minimum distance among elements belonging to the same q-ary coset is greater than or equal to $q - 1$, with equality in some cases.

Theorem 5.2.1 *Let $q \neq 2$ be a prime power and let $\mathbb{C}_s$ be a q-ary coset (modulo $q^m - 1$) with representative s. Define $L_s = \min\{| sq^j - sq^l |: 0 \leq j, l \leq m_s - 1, j \neq l\}$, where $| \cdot |$ is the absolute value function, and $sq^j - sq^l$ is considered modulo $q^m - 1$. Then it follows that $L_s \geq q - 1$, for all s, where s runs through the coset representatives. Moreover, there exists at least one coset $\mathbb{C}_{s^*}$ such that $L_{s^*} = q - 1$.*

Proof Assume without loss of generality (w. l. o. g.) that $j > l$ are integers such that $1 \leq j, l \leq m_s - 1$, and let $\mathbb{C}_s$ be an arbitrary coset. Applying the division algorithm for $sq^j - sq^l$ and $q^m - 1$ we have

$$sq^j - sq^l = (q^m - 1)b + r, 0 \leq r < q^m - 1 \Longrightarrow$$
$$\Longrightarrow sq^l(q^{j-l} - 1) - (q^m - 1)b = r.$$

Since $q - 1$ divides $q^{j-l} - 1$ and $q^m - 1$, it follows that $q - 1$ divides r, i.e., $L_s \geq q - 1$.

To complete the proof, it suffices to consider the coset containing the element $\frac{q^m - 1}{2} + 1$ since this coset also contains the element $\frac{q^m - 1}{2} + q$.   □

Corollary 5.2.3 can be utilized to compute the upper bound for the designed distance of certain classes of cyclic codes.

Corollary 5.2.3 *Let $q \neq 2$ be a prime power. If C is a q-ary cyclic code whose defining set Z contains i cosets, where $1 \leq i < q - 2$, then its designed distance δ is at most $i + 2$. In particular, if Z consists of only one q-ary coset, then $\delta = 2$.*

Proof It follows directly from Theorem 5.2.1.   □

In what follows, we introduce the concept of complementary coset, after showing some interesting properties of them.

Definition 5.2.1 Let $\mathbb{C}_s = \{s, qs, q^2s, q^3s, \ldots, q^{m_s-1}s\}$ be a q-coset with representative s. A complementary coset of $\mathbb{C}_s$ is a q-ary coset given by $\mathbb{C}_r = \{r, qr, q^2r, q^3r, \ldots, q^{m_r-1}r\}$ with representative r, containing an element $q^l r$, where $0 \leq l \leq m_r - 1$, such that $s + q^l r \equiv 0 \bmod (q^m - 1)$.

Proposition 5.2.2 establishes some properties of complementary cosets.

Proposition 5.2.2 *Let* $\mathbb{C}_s = \{s, qs, q^2s, q^3s, \ldots, q^{m_s-1}s\}$ *be a q-ary coset modulo* $n = q^m - 1$. *Then the following hold:*

(i) *For each q-ary coset $\mathbb{C}_s$ given, there exists only one complementary coset of* $\mathbb{C}_s$, *denoted by* $\overline{\mathbb{C}}_s$;

(ii) *The cyclotomic coset and its complementary coset have the same cardinality;*

(iii) *Defining the operation* $\mathbb{C}_s \oplus \overline{\mathbb{C}}_r = \{s + q^l r, sq + (q^l r)q, \ldots, sq^{m_s-1} + (q^l r)q^{m_s-1}\}$ *one has* $\mathbb{C}_s \oplus \overline{\mathbb{C}}_s = \mathbb{C}_0 = \{0\}$;

(iv) *If $\mathbb{C}_r$ is the complementary coset of $\mathbb{C}_s$ then $L_s = L_r$;*

(v) $\overline{\overline{\mathbb{C}}}_s = \mathbb{C}$.

Proof (i) Let $\mathbb{C}_s$ be a coset. Assume that

$$\mathbb{C}_{r_1} = \{r_1, qr_1, q^2r_1, q^3r_1, \ldots, q^{m_{r_1}-1}r_1\},$$

$$\mathbb{C}_{r_2} = \{r_2, qr_2, q^2r_2, q^3r_2, \ldots, q^{m_{r_2}-1}r_2\}$$

are two complementary cosets of $\mathbb{C}_s$ with representatives r_1 and r_2, respectively. From definition of complementary coset, there exist two elements $q^l r_1$, $0 \leq l \leq m_{r_1} - 1$ and $q^t r_2$, $0 \leq t \leq m_{r_2} - 1$ such that $s + q^l r_1 \equiv 0 \bmod n$ and $s + q^t r_2 \equiv 0 \bmod n$, and so $q^l r_1 \equiv q^t r_2 \bmod n$. Checking the latter equivalence we obtain one of the following three cases: $r_1 \equiv r_2 \bmod n$; $r_1 \equiv r_2 q^t \bmod n$, where $0 \leq t \leq m_{r_2} - 1$; $r_2 \equiv r_1 q^l \bmod n$, where $0 \leq t \leq m_{r_2} - 1$. In each of them it follows that $\mathbb{C}_{r_1} = \mathbb{C}_{r_2}$. Therefore each coset has only one complementary coset as well.

(ii) Let $\mathbb{C}_s$ be the coset containing s given by

$$\mathbb{C}_s = \{s, sq, sq^2, sq^3, \ldots, sq^{m_s-1}\}.$$

Suppose that $\mathbb{C}_s$ has cardinality m_s. Consider the q-ary coset containing the element $n - s$, where $n = q^m - 1$, and denote it by $\mathbb{C}_{[n-s]}$. Without considering the order one has

$$\mathbb{C}_{[n-s]} = \{n - s, [n - s]q, [n - s]q^2, \ldots, [n - s]q^{m_l-1}\}.$$

It is clear that $\mathbb{C}_{[n-s]}$ is the complementary coset of $\mathbb{C}_s$. We have to show that the equality $m_l = m_s$ is true. We first prove that $[n - s]q^l \not\equiv [n - s]q^t \bmod n$ holds for each $0 \leq t, l \leq m_s - 1$, that is, $m_l \geq m_s$. In fact, seeking a contradiction, we assume that $[n - s]q^l \equiv [n - s]q^t \bmod n$ holds, where $0 \leq t, l \leq m_s - 1$. Thus the congruence $sq^l \equiv sq^t \bmod n$ holds, where $0 \leq t, l \leq m_s - 1$, which is a contradiction.

On the other hand, seeking a contradiction, we assume that $sq^r \equiv sq^t \bmod n$ holds, for each $0 \leq r, t \leq m_l - 1$. Then one has $-sq^r \equiv -sq^t \bmod n$, that is, $[n - s]q^r \equiv [n - s]q^t \bmod n$ holds for each $0 \leq r, t \leq m_l - 1$, which is a contradiction. Thus $sq^r \not\equiv sq^t \bmod n$ for each $0 \leq r, t \leq m_l - 1$ and so, $m_l \leq m_s$. Therefore, one obtains $m_l = m_s$, as required.

(iii) Follows from direct computation.

(iv) Suppose that $\mathbb{C}_s$ is a q-ary coset with complementary $\mathbb{C}_r$ and $L_s = \mid sq^{t_1} - sq^{t_2} \mid$, where $0 \leq t_1, t_2 \leq m_s - 1, t_1 \neq t_2$. We may assume without loss of generality that

$$L_s = sq^{t_1} - sq^{t_2}. \tag{5.1}$$

From hypothesis, there exists $0 \leq l \leq m_r - 1$ such that $s \equiv -rq^l \bmod n$. Replacing the last equivalence in Equation (5.1) one obtains $L_s \equiv \mid rq^{t_2+l} - rq^{t_1+l} \mid \bmod n$. We may assume that $0 \leq t_1 + l, t_2 + l \leq m_r - 1$ and $t_2 > t_1$, so $L_s = rq^{t_2+l} - rq^{t_1+l}$. Therefore $L_s \leq L_r$.

On the other hand, assume $L_r = rq^{a_1} - rq^{a_2}$, where $0 \leq a_1, a_2 \leq m_r - 1, a_1 \neq a_2$. Since $r \equiv -sq^{m_r-l} \bmod n$ it follows that

$$L_r = rq^{a_1} - rq^{a_2} \equiv \mid sq^{m_r-l+a_2} - sq^{m_r-l+a_1} \mid \quad \bmod n.$$

As above, we can further assume w. l. o. g. that $0 \leq m_r - l + a_1, m_r - l + a_2 \leq m_s - 1$ and also $a_2 > a_1$. Thus

$$L_r = sq^{m_r-l+a_2} - sq^{m_r-l+a_1},$$

and so $L_r \leq L_s$. Therefore, the equality $L_r = L_s$ holds.

(v) Assume that $r \in \overline{\mathbb{C}}_s$. Then there exists $0 \leq t \leq m_s - 1$ such that $(s + q^t r) \equiv 0 \bmod n$ since, from Item (ii), $m_r = m_s$. Then one obtains $(r + q^{m_s-t}s) \equiv 0 \bmod n$. We can suppose that $0 \leq m_s - t \leq m_s - 1$ since for $t = 0$ the equivalences $(r + q^{m_s-t}s) \equiv (r + s) \equiv (s + q^t r) \bmod n$ hold. From these facts we conclude that $\mathbb{C}_s = \overline{\mathbb{C}}_s$ since, from Item (i), the complementary coset is unique.

$\square$

The following lemma already presented (see Lemma 5.1.3) gives necessary and sufficient condition under which a cyclic code contains its Euclidean dual.

Lemma 5.2.1 ([4, Lemma 1]) *Assume that* $\gcd(q, n) = 1$. *A cyclic code of length* n *over* F_q *with defining set* Z *contains its Euclidean dual code if and only if* $Z \cap Z^{-1} = \emptyset$, *where* $Z^{-1} = \{-z \bmod n \mid z \in Z\}$.

The next result characterizes Euclidean dual-containing cyclic codes in terms of complementary cosets.

Proposition 5.2.3 *Assume that* $\gcd(q, n) = 1$. *A cyclic codes of length* n *over* $\mathbb{F}_q$ *with defining set* $Z = \mathbb{C}_{r_1} \cup \mathbb{C}_{r_2} \cup \ldots \cup \mathbb{C}_{r_n}$ *contains its Euclidean dual code if and only if the union of the complementary cosets* $\overline{\mathbb{C}}_i$, *where* $i = r_1, r_2, \ldots, r_n$, *and* Z *does not have common elements.*

Proof Since the coset $\mathbb{C}_{-i}$ is the complementary coset of $\mathbb{C}_i$, for all $i = r_1, r_2, \ldots, r_n$, the result follows.   □

Remark 5.2.2 It is interesting to note that analogous result can be shown in the Hermitian case.

Let us consider the following result shown in [4].

Lemma 5.2.2 ([4, Lemmas 8 and 9]) *Let* $n \geq 1$ *be an integer and* q *be a power of a prime such that* $\gcd(n, q) = 1$ *and* $q^{\lfloor m/2 \rfloor} < n \leq q^m - 1$, *where* $m = ord_n(q)$.

(a) *The cyclotomic coset* $\mathbb{C}_x = \{xq^j \bmod n \mid 0 \leq j < m\}$ *has cardinality* m *for all* x *in the range* $1 \leq x \leq nq^{\lceil m/2 \rceil}/(q^m - 1)$.

(b) *If* x *and* y *are distinct integers in the range* $1 \leq x, y \leq \min\{\lfloor nq^{\lceil m/2 \rceil}/(q^m - 1) - 1 \rfloor, n - 1\}$ *such that* $x, y \neq 0$, *then the* q-ary cosets of x and y modulo n *are disjoint.*

We next improve the upper bound for the number of disjoint q-ary cosets modulo $q^m - 1$.

Theorem 5.2.2 *Let* $n = q^m - 1$, *where* q *is a prime power and* m *is even. If* x *and* y *are distinct integers in the range* $1 \leq x, y \leq 2q^{m/2}$, *such that* $x \neq 0 \bmod q$ *and* $y \neq 0 \bmod q$, *then the* q-ary cosets of x and y modulo n *are disjoint.*

Proof Let us consider the following result.

Theorem 5.2.3 ([165, Theorem 2.3]) *Let* $n = q^m - 1$, *where* q *is a prime power and* m *is even. Denote* $s^* = \min\{t : t \in \mathbb{C}_s\}$ *be the minimum coset representative. If* $0 \leq s \leq T$, *where* $T := 2q^{m/2}$, *and* $q \nmid s$ *then* $s = s^*$, *and* T *is the greatest value having this property.*

Applying Theorem 5.2.3, the result follows directly. To see this, note that if the inequalities $0 \leq s \leq T := 2q^{m/2}$ hold, it follows that $s = s^*$. Hence, there exist at least $2q^{m/2}$ q-cosets in the range $0 \leq s \leq T$. Because the minimum coset representatives belongs to disjoint cosets, there are exactly $2q^{m/2}$ disjoint coset in the range $0 \leq s \leq T := 2q^{m/2}$. The proof is complete.   □

Theorem 5.2.2 will be used in the construction of new families of quantum codes, (see Sect. 5.2.2).

Lemma 5.2.3 *Let $n = q^m - 1$, where $q \geq 3$ is a prime power and consider c be a positive integer. Then the c cosets given by*

$$\{\mathbb{C}_{q+1}, \mathbb{C}_{2q+1}, \mathbb{C}_{3q+1}, \ldots, \mathbb{C}_{cq+1}\}$$

are disjoint and each of them has m elements provided $cq + 1 < \lfloor q^{\lceil m/2 \rceil} - 1 \rfloor$. Moreover, each of these cosets are distinct of the cosets $\mathbb{C}_1, \mathbb{C}_2, \ldots, \mathbb{C}_c$.

Proof Apply Lemma 5.2.2.   □

Theorem 5.2.4 *Let $n = q^m - 1$, where $q \geq 3$, and assume that the inequality $cq + 1 < \lfloor q^{\lceil m/2 \rceil} - 1 \rfloor$ holds. Then the last elements in the c cosets given by*

$$\{\mathbb{C}_{q+1}, \mathbb{C}_{2q+1}, \mathbb{C}_{3q+1}, \ldots, \mathbb{C}_{cq+1}\},$$

form a sequence of c consecutive integers.

Proof Let us consider the q-ary cosets modulo $n = q^m - 1$ given by

$$\{\mathbb{C}_{q+1}, \mathbb{C}_{2q+1}, \mathbb{C}_{3q+1}, \ldots, \mathbb{C}_{cq+1}\}.$$

We show that the last elements in these cosets form a sequence of c consecutive positive integers. In fact, from Lemma 5.2.3, each of these cosets has cardinality m. Let $\mathbb{C}_s$ and $\mathbb{C}_{s+q}$ be two of them. Let u and v be the last elements in the cosets $\mathbb{C}_s$ and $\mathbb{C}_{s+q}$, respectively, where u and v are integers considered without using the modulo n operation. Then it follows that

$$u = sq^{m-1} = sq^t; \quad v = (s + q)q^{m-1} = (s + q)q^t.$$

Therefore, we know that $v = sq^t + q^{t+1}$; so, $v \equiv sq^t + 1 \mod (q^m - 1)$, that is,

$$v \equiv u + 1 \mod (q^m - 1).$$

Let $n = q^m - 1$. Applying the division algorithm for v and n and for $u + 1$ and n, there exist integers a, b, r_1 and r_2, where $0 \leq r_1, r_2 < n$ such that

$$v = an + r_1 \quad u + 1 = bn + r_2.$$

Since $v \equiv u + 1 \mod n$, it follows that $r_1 = r_2$. Denote this common number by v^*. We thus have

$$v = an + v^*; \quad u + 1 = bn + v^*,$$

where $0 \leq v^* < n$. Therefore,

$$u = bn + v^* - 1.$$

Let $u^* = v^* - 1$; it is clear that $0 \leq u^* < n$. Moreover, by the uniqueness of the remainder, u^* is the remainder of u when considering the modulo n. This means that the representatives v^* and u^* of v and u, respectively, are consecutive, i.e., $v^* = u^* + 1$. The proof is complete. $\square$

5.2.2 Code Constructions

In this subsection we apply some results of Sect. 5.2.1 in order to construct families of q-ary CSS quantum codes.

5.2.2.1 Construction I

In order to proceed further we need to prove Lemma 5.2.4.

Lemma 5.2.4 *Let p be a prime number, $p \geq 5$. Let $n = p^2 - 1$ and consider the first $2p - 2$ p-ary cosets modulo n given by*

$$\mathbb{C}_0 = \{0\},$$
$$\mathbb{C}_1 = \{1, \ p\},$$
$$\mathbb{C}_2 = \{2, \ 2p\},$$
$$\mathbb{C}_3 = \{3, \ 3p\},$$
$$\vdots$$
$$\mathbb{C}_{p-2} = \{p-2, \ (p-2)p\},$$
$$\mathbb{C}_{p+1} = \{p+1\},$$
$$\mathbb{C}_{p+2} = \{p+2, \ 1+2p\},$$
$$\vdots$$
$$\mathbb{C}_{2p-1} = \{2p-1, \ 1+(p-1)p\}.$$

Then, all these p-cosets modulo n are disjoint. Moreover, with exception of the cosets $\mathbb{C}_0$ and $\mathbb{C}_{p+1}$, which contain only one element, all of them have exactly two elements.

Proof We know that $p^2 - 1 > 1 + (p-1)p$ and $p^2 - 1 > (p-2)p$ are true. It is clear that $\mathbb{C}_0$ and $\mathbb{C}_{p+1}$ contain only one element. We next show that, except cosets $\mathbb{C}_0$ and $\mathbb{C}_{p+1}$, all of them have exactly two elements.

    Case 1: If $l = lp$, since $l = lp < p^2 - 1$ we obtain $p = 1$, a contradiction since p is a prime.

    Case 2: Assume that $p + l = 1 + lp$, where $2 \leq l \leq p - 1$ is an integer. Then one has $l - 1 = p(l - 1)$. Since $p + l = 1 + lp < p^2 - 1$ and $l - 1 \neq 0$, one obtains $p = 1$, which is a contradiction.

We now prove that all these cosets are disjoint.

Case 3: Assume that $1 \leq l \leq p - 1$ and $1 \leq k \leq p - 2$ and consider $1 + lp = kp$. Since $1 + lp = kp < p^2 - 1$ and $(k - l)p = 1$ hold, one has $p \mid 1$, which is a contradiction because p is a prime.

Case 4: Similarly, if $1 + lp = 1 + kp$, since $1 + lp = 1 + kp < p^2 - 1$ one has $l = k$.

Case 5: Moreover, if $kp = lp$, since $kp = lp < p^2 - 1$, one obtains $l = k$.

Case 6: If $p + j = kp$, where $1 \leq j \leq p - 1$ and $1 \leq k \leq p - 2$, we obtain $p(k - 1) = j$. Since $p + j = kp < p^2 - 1$, one can get $p \mid j$, which is a contradiction because $j < p$.

Case 7: If $p + j = 1 + lp$, where $2 \leq l, j \leq p - 1$, we conclude that $p(l - 1) = j - 1$. Since $p + j = 1 + lp < p^2 - 1$, it follows that $p \mid j - 1$, which is a contradiction because $j - 1 < p$.

Case 8: If $p - j = 1 + lp$, where $1 \leq l \leq p - 1$ and $2 \leq j \leq p$, since $0 < 1 + lp < p^2 - 1$, one obtains $p(1 - l) = j + 1$. If $l = 1$, $j = -1$, a contradiction; if $l > 1$, $j < -1$, which is a contradiction.

Case 9: If $p - j = kp$, where $1 \leq k \leq p - 2$ and $2 \leq j \leq p$, since $0 < p - j = kp < p^2 - 1$, one has $p(k - 1) = -j$. If $k = 1$ then $j = 0$, which is a contradiction; if $k > 1$ then $j < 0$, which is a contradiction.

Case 10: It is easy to see that the cyclotomic cosets $\mathbb{C}_0$ and $\mathbb{C}_{p+1}$ are disjoint from the other cosets and among them. $\square$

Applying Lemma 5.2.4 we obtain Theorem 5.2.5.

Theorem 5.2.5 *Let $p \geq 5$ be a prime number and $n = p^2 - 1$. Then there exists an $[[p^2 - 1, p^2 - 4p + 5, d \geq p]]_p$ quantum code.*

Proof Let C_1 be a cyclic code generated by the product of the minimal polynomials

$$C_1 = \langle g_1(x) \rangle = \langle M^{(0)}(x)M^{(1)}(x)\ldots M^{(p-2)}(x) \rangle,$$

and C_2 be the cyclic code generated by the product of the minimal polynomials

$$C_2 = \langle g_2(x) \rangle = \left\langle \prod_i M^{(i)}(x) \right\rangle,$$

where $M^{(i)}(x)$ are the minimal polynomials of α^i such that

$$i \notin \{p + 1, p + 2, \ldots, 2p - 1\}.$$

We know the minimum distance of the code C_1 is greater than or equal to p since its defining set contains the sequence of $p - 1$ consecutive integers given by $0, 1, \ldots, p - 2$. From the BCH bound, the minimum distance of C_1 is lower bounded by p. Similarly, the defining set of C, generated by the polynomial $h(x) = \frac{x^n - 1}{g_2(x)}$, contains the sequence of $p - 1$ consecutive integers given by $p + 1, p + 2, \ldots, p + (p - 1) = 2p - 1$; hence from the BCH bound, C also has minimum distance greater than or equal to p. Since the code $C_2^\perp$ is equivalent to C, $C_2^\perp$ also has minimum distance greater than or equal to p. Therefore, the resulting CSS code has minimum distance greater than or equal to p.

Next we compute the dimension of the new CSS codes. We know that the defining set Z_1 of C_1 has $p - 1$ disjoint cosets. Moreover, from Lemma 5.2.4, all of them (except coset) $\mathbb{C}_0$ have two elements. We know that the degree of the generator polynomial of a cyclic code is equal to the cardinality of its defining set. Hence, C_1 has dimension $k_1 = n - \partial(g_1(x))$, where $n = p^2 - 1$. From Lemma 5.2.4, Z_1 has $2(p-2) + 1 = 2p - 3$ elements, so $k_1 = p^2 - 2p + 2$. Similarly, the dimension k_2 of C_2 is equal to $k_2 = 2p - 3$. Thus, the CSS code has dimension $k_1 - k_2 = p^2 - 4p + 5$. Applying the CSS construction to C_1 and C_2, we have an $[[p^2 - 1, p^2 - 4p + 5, d \geq p]]_p$ CSS code. $\qquad\square$

Proceeding analogously as in the proof of Theorem 5.2.5, we obtain more families of quantum codes.

Corollary 5.2.4 *There exists an* $[[p^2 - 1, p^2 - 4c + 5, d \geq c]]_p$ *quantum code, where $c < p$, and p is prime.*

Proof Let C_1 and C_2 be two cyclic codes generated, respectively, by

$$M^{(0)}(x)M^{(1)}(x)M^{(2)}(x)\ldots M^{(c-2)}(x)$$

and

$$C_2 = \left\langle \prod_i M^{(i)}(x) \right\rangle,$$

where $M^{(i)}(x)$ are all minimal polynomials of α^i such that

$$i \notin \{p+1, p+2, \ldots, p+(c-1)\}.$$

Proceeding similarly as in the proof of Theorem 5.2.5, we have an

$$[[p^2 - 1, p^2 - 4c + 5, d \geq c]]_p$$

quantum code. $\qquad\square$

Remark 5.2.3 It is clear that the previous code constructions also hold when considering $q = p^m$ instead of considering p (p prime), and $\mathbb{F}_{q^l}$ ($l \geq 2$) instead of considering $\mathbb{F}_q$, since the properties of cosets and the minimal polynomials are the same when considered over $\mathbb{F}_p$ as well as over $\mathbb{F}_q$.

Based on Remark 5.2.3, Theorem 5.2.5 and Corollary 5.2.4 can be easily extended for all prime power q.

Theorem 5.2.6 *Let $q \geq 4$ be a prime power and $n = q^2 - 1$. Then, there exist CSS codes with parameters $[[q^2 - 1, q^2 - 4q + 5, d \geq q]]_q$.*

Proof It suffices to consider C_1 and C_2 as the cyclic codes generated, respectively, by

$$C_1 = \langle M^{(0)}(x)M^{(1)}(x)M^{(2)}(x)\ldots M^{(q-2)}(x)\rangle$$

and

$$C_2 = \left\langle \prod_i M^{(i)}(x)\right\rangle,$$

where $M^{(i)}(x)$ are all minimal polynomials of α^i such that

$$i \notin \{q+1, q+2, \ldots, 2q-1\}.$$

Proceeding similarly as in the proof of Theorem 5.2.5 changing p for q, the result follows. □

Corollary 5.2.5 *There exists an* $[[q^2-1, q^2-4c+5, d \geq c]]_q$ *quantum code, where* $c < q$, $q = p^m$, *and* $q \geq 4$.

Proof Let C_1 and C_2 be cyclic code generated, respectively, by

$$C_1 = \langle M^{(0)}(x)M^{(1)}(x)M^{(2)}(x)\ldots M^{(c-2)}(x)\rangle$$

and

$$C_2 = \left\langle \prod_i M^{(i)}(x)\right\rangle,$$

where $M^{(i)}(x)$ are all minimal polynomials of α^i such that

$$i \notin \{q+1, q+2, \ldots, q+(c-1)\}.$$

Proceeding similarly as in the proof of Corollary 5.2.4 changing p for q, the result follows. □

5.2.2.2   Construction II

We need to utilize the following result shown in [166] in order to obtain our families of quantum codes.

Theorem 5.2.7 *[166]* $|\mathbb{C}_s| = m$ *for all* $0 < s < T := 2q^{m/2}$ *except* $|\mathbb{C}_{q^{m/2}+1}| = m/2$ *when m is even.*

Once the upper bound for the number of distinct q-ary cosets has been improved (see Theorem 5.2.2), we are able to construct families q-ary CSS codes with good parameters. Theorem 5.2.8 asserts the existence of such codes.

Theorem 5.2.8 *Let $n = q^m - 1$, where $q \geq 4$ is a prime power and $m \geq 2$ is an even integer. Then there exists an $[[n, n - 2m(c - 2) - m/2 - 1, d \geq c]]_q$ quantum code, where $2 \leq c \leq q$.*

Proof Let C_1 be the cyclic code generated by

$$M^{(0)}(x)M^{(1)}(x)\ldots M^{(c-2)}(x),$$

and C_2 be the cyclic code generated by $g_2(x)$, that is the product of the minimal polynomials

$$g_2(x) = \prod_i M^{(i)}(x),$$

where $M^{(i)}(x)$ are the minimal polynomials of α^i such that

$$i \notin \{q^{m/2} + 1, q^{m/2} + 2, \ldots, q^{m/2} + c - 1\}.$$

From the BCH bound, the minimum distance of C_1 is greater than or equal to c, since its defining set contains the sequence $0, 1, \ldots, c - 2$. Again, from the BCH bound, the minimum distance of $C_2^{\perp}$ is also greater than or equal to c, because $C_2^{\perp}$ is equivalent to $C = \langle (x^n - 1)/g_2(x) \rangle$ and C contains the sequence $q^{m/2} + 1, q^{m/2} + 2, \ldots, q^{m/2} + c - 1$. From the CSS construction, the resulting quantum code has minimum distance greater than or equal to c. Moreover, we have $C_2 \subset C_1$.

The degree of $g_1(x)$ is equal to the cardinality of the defining set Z_1 of C_1. Moreover, from Theorem 5.2.7, Z_1 has $m(c - 2) + 1$ elements, so the dimension k_1 of C_1 is

$$k_1 = n - m(c - 2) - 1.$$

Similarly, by applying Theorem 5.2.7, since $q^{m/2} + c - 1 < T := 2q^{m/2}$, then C_2 has dimension

$$k_2 = n - [n - (m(c - 2) + m/2)] = m(c - 2) + m/2.$$

Thus, the CSS code has dimension $n - 2m(c - 2) - m/2 - 1$. Therefore, an

$$[[n, n - 2m(c - 2) - m/2 - 1, d \geq c]]_q,$$

quantum code can be constructed. The proof is complete.   □

Quantum codes with parameters $[[15, 9, d \geq 3]]_4$, $[[15, 5, d \geq 4]]_4$, $[[24, 18, d \geq 3]]_5$ and $[[24, 14, d \geq 4]]_5$ can be constructed by applying Construction I.

5.2.2.3   Construction III

We start by showing Theorem 5.2.9, a particular case of Theorem 5.2.10.

Theorem 5.2.9 *Let $q \geq 5$ be an odd prime power. Then there exists an $[[q^3 - 1, q^3 -21, d \geq 5]]_q$ quantum code.*

Proof It is easy to check that $\mathbb{C}_{(\frac{q^3-1}{2})}$ contains only one element. Moreover, the cosets $\mathbb{C}_1$, $\mathbb{C}_2$ and $\mathbb{C}_3$ are disjoint and each of them has three elements.

We next show that the q-cosets $\mathbb{C}_{(\frac{q^3-1}{2})}$, $\mathbb{C}_{(\frac{q^3-1}{2}+1)}$, $\mathbb{C}_{(\frac{q^3-1}{2}+2)}$ and $\mathbb{C}_{(\frac{q^3-1}{2}+3)}$ are distinct of the cosets $\mathbb{C}_1$, $\mathbb{C}_2$ and $\mathbb{C}_3$. Clearly the coset $\mathbb{C}_{(\frac{q^3-1}{2})}$ is distinct of such cosets. Consider the coset $\mathbb{C}_{(\frac{q^3-1}{2}+1)}$ and suppose w.l.o.g. that $\frac{q^3-1}{2} + 1$ is even. Then, from Corollary 5.2.2, $\mathbb{C}_{(\frac{q^3-1}{2}+1)}$ is disjoint from $\mathbb{C}_1$ and $\mathbb{C}_3$. If $(\frac{q^3-1}{2} + 1)q^i \equiv 2 \mod n$, then it follows that

$$[(q^3 - 1)q^i + 2q^i] \equiv 4 \mod n$$
$$\Longrightarrow [(q^3 - 1)q^i + 2q^i] \equiv 2q^i \equiv 4 \mod n.$$

Since $2q^i - 4 < q^3 - 1$ and because the equality $2q^i = 4$ is not satisfied, the coset $\mathbb{C}_{(\frac{q^3-1}{2}+1)}$ is disjoint of $\mathbb{C}_2$. The other cases are similar to the previous one.

In what follows, we show that each of cosets $\mathbb{C}_{(\frac{q^3-1}{2}+1)}$, $\mathbb{C}_{(\frac{q^3-1}{2}+2)}$ and $\mathbb{C}_{(\frac{q^3-1}{2}+3)}$ are disjoint among them. From Corollary 5.2.2,

$$\mathbb{C}_{(\frac{q^3-1}{2}+1)} \neq \mathbb{C}_{(\frac{q^3-1}{2}+2)}$$

and

$$\mathbb{C}_{(\frac{q^3-1}{2}+2)} \neq \mathbb{C}_{(\frac{q^3-1}{2}+3)}.$$

To show that $\mathbb{C}_{(\frac{q^3-1}{2}+1)}$ is disjoint from $\mathbb{C}_{(\frac{q^3-1}{2}+3)}$, let us consider the following congruence:

$$\left(\frac{q^3 - 1}{2} + 1\right) \equiv \left(\frac{q^3 - 1}{2} + 3\right)q^i \mod n.$$

We then obtain

$$(q^3 - 1 + 2) \equiv [(q^3 - 1)q^i + 6q^i] \mod n$$
$$\Longrightarrow 6q^i \equiv 2 \mod n.$$

Since $6q^i \equiv 2 \mod n$ does not hold, the result follows.

It is easy to see that each one of the cosets $\mathbb{C}_{(\frac{q^3-1}{2}+1)}$, $\mathbb{C}_{(\frac{q^3-1}{2}+2)}$ and $\mathbb{C}_{(\frac{q^3-1}{2}+3)}$ has three elements.

Let C_1 be the cyclic code generated by

$$M^{(0)}(x)M^{(1)}(x)M^{(2)}(x)M^{(3)}(x)$$

and C_2 be the cyclic code generated by

$$\prod_i M^{(i)}(x),$$

where $M^{(i)}(x)$ are the minimal polynomials such that $i \notin \{b, \ b+1, \ b+2, \ b+3\}$ and i runs through the coset representatives mod $(q^3 - 1)$ and $b = \frac{q^3-1}{2}$. Proceeding similarly as in the proof of Theorem 5.2.5 and applying the CSS construction, the code follows. $\square$

Applying the previous theorem one can construct quantum codes with parameters $[[124, 104, d \geq 5]]_5$, $[[342, 322, d \geq 5]]_7$, $[[1330, 1310, d \geq 5]]_{11}$, and so on.

We can also construct CSS codes with minimum distance greater than three and four, as states the next result.

Corollary 5.2.6 *Let $q \geq 5$ be an odd prime power. Then there exist CSS codes with parameters $[[q^3 - 1, q^3 - 9, d \geq 3]]_q$ and $[[q^3 - 1, q^3 - 15, d \geq 4]]_q$.*

Proof For the first construction, consider C_1 be the cyclic code generated by

$$M^{(0)}(x)M^{(1)}(x)$$

and C_2 be the cyclic code generated by

$$\prod_i M^{(i)}(x),$$

where $M^{(i)}$ are the minimal polynomials such that $i \notin \{b, \ b+1\}$ and i runs through the coset representatives mod $(q^3 - 1)$ and $b = \frac{q^3-1}{2}$.

For the second, let us consider C_1 as the cyclic code generated by

$$M^{(0)}(x)M^{(1)}(x)M^{(2)}(x)$$

and C_2 be the cyclic code generated by

$$\prod_i M^{(i)}(x),$$

where $M^{(i)}$ are the minimal polynomials such that $i \notin \{b, b+1, b+2\}$ and i runs through the coset representatives mod $(q^3 - 1)$ and $b = \frac{q^3-1}{2}$. Proceeding similarly as in the proof of Theorem 5.2.5 and applying the CSS construction to these code, the results follows.   □

Applying Corollary 5.2.6 one can construct quantum codes with parameters $[[124, 116, d \geq 3]]_5$, $[[124, 110, d \geq 4]]_5$, $[[342, 334, d \geq 3]]_7$, $[[342, 328, d \geq 4]]_7$, and so on.

Theorem 5.2.9 can be generalized in the following way.

Theorem 5.2.10 *Let $q \geq 5$ be an odd prime power. Then there exists an* $[[q^m - 1, q^m - 6m - 3, d \geq 5]]_q$ *CSS code.*

Proof The coset $\mathbb{C}_{(\frac{q^m-1}{2})}$ contains only the element $\left(\frac{q^m-1}{2}\right)$. Moreover, proceeding similarly as in the proof of Theorem 5.2.9, it is easy to see that the cosets $\mathbb{C}_{(\frac{q^m-1}{2})}$, $\mathbb{C}_{(\frac{q^m-1}{2}+1)}$, $\mathbb{C}_{(\frac{q^m-1}{2}+2)}$ and $\mathbb{C}_{(\frac{q^m-1}{2}+3)}$ are disjoint from the cosets $\mathbb{C}_1$, $\mathbb{C}_2$ and $\mathbb{C}_3$ and also disjoint among them. Moreover, it is easy to see that each of these cosets has m elements. Let C_1 be the cyclic code generated by

$$M^{(0)}(x)M^{(1)}(x)M^{(2)}(x)M^{(3)}(x),$$

and C_2 be the cyclic code generated by

$$\prod_i M^{(i)}(x),$$

where $M^{(i)}$ are the minimal polynomials of α^i such that $i \notin \{b, b+1, b+2, b+3\}$ and i runs through the coset representatives mod $(q^m - 1)$ and $b = \frac{q^m-1}{2}$. Applying the CSS construction to these codes, we are done.   □

Corollary 5.2.7 *Let $q \geq 5$ be an odd prime power. Then there exist CSS codes with parameters $[[q^m - 1, q^m - 2m - 3, d \geq 3]]_q$ and $[[q^m - 1, q^m - 4m - 3, d \geq 4]]_q$.*

Applying Theorem 5.2.10 and Corollary 5.2.7 one can construct quantum codes with parameters $[[80, 74, d \geq 3]]_9$, $[[80, 70, d \geq 4]]_9$, $[[80, 66, d \geq 5]]_9$, $[[624, 614, d \geq 3]]_5$, $[[624, 606, d \geq 4]]_5$, $[[624, 598, d \geq 5]]_5$, $[[728, 720, d \geq 3]]_9$, $[[728, 714, d \geq 4]]_9$, $[[728, 708, d \geq 5]]_9$ and so on.

Theorem 5.2.11 *Let $n = q^m - 1$, where $q \geq 3$ is a prime power and $m \geq 3$. Then there exists an $[[n, n - m(2c - 3) - 1, d \geq c]]_q$ quantum code, where $2 \leq c \leq q$ and $(c - 1)q + 1 < \lfloor q^{\lceil m/2 \rceil} - 1 \rfloor$.*

Proof Let C_1 be the cyclic code generated by

$$M^{(0)}(x)M^{(1)}(x)\ldots M^{(c-2)}(x),$$

where $2 \leq c \leq q$, and C_2 be the cyclic code generated by

$$\prod_i M^{(i)}(x),$$

where $M^{(i)}(x)$ are the minimal polynomials such that $i \notin \{q+1, 2q+1, \ldots, (c-1)q+1\}$. Applying Lemma 5.2.3 and Theorem 5.2.4, and proceeding similarly as in the proof of Theorem 5.2.5 the result follows. $\square$

Example 5.2.1 Applying Theorem 5.2.11 with $q = 3$, $n = 3^3 - 1 = 26$ and $d \geq 3$ one can construct an $[[26, 16, d \geq 3]]_3$ quantum code. If $q = 4$, $n = 63$ and $d \geq 3$ one has an $[[63, 53, d \geq 3]]_4$ quantum code. Similarly, if we take $q = 5$, $n = 124$ and $d \geq 4$ one has an $[[124, 108, d \geq 4]]_5$ quantum code and so on.

5.2.3 Code Comparison

In this section we compare the parameters of our CSS codes with the parameters of the best CSS codes shown in [4]. As was said, we utilize the code comparison described in Remark 5.1.2. Such criterion is usual in the literature.

In Table 5.8, the parameters $[[n, k, d \geq c]]_q$ assume the values $[[q^2 - 1, q^2 - 4c + 5, d \geq c]]_q$, where $2 \leq c \leq q$ and $q \geq 4$ is a prime power.

In Table 5.9, $n = q^m - 1$ is the code length, where $q \geq 4$ and $m \geq 2$ is an even integer; $k = n - 2m(c - 2) - m/2 - 1$, where $2 \leq c \leq q$ and d is the minimum distance of the respective code. The parameters of our codes are obtained from Construction I. The parameters $[[n', k', d']]_q = [[n, n - 2m(\lceil (\delta - 1)(1 - 1/q)\rceil), d \geq \delta]]_q$. are the parameters of the codes shown in [4].

In Table 5.10, our codes are derived from Construction II, where $q \geq 5$ is an odd prime power and $[[n', k', d']]_q$ assumes the values mentioned above. In Table 5.11, the codes are derived from Construction III, where $n = q^m - 1$, $q \geq 3$ is a prime power, $m \geq 3$ and $2 \leq c \leq q$. The parameters $[[n', k', d']]_q$ assume the values mentioned above.

As can be seen in Tables 5.8, 5.9, 5.10 and 5.11, our CSS codes have parameters better than the ones available in [4]. More precisely, fixing n and d, the codes constructed here achieve greater values of the number of qudits than the codes shown in [4].

5.3 BCH Codes—Part III

In this section, we construct more families of quantum codes derived from BCH codes. The codes constructed here can be found in our paper [89].

Table 5.8 Comparisons

Our CSS codes	CSS Codes in [4]
$[[n, k, d \geq c]]_q$	$[[n', k', d']]_q$
$[[15, 9, d \geq 3]]_4$	$[[15, 7, d' \geq 3]]_4$
$[[15, 5, d \geq 4]]_4$	—
$[[24, 18, d \geq 3]]_5$	$[[24, 16, d' \geq 3]]_4$
$[[24, 10, d \geq 5]]_5$	—
$[[48, 42, d \geq 3]]_7$	$[[48, 40, d' \geq 3]]_7$
$[[48, 38, d \geq 4]]_7$	$[[48, 36, d' \geq 4]]_7$
$[[48, 34, d \geq 5]]_7$	$[[48, 32, d' \geq 5]]_7$
$[[48, 30, d \geq 6]]_7$	$[[48, 28, d' \geq 6]]_7$
$[[48, 26, d \geq 7]]_7$	—
$[[63, 57, d \geq 3]]_8$	$[[63, 55, d' \geq 3]]_8$
$[[63, 53, d \geq 4]]_8$	$[[63, 51, d' \geq 4]]_8$
$[[63, 49, d \geq 5]]_8$	$[[63, 47, d' \geq 5]]_8$
$[[63, 45, d \geq 6]]_8$	$[[63, 43, d' \geq 6]]_8$
$[[63, 41, d \geq 7]]_8$	$[[63, 39, d' \geq 7]]_8$
$[[63, 37, d \geq 8]]_8$	—
$[[80, 74, d \geq 3]]_9$	$[[80, 72, d' \geq 3]]_9$
$[[80, 70, d \geq 4]]_9$	$[[80, 68, d' \geq 4]]_9$
$[[80, 66, d \geq 5]]_9$	$[[80, 64, d' \geq 5]]_9$
$[[80, 54, d \geq 8]]_9$	$[[80, 52, d' \geq 8]]_9$
$[[80, 50, d \geq 9]]_9$	—
$[[120, 114, d \geq 3]]_{11}$	$[[120, 112, d' \geq 3]]_{11}$
$[[120, 106, d \geq 5]]_{11}$	$[[120, 104, d' \geq 5]]_{11}$
$[[120, 98, d \geq 7]]_{11}$	$[[120, 96, d' \geq 7]]_{11}$
$[[120, 90, d \geq 9]]_{11}$	$[[120, 88, d' \geq 9]]_{11}$
$[[120, 82, d \geq 11]]_{11}$	—
$[[168, 162, d \geq 3]]_{13}$	$[[168, 160, d' \geq 3]]_{13}$
$[[168, 154, d \geq 5]]_{13}$	$[[168, 152, d' \geq 5]]_{13}$
$[[168, 146, d \geq 7]]_{13}$	$[[168, 144, d' \geq 7]]_{13}$
$[[168, 138, d \geq 9]]_{13}$	$[[168, 136, d' \geq 9]]_{13}$
$[[168, 130, d \geq 11]]_{13}$	$[[168, 128, d' \geq 11]]_{13}$
$[[168, 122, d \geq 13]]_{13}$	—

Table 5.9 Code comparison

Our CSS codes—Construction I	CSS codes in [4]
$[[n, n - 2m(c - 2) - m/2 - 1, d \geq c]]_q$	$[[n', k', d']]_q$
$m = 2$	
$[[15, 9, d \geq 3]]_4$	$[[15, 7, d' \geq 3]]_4$
$[[15, 5, d \geq 4]]_4$	$[[15, 3, d' \geq 4]]_4$
$[[24, 18, d \geq 3]]_5$	$[[24, 16, d' \geq 3]]_7$
$[[24, 14, d \geq 4]]_5$	$[[24, 12, d' \geq 4]]_7$
$[[24, 10, d \geq 5]]_5$	$[[24, 8, d' \geq 5]]_7$
$[[63, 57, d \geq 3]]_8$	$[[63, 55, d' \geq 3]]_8$
$[[63, 53, d \geq 4]]_8$	$[[63, 51, d' \geq 4]]_8$
$[[63, 49, d \geq 5]]_8$	$[[63, 47, d' \geq 5]]_8$
$[[63, 45, d \geq 6]]_8$	$[[63, 43, d' \geq 6]]_8$
$[[63, 41, d \geq 7]]_8$	$[[63, 39, d' \geq 7]]_8$
$[[63, 37, d \geq 8]]_8$	$[[63, 35, d' \geq 8]]_8$
$m = 4$	
$[[255, 244, d \geq 3]]_4$	$[[255, 239, d' \geq 3]]_4$
$[[255, 236, d \geq 4]]_4$	$[[255, 231, d' \geq 4]]_4$
$[[624, 613, d \geq 3]]_5$	$[[624, 608, d' \geq 3]]_5$
$[[624, 605, d \geq 4]]_5$	$[[624, 600, d' \geq 4]]_5$
$[[624, 597, d \geq 5]]_5$	$[[624, 592, d' \geq 5]]_5$

The first construction generates quantum codes with parameters

(i) $[[n, n - 4(c - 2) - 2, d \geq c]]_q$, where $n = q^4 - 1$, and $3 \leq c \leq q^2$;

The second one produces codes with parameters

(ii) $[[n, n - 2mc - 2, d \geq c + 2]]_q$, for all $1 \leq c \leq q^2 - 2$;
(iii) $[[n, n - 2m(q^2 - 1) - 2, d \geq q^2 + 2]]_q$;
(iv) $[[n, n - 2m(c - 1) - 2, d \geq c + 2]]_q$, for all $q^2 + 1 \leq c \leq 2q^2 + 2$;
(v) $[[n, n - 4m(q^2 - 1) - 2, d \geq 2q^2 + 2]]_q$, for all $q^2 + 1 \leq c \leq 2q^2 + 2$, where $n = q^{2m} - 1$, $q \geq 4$ is a prime power, and $m = \mathrm{ord}_n(q^2) \geq 3$.

The third one generates families of quantum codes with parameters

(vi) $[[n, n - m(2c - 1) - 2, d \geq c + 2]]_q$, for all $1 \leq c \leq q - 2$;
(vii) $[[n, n - m(2q - 3) - 2, d \geq q + 1]]_q$;
(viii) $[[n, n - m(2q - 1) - 1, d \geq q + 3]]_q$;
(ix) $[[n, n - m(2c - 4) - 2, d \geq c + 2]]_q$;
(x) $[[n, n - m(4q - 8) - 2, d \geq 2q]]_q$;
(xi) $[[n, n - m(4q - 5) - 2, d \geq 2q + 2]]_q$, where $q + 1 < c < 2q - 2$; where $n = q^m - 1$, $q \geq 4$ is a prime power, and $m = \mathrm{ord}_n(q) \geq 3$.

Table 5.10 Code comparison

Our CSS codes—Construction II	CSS Codes in [4]
$[[q^m - 1, q^m - 2m - 3, d \geq 3]]_q$	$[[n', k', d']]_q$
$[[q^m - 1, q^m - 4m - 3, d \geq 4]]_q$	$[[n', k', d']]_q$
$[[q^m - 1, q^m - 6m - 3, d \geq 5]]_q$	$[[n', k', d']]_q$
$m = 2$	
$[[80, 74, d \geq 3]]_9$	$[[80, 72, d' \geq 3]]_9$
$[[80, 70, d \geq 4]]_9$	$[[80, 68, d' \geq 4]]_9$
$[[80, 66, d \geq 5]]_9$	$[[80, 66, d' \geq 5]]_9$
$m = 3$	
$[[124, 116, d \geq 3]]_5$	$[[124, 112, d' \geq 3]]_5$
$[[124, 110, d \geq 4]]_5$	$[[124, 106, d' \geq 4]]_5$
$[[124, 104, d \geq 5]]_5$	$[[124, 100, d' \geq 5]]_5$
$[[342, 334, d \geq 3]]_7$	$[[342, 330, d' \geq 3]]_7$
$[[342, 328, d \geq 4]]_7$	$[[342, 324, d' \geq 4]]_7$
$[[342, 322, d \geq 5]]_7$	$[[342, 318, d' \geq 5]]_7$
$m = 4$	
$[[624, 614, d \geq 3]]_5$	$[[624, 608, d' \geq 3]]_5$
$[[624, 606, d \geq 4]]_5$	$[[624, 600, d' \geq 4]]_5$
$[[624, 598, d \geq 5]]_5$	$[[624, 592, d' \geq 5]]_5$

We need to recall three useful lemmas from [4].

Lemma 5.3.1 ([4, Lemma 1]) *Assume that* $\gcd(q, n) = 1$. *A cyclic code of length n over* $\mathbb{F}_q$ *with defining set Z contains its Euclidean dual code if and only if* $Z \cap Z^{-1} = \emptyset$, *where* $Z^{-1} = \{-z \bmod n | z \in Z\}$.

Lemma 5.3.2 *Assume that* $\gcd(q, n) = 1$. *A cyclic code of length n over* $\mathbb{F}_{q^2}$ *with defining set Z contains its Hermitian dual code if and only if* $Z \cap Z^{-q} = \emptyset$, *where* $Z^{-q} = \{-qz \bmod n | z \in Z\}$.

Lemma 5.3.3 (Hermitian Construction) *If there exists a classical linear* $[n, k, d]_{q^2}$ *code D such that* $D^{\perp_H} \subset D$, *then there exists an* $[[n, 2k - n, \geq d]]_q$ *stabilizer code that is pure to d. If the minimum distance* $d^{\perp_H}$ *of* $D^{\perp_H}$ *exceeds d, then the stabilizer code is pure and has minimum distance d.*

We utilize the notation $\mathbb{C}_{[a]}$ to denote the cyclotomic coset containing a, where a is not necessarily the smallest number in $\mathbb{C}_{[a]}$.

Table 5.11 Code comparison

Our CSS codes—Construction III	CSS Codes in [4]
$[[n, n - m(2c - 3) - 1, d \geq c]]_q, 2 \leq c \leq q$	$[[n', k', d']]_q$
$m = 3$	
$[[26, 16, d \geq 3]]_3$	$[[26, 14, d' \geq 3]]_3$
$[[63, 53, d \geq 3]]_4$	$[[63, 51, d' \geq 3]]_4$
$[[63, 47, d \geq 4]]_4$	$[[63, 45, d' \geq 4]]_4$
$[[124, 114, d \geq 3]]_5$	$[[124, 112, d' \geq 3]]_5$
$[[124, 108, d \geq 4]]_5$	$[[124, 106, d' \geq 4]]_5$
$[[124, 102, d \geq 5]]_5$	$[[124, 100, d' \geq 5]]_5$
$[[342, 332, d \geq 3]]_7$	$[[342, 330, d' \geq 3]]_7$
$[[342, 326, d \geq 4]]_7$	$[[342, 324, d' \geq 4]]_7$
$[[342, 320, d \geq 5]]_7$	$[[342, 318, d' \geq 5]]_7$
$[[342, 314, d \geq 6]]_7$	$[[342, 312, d' \geq 6]]_7$
$[[342, 308, d \geq 7]]_7$	$[[342, 306, d' \geq 7]]_7$
$m = 4$	
$[[255, 242, d \geq 3]]_4$	$[[255, 239, d' \geq 3]]_4$
$[[255, 234, d \geq 4]]_4$	$[[255, 231, d' \geq 4]]_4$
$[[624, 611, d \geq 3]]_5$	$[[624, 608, d' \geq 3]]_5$
$[[624, 603, d \geq 4]]_5$	$[[624, 600, d' \geq 4]]_5$
$[[624, 595, d \geq 5]]_5$	$[[624, 592, d' \geq 5]]_5$

5.3.1 Construction I

Let us prove the first result.

Lemma 5.3.4 *Let $n = q^4 - 1$, where $q \geq 3$ is a prime power, and consider the first $q^2 - 1$ q^2-ary cosets modulo n given by*

$$\mathbb{C}_{[q^2+1]},$$
$$\mathbb{C}_{[q^2+2]} = \{q^2 + 2, \ 1 + 2q^2\},$$
$$\vdots$$
$$\mathbb{C}_{[2q^2-1]} = \{2q^2 - 1, \ 1 + (q^2 - 1)q^2\}.$$

Then the following hold:

(a) $\mathbb{C}_{[q^2+1]}$ contains only one element;
(b) each of the other cosets contains two elements;
(c) each of these cosets are mutually disjoint.

Proof Note first that the inequality $n > 1 + (q^2 - 1)q^2$ holds.

(a) This follows from the fact that $(q^2 + 1)q^2 \equiv q^2 + 1 \bmod n$.

(b) We prove that each of the cosets $\mathbb{C}_{[q^2+2]}$, $\mathbb{C}_{[2q^2-1]}$ has exactly two elements. To do this, assume that $q^2 + j \equiv 1 + jq^2 \bmod n$, where $j = 2, \ldots, q^2 + 1$. Because $1 + jq^2 < n$, we have $q^2 + j = 1 + jq^2$; hence, $j - 1 = (j - 1)q^2$, which is a contradiction.

(c) It is clear that coset $\mathbb{C}_{[q^2+1]}$ is disjoint from the other cosets, since it has only one element. Assume next that $\mathbb{C}_{[q^2+i]} = \mathbb{C}_{[q^2+j]}$, where $2 \leq i, j \leq q^2 - 1$, where $i \neq j$, Thus either $q^2 + i \equiv q^2 + j \bmod n$ or $q^2 + i \equiv (q^2 + j)q^2 \bmod n$, where $2 \leq i, j \leq q^2 - 1$. Since $2q^2 + 1 < q^4 - 1$ and $1 + (q^2 - 1)q^2 < q^4 - 1$ hold, such inequalities imply that $q^2 + i = q^2 + j$ or $q^2 + i = 1 + jq^2$. The first case implies $i = j$, a contradiction, and the second implies $q^2 | (i - 1)$, which is also a contradiction. Therefore, all these cosets are mutually disjoint. The proof is complete.

$\square$

In the sequence we use Lemma 5.3.4 to show how to construct quantum codes of length $n = q^4 - 1$.

Theorem 5.3.1 *Let $q \geq 3$ be a prime power and $n = q^4 - 1$. Then there exists an $[[n, n - 4(q^2 - 2) - 2, d \geq q^2]]_q$ quantum error-correcting code.*

Proof Let us consider C as the cyclic code generated by the product of the minimal polynomials

$$g(x) = M^{(q^2+1)}(x)M^{(q^2+2)}(x) \cdot \ldots \cdot M^{(q^2+j)}(x),$$

where $1 \leq j \leq q^2 - 1$. We show first that C is Hermitian dual-containing. Seeking a contradiction, we suppose $Z \cap Z^{-q} \neq \emptyset$. Thus there exist i, j, where $1 \leq i, j \leq q^2 - 1$ such that $\mathbb{C}_{[q^2+j]} = \mathbb{C}_{[-q(q^2+i)]}$. Hence, $q^2 + j \equiv -q(q^2 + i)q^{2k}$, where $k = 0$ or $k = 1$. If $k = 0$, we have $q^3 + qi + q^2 + j < q^4 - 1$, so $q^2 + j = -q^3 - qi$, a contradiction. If $k = 1$, since $\gcd(q^2, n) = 1$ and $q^4 \equiv 1 \bmod n$, we have

$$q^2 + j \equiv -q^3(q^2 + i) \bmod \ n \Longrightarrow$$
$$q^5 + q^3 i \equiv -(q^2 + j) \bmod \ n \Longrightarrow$$
$$q + q^3 i \equiv -(q^2 + j) \bmod n,$$

where $1 \leq i, j, q^2 - 1$.

If $i < q$ then $iq^3 + q + q^2 + j < q^4 - 1$, so $q + q^3 i = -(q^2 + j)$, a contradiction. On the other hand, if $i \geq q$, from the division algorithm we write $i = lq + r$, where r, l are integers such that $0 \leq r \leq q - 1$. We also have $0 \leq l \leq q - 1$; hence,

$$q + q^3 i = q + q^3(lq + r) \equiv q + l + q^3 r \bmod n.$$

Computing $q + l + q^3 r + q^2 + j$ we obtain

$$q + l + q^3 r + q^2 + j < q^3(q-1) + 2q + 2q^2 = q^4 - q^3 + 2q + 2q^2.$$

Since $q^3 > 2q^2 + 2q + 1$, it follows that $q + l + q^3 r + q^2 + j < q^4 - 1$; hence, $q + l + q^3 r = -q^2 - j$, a contradiction. Consequently, C is Hermitian dual-containing.

We next compute the minimum distance and the dimension of C. Since the defining set of C contains the sequence $q^2 + 1, q^2 + 2, \ldots, 2q^2 - 1$, it follows from the BCH bound that C has minimum distance greater than or equal to q^2. On the other hand, from Lemma 5.3.4, the defining set of C has $2(q^2 - 2) + 1$ elements. Hence, $g(x)$ has degree $\deg(g(x)) = 2(q^2 - 2) + 1$, so C has dimension $n - 2(q^2 - 2) - 1$, i.e., C is an $[n, n - 2(q^2 - 2) - 1, d \geq q^2]_{q^2}$ code. Applying Lemma 5.1.5, there exists an $[[n, n - 4(q^2 - 2) - 2, d \geq q^2]]_q$ quantum code. The proof is complete. $\qquad\square$

Corollary 5.3.1 *Let $q \geq 3$ be a prime power and $n = q^4 - 1$. Then there exists an $[[n, n - 4(c-2) - 2, d \geq c]]_q$, where $3 \leq c \leq q^2 - 1$*

Proof It suffices to consider C as the cyclic code generated by

$$M^{(q^2+1)}(x) M^{(q^2+2)}(x) \cdot \ldots \cdot M^{(q^2+c-1)}(x),$$

after proceeding similarly to the proof of Theorem 5.3.1. $\qquad\square$

Example 5.3.1 As an example, let us consider $m = 2$ and $q = 3$. Let C be generated by $M^{(10)}(x) M^{(11)}(x)$. Applying Theorem 5.3.1 we have an $[[80, 74, d \geq 3]]_3$ code.

5.3.2 Construction II

In this subsection, we construct suitable Hermitian dual-containing non-narrow-sense BCH codes with good parameters in order to obtain good quantum codes derived from them.

We start with the following result.

Lemma 5.3.5 *Let $q \neq 2$ be a prime power and $n = q^{2m} - 1$, where $m = \mathrm{ord}_n(q^2) \geq 3$. If $s = \sum_{i=0}^{m-1} (q^2)^i$, then the q^2-coset $\mathbb{C}_{[s]}$ has only one element.*

Proof We know that $\gcd(q^2, n) = 1$ and $q^{2m} \equiv 1 \bmod n$. The result follows from direct computation.

$$sq^{2j} = \left(\sum_{i=0}^{m-1} (q^2)^i \right) q^{2j}$$

$$= q^{2j}(q^2)^{m-1} + q^{2j}(q^2)^{m-2} + \cdots + q^{2j}q^2 + q^{2j} =$$

$$
\begin{aligned}
&= q^{2j}q^{2m}q^{-2} + q^{2j}q^{2m}q^{-4} + \cdots + q^{2j}q^{2m}q^{-2j+2} + \\
&\quad + q^{2j}q^{2m}q^{-2j} + q^{2j}q^{2m}q^{-2j-2} + \cdots + q^{2j}q^2 + q^{2j} \\
&\equiv (\bmod\ n) q^{2j}q^{-2} + q^{2j}q^{-4} + \cdots + q^{2j}q^{-2j+2} + \\
&\quad + q^{2j}q^{-2j} + q^{2m-2} + q^{2m-4} + \cdots + q^{2j}q^2 + q^{2j} = \\
&= (q^2)^{m-1} + (q^2)^{m-2} + \cdots + (q^2)^{j+1} + (q^2)^j + \\
&\quad + (q^2)^{j-1} + (q^2)^{j-2} + \cdots + q^2 + 1 = \\
&= \sum_{i=0}^{m-1} (q^2)^i = s
\end{aligned}
$$

$\square$

Lemma 5.3.6 *Let* $q \neq 2$ *be a prime power and* $n = q^{2m} - 1$, *where* $m = \mathrm{ord}_n(q^2) \geq 3$. *Let* $s = \displaystyle\sum_{i=0}^{m-1} (q^2)^i$. *Then the following results hold:*

(a) *the* q^2*-ary cosets of the form* $\mathbb{C}_{[s+i]}$ *are mutually disjoints, where* $1 \leq i \leq q^2 - 1$;
(b) *the* q^2*-ary cosets of the form* $\mathbb{C}_{[s-j]}$ *are mutually disjoints, where* $1 \leq j \leq q^2 - 1$;
(c) *the* q^2*-ary cosets of the form* $\mathbb{C}_{[s+i]}$ *are mutually disjoints from the* q^2*-ary cosets of the form* $\mathbb{C}_{[s-j]}$, *where* $1 \leq i, j \leq q^2 - 1$.

Proof We only show Item (a). Items (b) and (c) are left to the reader.

(a) Assume that there exist $i \neq j$, where $1 \leq i, j \leq q^2 - 1$ such that $\mathbb{C}_{[s+i]} = \mathbb{C}_{[s+j]}$. Then there exists $0 \leq t \leq m - 1$ such that $s + i \equiv (s + j)q^{2t} \bmod n$. From Lemma 5.3.5, we know that $sq^{2t} \equiv s \bmod n$. Since $\gcd(q^2, n) = 1$ and $q^{2m} \equiv 1 \bmod n$, then one has

$$
\begin{aligned}
s + i &\equiv (s + j)q^{2t} \equiv s + jq^{2t} \bmod n \\
&\Longrightarrow i \equiv jq^{2t} \bmod n.
\end{aligned}
$$

Because $1 \leq i, j \leq q^2 - 1$, it follows that

$$
i \equiv jq^{2t} \bmod n \Longrightarrow i = jq^{2t}.
$$

If $t = 0$, then $i = j$, a contradiction; if $t \geq 1$, the equality $i = jq^{2t}$ does not hold. Therefore, it follows that the cosets $\mathbb{C}_{[s+i]}$ and $\mathbb{C}_{[s+j]}$ are disjoint. The proof is complete. $\square$

Exercise 5.3.1 Show Items (b) and (c) of Lemma 5.3.6.

Lemma 5.3.7 *Let* $q \geq 4$ *be a prime power and* $n = q^{2m} - 1$, *where* $m = \mathrm{ord}_n(q^2) \geq 3$. *Let* $s = \displaystyle\sum_{i=0}^{m-1} (q^2)^i$. *Then the following hold:*

(a) *the cosets of the form* $\mathbb{C}_{[s+i]}$, *where* $1 \leq i \leq q^2 - 1$, *contain* m *elements;*
(b) *the cosets of the form* $\mathbb{C}_{[s-j]}$, $1 \leq j \leq q^2 - 1$, *contain* m *elements.*

Proof We prove Item (a). Item (b) is left as exercise.

(a) The elements of $\mathbb{C}_{[s+i]}$ are of the form $(s+i)q^{2t}$, where $0 \leq t \leq m - 1$ for all $1 \leq i \leq q^2 - 1$. Since $\gcd(q^2, n) = 1$, $q^{2m} \equiv 1 \bmod n$ and $sq^{2t} \equiv s \bmod n$, it follows that

$$(s+i)q^{2t} \equiv s + iq^{2t} \bmod n.$$

Let us consider that $0 \leq t \leq m - 2$. We then have

$$
\begin{aligned}
s + iq^{2t} \\
< (q^{2m} - 1)/(q^2 - 1) + q^{2m-2} \\
\leq (q^{2m} - 1)/15 + (q^{2m} - 1)/15 \\
< q^{2m} - 1.
\end{aligned}
$$

Hence, the first $m - 1$ elements belonging to $\mathbb{C}_{[s+i]}$ are distinct, for all $1 \leq i \leq q^2 - 1$, i.e., the cosets $\mathbb{C}_{[s+i]}$ contain m elements, because $m - 1 > m/2$. $\qquad\square$

Lemma 5.3.8 *Let* $q \geq 4$ *be a prime power and* $n = q^{2m} - 1$, *where* $m = \mathrm{ord}_n(q^2) \geq 3$. *Let* $s = \displaystyle\sum_{i=0}^{m-1} (q^2)^i$. *If* C *is the cyclic code generated by the product of the minimal polynomials*

$$M^{(s)}(x)M^{(s+1)}(x)\cdots M^{(s+i)}(x)M^{(s-1)}(x)\cdots M^{(s-j)}(x),$$

where $1 \leq i, j \leq q^2 - 1$, *then* C *is Hermitian dual-containing.*

Proof According to Lemma 5.1.4, we have to show that $Z \cap Z^{-q} = \emptyset$. Forcing a contradiction, we assume that $Z \cap Z^{-q} \neq \emptyset$. The cases concerning the coset $\mathbb{C}_{[s]}$ are immediate. Assume first that $\mathbb{C}_{[s+j]} = \mathbb{C}_{[-q(s+i)]}$, $1 \leq i, j \leq q^2 - 1$. Then there exists $0 \leq h \leq m - 1$ such that

$$s + j \equiv -q(s+i)q^{2h} \bmod n.$$

Because $\gcd(q^2, n) = 1$, $q^{2m} \equiv 1 \bmod n$ and $sq^{2t} \equiv s \bmod n$ for all $0 \leq t \leq m - 1$, we obtain

$$s + j \equiv -qs - qiq^{2h} \bmod n,$$

where $0 \leq h \leq m - 1$. We now compute the expression $s + j + q(s + iq^{2h})$, $0 \leq h \leq m - 1$. If $h \leq m - 2$, one has

$$s + j + q(s + iq^{2h})$$

$$\leq \frac{q^{2m} - 1}{q^2 - 1} + j + q\frac{q^{2m} - 1}{q^2 - 1} + iq^{2m-3}$$

$$\leq \frac{q^{2m} - 1}{q - 1} + (q^2 - 1)(1 + q^{2m-3}).$$

It is easy to see that

$$\frac{q^{2m} - 1}{q - 1} + (q^2 - 1)(1 + q^{2m-3}) < q^{2m} - 1.$$

Since $s + j = -qs - iq^{2h+1}$ does not hold, we have a contradiction.

If $h = m - 1$, we will verify the equivalence $s + j \equiv -q(s + i)q^{2m-2} \bmod n$:

$$s + j \equiv -q(s + i)q^{2m-2} \bmod n$$

$$\Longrightarrow j(q^2 - 1) \equiv -iq^{2m-1}(q^2 - 1) \bmod n$$

$$\Longrightarrow (j + iq^{2m-1})(q^2 - 1) \equiv 0 \bmod n.$$

Applying the algorithm of division for i and q we have $i = aq + r$, where $0 \leq r < q$. Because $1 \leq i \leq q^2 - 1$ we also have $0 \leq a < q$; hence,

$$(j + iq^{2m-1})(q^2 - 1)$$

$$\equiv [j + (aq + r)q^{2m-1}](q^2 - 1)$$

$$\equiv (j + a)(q^2 - 1) + r(q^2 - 1)q^{2m-1}$$

$$\equiv (j + a)(q^2 - 1) + rq - rq^{2m-1} \equiv 0 \bmod n$$

$$\Longrightarrow rq^{2m-1} - rq - (j + a)(q^2 - 1) \equiv 0 \bmod n.$$

If $r = 0$, it follows that $(j + a)(q^2 - 1) < q^{2m} - 1$, so $(j + a)(q^2 - 1) \not\equiv 0 \bmod n$. If $r > 0$, then $0 < rq^{2m-1} - rq - (j + a)(q^2 - 1) < q^{2m} - 1$, which is a contradiction.

The cases $\mathbb{C}_{[s+j]} = \mathbb{C}_{[-q(s-i)]}$, $\mathbb{C}_{[s-j]} = \mathbb{C}_{[-q(s+i)]}$ and $\mathbb{C}_{[s-j]} = \mathbb{C}_{[-q(s-i)]}$ are analogous to the previous one, so the proof is omitted. Therefore, C is Hermitian dual-containing, as required. $\qquad\square$

Theorem 5.3.2 given in the following is the main result of this subsection.

Theorem 5.3.2 *Let* $q \geq 4$ *be a prime power and* $n = q^{2m} - 1$, *where* $m = \mathrm{ord}_n(q^2) \geq 3$. *Then there exists an* $[[n, n - 4m(q^2 - 1) - 2, d \geq q^2 + 2]]_q$ *quantum error-correcting code.*

Proof Let C be the cyclic code generated by

$$M^{(s)}(x)M^{(s+1)}(x) \cdot \ldots \cdot M^{(s+q^2-1)}(x)M^{(s-1)}(x) \cdot \ldots \cdot M^{(s-q^2+1)}(x).$$

From Lemmas 5.3.6 and 5.3.7, it is easy to see that C is an $[n, n - 2m(q^2 - 1) - 1, d \geq 2q^2 + 2]_{q^2}$ code. From Lemma 5.3.8, C is Hermitian dual-containing. Applying the Hermitian construction, an $[[n, n - 4m(q^2 - 1) - 2, d \geq 2q^2 + 2]]_q$ quantum code can be constructed. The proof is complete. $\qquad\square$

Corollary 5.3.2 *Let $q \geq 4$ be a prime power and $n = q^{2m} - 1$, where $m = \mathrm{ord}_n(q^2) \geq 3$. Then there exist quantum codes with parameters*

- $[[n, n - 2mc - 2, d \geq c + 2]]_q$, where $1 \leq c < q^2 - 1$;
- $[[n, n - 2m(q^2 - 1) - 2, d \geq q^2 + 2]]_q$;
- $[[n, n - 2m(c - 1) - 2, d \geq c + 2]]_q$, where $q^2 + 1 \leq c \leq 2q^2 - 2$.

Exercise 5.3.2 Prove Corollary 5.3.2.

5.3.3 Construction III

The result given in the sequence is analogous to Lemma 5.3.5.

Lemma 5.3.9 *Let $q \neq 2$ be a prime power and $n = q^m - 1$, where $m = \mathrm{ord}_n(q) \geq 3$. If $s = \sum_{i=0}^{m-1} q^i$, then the q-coset $\mathbb{C}_{[s]}$ has only one element.*

Proof Left to exercise. $\qquad\square$

Exercise 5.3.3 Prove Lemma 5.3.9.

The following two results are analogous to Lemmas 5.3.6 and 5.3.7, respectively.

Lemma 5.3.10 *Let $q \geq 3$ be a prime power and $n = q^m - 1$, where $m = \mathrm{ord}_n(q) \geq 3$. Let $s = \sum_{i=0}^{m-1} q^i$. Then the following are true:*

(a) *the q-cosets of the form $\mathbb{C}_{[s+i]}$ are mutually disjoints, where $1 \leq i \leq q - 1$;*
(b) *the q-cosets of the form $\mathbb{C}_{[s-j]}$ are mutually disjoints, where $1 \leq j \leq q - 1$;*
(c) *the q-cosets of the form $\mathbb{C}_{[s+i]}$ are mutually disjoints to the q-cosets of the form $\mathbb{C}_{[s-j]}$, where $1 \leq i, j \leq q - 1$.*

Lemma 5.3.11 *Let $q \geq 4$ be a prime power and $n = q^m - 1$, where $m = \mathrm{ord}_n(q) \geq 3$. Let $s = \sum_{i=0}^{m-1} q^i$. Then the following hold:*

(a) *the cosets of the form $\mathbb{C}_{[s+i]}$, where $1 \leq i \leq q - 1$, contain m elements;*
(b) *the cosets of the form $\mathbb{C}_{[s-j]}$, where $1 \leq j \leq q - 1$, contain m elements.*

Exercise 5.3.4 Show Lemmas 5.3.10 and 5.3.11.

Lemma 5.3.12 *Let $q \geq 4$ be a prime power and $n = q^m - 1$, where $m = \text{ord}_n(q) \geq 3$. Let $s = \sum_{i=0}^{m-1} q^i$. Let C be the cyclic code generated by*

$$M^{(s)}(x) M^{(s+1)}(x) \cdot \ldots \cdot M^{(s+j)}(x) M^{(s-1)}(x) \cdot \ldots \cdot M^{(s-j)}(x),$$

where $1 \leq j \leq q - 1$. Then C is Euclidean dual-containing.

Proof From Lemma 5.1.3, it is sufficient to prove that $Z \cap Z^{-1} = \emptyset$. Forcing a contradiction, we assume the $Z \cap Z^{-1} \neq \emptyset$. The cases concerning the coset $\mathbb{C}_{[s]}$ are trivial.

(1) Assume first that $\mathbb{C}_{[s+i]} = \mathbb{C}_{[-(s+j)]}$, where $1 \leq i, j \leq q - 1$. Then there exists $0 \leq t \leq m - 1$ such that $s + i \equiv -(s + j) \bmod n$. Since $\gcd(q, n) = 1$, $q^m \equiv 1 \bmod n$ and $sq^t \equiv s \bmod n$, $0 \leq t \leq m - 1$, we have

$$s + i \equiv -s - jq^t \bmod n \Longrightarrow 2s \equiv -(i + jq^t) \quad \bmod n.$$

If $0 \leq t \leq m - 2$ and because $q \geq 4$, it follows that

$$2s + i + jq^t$$
$$\leq \frac{2q^m - 2}{q - 1} + (q - 1)(1 + q^{m-2})$$
$$< q^m - 1$$
$$\Longrightarrow 2s + i + jq^t < q^m - 1.$$

Hence $s + i = -s - jq^t$, a contradiction.

Let us next consider the case $t = m - 1$. We know that for each $1 \leq i, j \leq q - 3$ we have $2s + i + jq^{m-1} < q^m - 1$; since $s + i = -s - jq^{m-1}$ does not hold, this implies in a contradiction. Analogously, if $j = q - 3$ and $1 \leq i \leq q - 1$, it follows that $2s + i + jq^{m-1} < q^m - 1$, and because $s + i \neq -s - jq^{m-1}$, we have a contradiction.

If $j \geq q - 2$, we have $2s + i + jq^t > q^m - 1$. Let us compute the equivalence $2s \equiv -(i + jq^{m-1}) \bmod n$ for $j = q - 2$ and $1 \leq i \leq q - 1$.

$$2s \equiv -[i + (q - 2)q^{m-1}] \bmod n \Longrightarrow 2s \equiv -i - 1 + 2q^{m-1} \bmod n.$$

As $0 < 2s + i + 1 - 2q^{m-1} < q^m - 1$ and also $2s \neq -i - 1 + 2q^{m-1}$ are true, one has a contradiction.

Let $j = q - 1$ and $1 \leq i \leq q - 1$. Computing the equivalence $2s \equiv -(i + jq^{m-1}) \bmod n$ we obtain

$$2s \equiv -[i + (q - 1)q^{m-1}] \bmod n \Longrightarrow 2s \equiv -i - 1 + q^{m-1} \bmod n.$$

Because $0 < 2s + i + 1 - q^{m-1} < q^m - 1$ and $2s \neq -i - 1 + q^{m-1}$ hold, then the equivalence $2s \equiv -[i + (q-1)q^{m-1}] \bmod n$ does not hold, a contradiction.

(2) Suppose $\mathbb{C}_{[s-i]} = \mathbb{C}_{[-(s-j)]}$, where $1 \leq i, j \leq q - 1$. Then there exists $0 \leq t \leq m - 1$ such that $s - i \equiv -(s-j)q^t \bmod n$. We have

$$s - i \equiv -s + jq^t \bmod n \implies 2s \equiv i + jq^t \bmod n.$$

If $0 \leq t \leq m - 2$ then the inequalities $2s < q^m - 1$ and $2s > i + jq^t$ hold, which is a contradiction. If $t = m - 1$, then

$$2s \equiv i + jq^{m-1} \bmod n$$

$$\implies 2(q^m - 1) \equiv (q-1)(i + jq^{m-1}) \bmod n$$
$$\implies (q-1)i + (q-1)jq^{m-1} \equiv 0 \bmod n$$
$$\implies (q-1)i + j - jq^{m-1} \equiv 0 \bmod n$$
$$\implies jq^{m-1} - i(q-1) - j \equiv 0 \bmod n.$$

Since $0 < jq^{m-1} - i(q-1) - j < q^m - 1$, the equivalence $2s \equiv i + jq^t \bmod n$ does not hold, a contradiction.

(3) Assume that $\mathbb{C}_{[s+i]} = \mathbb{C}_{[-(s-j)]}$, where $1 \leq i, j \leq q - 1$. Then there exists $0 \leq t \leq m - 1$ such that $s + i \equiv -(s-j)q^t \bmod n$, so $2s \equiv jq^t - i \bmod n$. If $t = 0$ and $i = j$ we have $2s \equiv 0 \bmod n$, a contradiction. If $0 \leq t \leq m - 2$ and $i \neq j$, we know that

$$2s \equiv jq^t - i \bmod n \implies (q-1)(jq^t - i) \equiv 0 \bmod n;$$

hence, $-(q^m - 1) < (q-1)(jq^t - i) < q^m - 1$ and $(q-1)(jq^t - i) \neq 0$, a contradiction.

If $t = m - 1$, we obtain

$$(q-1)(jq^{m-1} - i) \equiv 0 \bmod n \implies jq^{m-1} + i(q-1) - j \equiv 0 \bmod n.$$

Since $0 < jq^{m-1} + i(q-1) - j < q^m - 1$, the equivalence $s + i \equiv -(s-j)q^t \bmod n$ does not hold, a contradiction.

(4) Suppose finally that $\mathbb{C}_{[s-i]} = \mathbb{C}_{[-(s+j)]}$; then $s - i \equiv -(s+j)q^t \bmod n$ for some $0 \leq t \leq m - 1$. Thus, $2s \equiv i - jq^t \bmod n$. As in the previous case, if $t = 0$ and $i = j$ we have $2s \equiv 0 \bmod n$, a contradiction. If $0 \leq t \leq m - 2$ and $i \neq j$ we then know that

$$2s \equiv i - jq^t \bmod n \implies (q-1)(i - jq^t) \equiv 0 \bmod n,$$

which is a contradiction. Moreover, it is easy to see that the last equivalence does not hold, a contradiction.

Therefore, C is Euclidean dual-containing code, as required. The proof is complete.   □

We next recall Corollary 5.1.2 shown in [62].

Corollary 5.3.3 *Assume that we have an $[N_0, K_0]$ linear code L which contains its Euclidean dual, $L^\perp \leq L$, and which can be enlarged to an $[N_0, K_0']$ linear code L', where $K_0' \geq K_0 + 2$. Then there exists a quantum code with parameters $[[N_0, K_0 + K_0' - N_0, d \geq min\{d, \lceil \frac{q+1}{q} d' \rceil\}]]$, where $d = w(L \backslash L'^\perp)$ and $d' = w(L' \backslash L'^\perp)$.*

We are now ready to state the main results of this subsection.

Theorem 5.3.3 *Let $q \geq 4$ be a prime power and $n = q^m - 1$, where $m = ord_n(q) \geq 3$. Then there exists an $[[n, n - m(2c - 1) - 2, d \geq c + 2]]_q$ quantum code, for all $1 \leq c \leq q - 2$.*

Proof Let C be the cyclic code generated by

$$M^{(s)}(x)M^{(s+1)}(x) \cdot \ldots \cdot M^{(s+i)}(x)M^{(s-1)}(x) \cdot \ldots \cdot M^{(s-j)}(x),$$

where $1 \leq i + j = c \leq q - 2$. It is easy to see that C is an $[n, n - mc - 1, d \geq c + 2]_q$ code, where $1 \leq c \leq q - 2$. Moreover, from Lemma 5.1.3, C is Euclidean dual-containing.

Let C' be the code generated by

$$M^{(s)}(x)M^{(s+1)}(x) \cdot \ldots \cdot M^{(s+i)}(x)M^{(s-1)}(x) \cdot \ldots \cdot M^{(s-j+1)}(x).$$

We know that C' is an enlargement of C and has parameters $[n, n - m(c - 1) - 1, d' \geq c + 1]_q$. Applying Corollary 5.1.2 to C and C' we obtain an $[[n, n - m(2c - 1) - 2, d \geq c + 2]]_q$ code, as required. The proof is complete.   □

Theorem 5.3.4 *Let $q \geq 4$ be a prime power and $n = q^m - 1$, where $m = ord_n(q) \geq 3$. Then there exist quantum codes with parameters*

- $[[n, n - m(2q - 3) - 2, d \geq q + 1]]_q$;
- $[[n, n - m(2q - 1) - 1, d \geq q + 3]]_q$;
- $[[n, n - m(2c - 4) - 2, d \geq c + 2]]_q$;
- $[[n, n - m(4q - 8) - 2, d \geq 2q]]_q$;
- $[[n, n - m(4q - 5) - 2, d \geq 2q + 2]]_q$, where $q + 1 < c < 2q - 2$.

Exercise 5.3.5 Show Theorem 5.3.4.

5.3.4 Code Comparison

As usual, we compare here the parameters of our codes with the codes displayed in the literature. In Table 5.12, our Hermitian quantum codes have parameters $[[n, n - 4(c - 2) - 2, d \geq c]]_q$, where $3 \leq c \leq q^2$ and $n = q^4 - 1$; $[[n', k', d']]_q = [[n', n' - 2m\lceil(\delta - 1)(1 - 1/q^2)\rceil, d' \geq \delta]]_q$ are the parameters of the Hermitian quantum codes shown in Theorem 21 in [4], where $m = \text{ord}_n(q^2) = 2$ and $2 \leq \delta \leq \lfloor n(q^m - 1)/(q^{2m} - 1)\rfloor$.

In Table 5.13, our quantum codes are obtained from Sect. 5.3.2 and have parameters $[[n, k, d]]_q$ given by

- $[[n, n - 2mc - 2, d \geq c + 2]]_q$, where $1 \leq c < q^2 - 1$;
- $[[n, n - 2m(q^2 - 1) - 2, d \geq q^2 + 2]]_q$;
- $[[n, n - 2m(c - 1) - 2, d \geq c + 2]]_q$, where $q^2 + 1 \leq c \leq 2q^2 - 2$.
- $[[n, n - 4m(q^2 - 1) - 2, d \geq 2q^2 + 2]]_q$, where $n = q^{2m} - 1$, $q \geq 4$ is a prime power, $m = \text{ord}_n(q^2) \geq 3$;

$[[n', k', d']]_q = [[n', n' - 2m\lceil(\delta - 1)(1 - 1/q^2)\rceil, d' \geq \delta]]_q$ are the parameters of the Hermitian quantum codes shown in Theorem 21 in [4], where $m = \text{ord}_n(q^2) \geq 3$ and $2 \leq \delta \leq \lfloor n(q^m - 1)/(q^{2m} - 1)\rfloor$.

In Table 5.14, our codes are derived from Sect. 5.3.3 and have parameters $[[n, k, d]]_q$ given by

- $[[n, n - m(2c - 1) - 2, d \geq c + 2]]_q$, where $1 \leq c \leq q - 2$;
- $[[n, n - m(2q - 3) - 2, d \geq q + 1]]_q$;
- $[[n, n - m(2q - 1) - 1, d \geq q + 3]]_q$;
- $[[n, n - m(2c - 4) - 2, d \geq c + 2]]_q$;
- $[[n, n - m(4q - 8) - 2, d \geq 2q]]_q$;
- $[[n, n - m(4q - 5) - 2, d \geq 2q + 2]]_q$, where $q + 1 < c < 2q - 2$.

The parameters $[[n'', k'', d'']]_q$ are the parameters of quantum BCH codes derived from q-ary Steane's construction (see Corollary 5.1.2) applied to narrow-sense BCH codes. These codes were obtained by the same method presented in Table I in [148] by considering the criterion for classical Euclidean dual-containing BCH codes of Theorems 3 and 5 in [4].

As we can see in Tables 5.12, 5.13 and 5.14, according to the procedure described in Remark 5.1.2, our quantum codes have parameters better than the ones exhibited in the literature.

5.4 Algebraic Geometry Codes

In this section, we construct several families of quantum codes with good and asymptotically good parameters. These quantum codes are derived from (classical) t-point $(t \geq 1)$ algebraic geometry (AG) codes by applying the CSS construction. More

Table 5.12 Code comparison

Our Hermitian codes		Hermitian codes in [4]
$[[n, n - 4(c - 2) - 2, d \geq c]]_q$		$[[n', k', d']]_q$
	$m = 2, q = 3$	
$[[80, 74, d \geq 3]]_3$		$[[80, 72, d' \geq 3]]_3$
$[[80, 70, d \geq 4]]_3$		$[[80, 68, d' \geq 4]]_3$
$[[80, 66, d \geq 5]]_3$		$[[80, 64, d' \geq 5]]_3$
$[[80, 62, d \geq 6]]_3$		$[[80, 60, d' \geq 6]]_3$
$[[80, 58, d \geq 7]]_3$		$[[80, 56, d' \geq 7]]_3$
$[[80, 54, d \geq 8]]_3$		$[[80, 52, d' \geq 8]]_3$
$[[80, 50, d \geq 9]]_3$		
	$m = 2, q = 4$	
$[[255, 249, d \geq 3]]_4$		$[[255, 247, d' \geq 3]]_4$
$[[255, 245, d \geq 4]]_4$		$[[255, 243, d' \geq 4]]_4$
$[[255, 241, d \geq 5]]_4$		$[[255, 239, d' \geq 5]]_4$
$[[255, 237, d \geq 6]]_4$		$[[255, 235, d' \geq 6]]_4$
$[[255, 233, d \geq 7]]_4$		$[[255, 231, d' \geq 7]]_4$
$[[255, 229, d \geq 8]]_4$		$[[255, 227, d' \geq 8]]_4$
$[[255, 225, d \geq 9]]_4$		$[[255, 223, d' \geq 9]]_4$
$[[255, 221, d \geq 10]]_4$		$[[255, 219, d' \geq 10]]_4$
$[[255, 217, d \geq 11]]_4$		$[[255, 215, d' \geq 11]]_4$
$[[255, 213, d \geq 12]]_4$		$[[255, 211, d' \geq 12]]_4$
$[[255, 209, d \geq 13]]_4$		$[[255, 207, d' \geq 13]]_4$
$[[255, 205, d \geq 14]]_4$		$[[255, 203, d' \geq 14]]_4$
$[[255, 201, d \geq 15]]_4$		$[[255, 199, d' \geq 15]]_4$
$[[255, 197, d \geq 16]]_4$		
	$m = 2, q = 5$	
$[[624, 618, d \geq 3]]_5$		$[[624, 616, d' \geq 3]]_5$
$[[624, 614, d \geq 4]]_5$		$[[624, 612, d' \geq 4]]_5$
$[[624, 610, d \geq 5]]_5$		$[[624, 608, d' \geq 5]]_5$
$[[624, 606, d \geq 6]]_5$		$[[624, 604, d' \geq 6]]_5$
$[[624, 602, d \geq 7]]_5$		$[[624, 600, d' \geq 7]]_5$
$[[624, 598, d \geq 8]]_5$		$[[624, 596, d' \geq 8]]_5$
$[[624, 594, d \geq 9]]_5$		$[[624, 592, d' \geq 9]]_5$
$[[624, 590, d \geq 10]]_5$		$[[624, 588, d' \geq 10]]_5$
$[[624, 586, d \geq 11]]_5$		$[[624, 584, d' \geq 11]]_5$
$[[624, 582, d \geq 12]]_5$		$[[624, 580, d' \geq 12]]_5$

Table 5.13 Code comparison

Our Hermitian codes		Hermitian codes in [4]
$[[n, k, d]]_q$		$[[n', k', d']]_q$
	$m = 3, q = 4$	
$[[4095, 4087, d \geq 3]]_4$		$[[4095, 4083, d' \geq 3]]_4$
$[[4095, 4081, d \geq 4]]_4$		$[[4095, 4077, d' \geq 4]]_4$
$[[4095, 4075, d \geq 5]]_4$		$[[4095, 4071, d' \geq 5]]_4$
$[[4095, 4069, d \geq 6]]_4$		$[[4095, 4065, d' \geq 6]]_4$
$[[4095, 4063, d \geq 7]]_4$		$[[4095, 4059, d' \geq 7]]_4$
$[[4095, 4057, d \geq 8]]_4$		$[[4095, 4053, d' \geq 8]]_4$
$[[4095, 4051, d \geq 9]]_4$		$[[4095, 4047, d' \geq 9]]_4$
$[[4095, 4045, d \geq 10]]_4$		$[[4095, 4041, d' \geq 10]]_4$
$[[4095, 4039, d \geq 11]]_4$		$[[4095, 4035, d' \geq 11]]_4$
$[[4095, 4033 d \geq 12]]_4$		$[[4095, 4029, d' \geq 12]]_4$
$[[4095, 4027, d \geq 13]]_4$		$[[4095, 4023, d' \geq 13]]_4$
$[[4095, 4021, d \geq 14]]_4$		$[[4095, 4017, d' \geq 14]]_4$
$[[4095, 4015, d \geq 15]]_4$		$[[4095, 4011, d' \geq 15]]_4$
$[[4095, 4009, d \geq 16]]_4$		$[[4095, 4005, d' \geq 16]]_4$
$[[4095, 4003, d \geq 18]]_4$		$[[4095, 3999, d' \geq 18]]_4$
$[[4095, 3997, d \geq 19]]_4$		$[[4095, 3993, d' \geq 19]]_4$
$[[4095, 3949, d \geq 27]]_4$		$[[4095, 3945, d' \geq 27]]_4$
$[[4095, 3925, d \geq 31]]_4$		$[[4095, 3921, d' \geq 31]]_4$
$[[4095, 3919, d \geq 32]]_4$		$[[4095, 3915, d' \geq 32]]_4$
$[[4095, 3913, d \geq 34]]_4$		$[[4095, 3909, d' \geq 34]]_4$

specifically, we construct two classical nested AG codes $C_1 \subset C_2$, applying after the CSS construction. Many of these codes have large minimum distances when compared with their code lengths, as well as they also have small Singleton defects. As an example, we construct a family $[[46, 2(t_2 - t_1), d]]_{25}$ of quantum codes, where t_1, t_2 are positive integers where $1 < t_1 < t_2 < 23$ and $d \geq \min\{46 - 2t_2, 2t_1 - 2\}$, of length $n = 46$, with minimum distance in the range $2 \leq d \leq 20$, having Singleton defect at most four. Furthermore, by applying the CSS construction to sequences of t-point classical AG codes constructed here, we obtain sequences of asymptotically good quantum codes. The content presented here can be found in our paper [106].

As we know, methods and techniques of construction of quantum codes with good parameters were extensively investigated in the literature [4, 25–27, 71, 73, 80, 89, 91, 94, 97, 100, 119, 147, 148]. Many of these works were performed by applying one (or two or all of them) of the following techniques:

(1) the CSS construction based on linear Euclidean self-orthogonal codes or even based on two nested linear codes [4, 25, 73, 80, 100];

Table 5.14   Code comparison

Our codes - construction III		Codes derived from [62]
$[[n, k, d]]_q$		$[[n'', k'', d'']]_q$
	$m = 3, q = 4$	
$[[63, 58, d \geq 3]]_4$		$[[63, 54, d' \geq 4]]_4$
$[[63, 52, d \geq 4]]_4$		$[[63, 48, d' \geq 4]]_4$
$[[63, 41, d \geq 7]]_4$		$[[63, 39, d' \geq 7]]_4$
$[[63, 28, d \geq 10]]_4$		$[[63, 24, d' \geq 10]]_4$
	$m = 3, q = 5$	
$[[124, 119, d \geq 3]]_5$		$[[124, 115, d' \geq 3]]_5$
$[[124, 113, d \geq 4]]_5$		$[[124, 109, d' \geq 4]]_5$
$[[124, 107, d \geq 5]]_5$		$[[124, 103, d' \geq 5]]_5$
$[[124, 96, d \geq 8]]_5$		$[[124, 94, d' \geq 8]]_5$
$[[124, 92, d \geq 9]]_5$		$[[124, 88, d' \geq 9]]_5$
$[[124, 86, d \geq 10]]_5$		$[[124, 82, d' \geq 10]]_5$
$[[124, 77, d \geq 12]]_5$		$[[124, 73, d' \geq 12]]_5$
	$m = 3, q = 7$	
$[[342, 337, d \geq 3]]_7$		$[[342, 333, d' \geq 3]]_7$
$[[342, 331, d \geq 4]]_7$		$[[342, 327, d' \geq 4]]_7$
$[[342, 325, d \geq 5]]_7$		$[[342, 321, d' \geq 5]]_7$
$[[342, 319, d \geq 6]]_7$		$[[342, 315, d' \geq 6]]_7$
$[[342, 313, d \geq 7]]_7$		$[[342, 309, d' \geq 7]]_7$
$[[342, 302, d \geq 10]]_7$		$[[342, 300, d' \geq 10]]_7$
$[[342, 298, d \geq 11]]_7$		$[[342, 294, d' \geq 11]]_7$
$[[342, 292, d \geq 12]]_7$		$[[342, 288, d' \geq 12]]_7$
$[[342, 286, d \geq 13]]_7$		$[[342, 282, d' \geq 13]]_7$
$[[342, 271, d \geq 16]]_7$		$[[342, 267, d' \geq 16]]_7$
	$m = 4, q = 4$	
$[[255, 249, d \geq 3]]_4$		$[[255, 243, d' \geq 3]]_4$
$[[255, 241, d \geq 4]]_4$		$[[255, 235, d' \geq 4]]_4$
$[[255, 226, d \geq 7]]_4$		$[[255, 223, d' \geq 7]]_4$
$[[255, 209, d \geq 10]]_4$		$[[255, 203, d' \geq 10]]_4$
	$m = 4, q = 5$	
$[[624, 618, d \geq 3]]_5$		$[[624, 612, d' \geq 3]]_5$
$[[624, 610, d \geq 4]]_5$		$[[624, 604, d' \geq 4]]_5$
$[[624, 602, d \geq 5]]_5$		$[[624, 596, d' \geq 5]]_5$
$[[624, 587, d \geq 8]]_5$		$[[624, 584, d' \geq 8]]_5$
$[[624, 582, d \geq 9]]_5$		$[[624, 576, d' \geq 9]]_5$
$[[624, 574, d \geq 10]]_5$		$[[624, 568, d' \geq 10]]_5$
$[[624, 562, d \geq 12]]_5$		$[[624, 556, d' \geq 12]]_5$

(2) the Hermitian construction applied to Hermitian self-orthogonal codes [4, 25, 71, 73, 80, 91, 100];

(3) the Steane enlargement of CSS construction applied to Euclidean self-orthogonal codes [89, 100, 147, 148].

In particular, the CSS construction was also utilized in chains of nested linear codes to construct quantum codes whose parameters are asymptotically good [12, 26, 27, 81, 117]. All these latter asymptotically good quantum codes were constructed by applying the CSS construction to families of AG codes. In fact, the class of AG codes is a good source in order to obtain asymptotically good codes (see for example [47, 151]). In Refs. [12, 26, 27, 117], the authors constructed asymptotically good binary quantum codes and, in Ref. [81], the authors presented families of nonbinary asymptotically good quantum codes by means of one-point AG codes.

The aim here is to construct classical t-point ($t \geq 1$) AG codes (which are a generalization of one-point AG codes) as well as AG codes whose divisor G is not a rational place, after applying the CSS construction to these codes, in order to generate nonbinary quantum codes with good parameters. Additionally, we also construct sequences of classical t-point AG codes to obtain sequences of asymptotically good quantum codes by means of the CSS construction.

The constructions performed here are natural generalizations of the works dealing with constructions of quantum codes derived from one-point AG codes (see for example [12, 26, 27, 117]).

5.4.1 Preliminaries

Recall that a q-ary quantum code $\mathbb{Q}$ of length n is a K-dimensional subspace of the q^n-dimensional Hilbert space $(\mathbb{C}^q)^{\otimes n}$, where $\otimes n$ denotes the tensor product of vector spaces. If $K = q^k$ we write $[[n, k, d]]_q$ to denote a q-ary quantum code of length n and minimum distance d. Let $[[n, k, d]]_q$ be a quantum code. The quantum Singleton bound (QSB) asserts that $k + 2d \leq n + 2$. If the equality holds then the code is MDS.

5.4.2 Our Codes

This subsection is divided into three parts. The first part deals with constructions of quantum t-point algebraic geometry codes. In the second, we construct AG codes where the divisor G is a sum of non-rational places and, in the third part, we construct sequences of asymptotically good quantum codes derived from AG codes.

5.4.2.1 Construction I

In the first result we utilize two t-point ($t \geq 1$) AG codes to construct quantum codes with good parameters.

Theorem 5.4.1 (General t-point construction, $t \geq 1$) *Let q be a prime power and $F/\mathbb{F}_q$ be an algebraic function field of genus g, with $n + t$ pairwise distinct rational places. Assume that a_i, b_i, $i = 1, \ldots, t$, are positive integers such that $a_i \leq b_i$ for all i, and $2g - 2 < \sum_{i=1}^{t} a_i < \sum_{i=1}^{t} b_i < n$. Then there exists an $[[n, k, d]]_q$ quantum code, where $k = \sum_{i=1}^{t} b_i - \sum_{i=1}^{t} a_i$ and $d \geq \min \left\{ n - \sum_{i=1}^{t} b_i, \sum_{i=1}^{t} a_i - (2g - 2) \right\}$.*

Proof Let $\{P_1, P_2, \ldots, P_n, P_{n+1}, \ldots, P_{n+t}\}$ be the set of places of $F/\mathbb{F}_q$ of degree one. Let $D = P_1 + \ldots + P_n$ be a divisor of $F/\mathbb{F}_q$. Assume also that G_1 and G_2 are two divisors of $F/\mathbb{F}_q$ given, respectively, by $G_1 = a_1 P_{n+1} + \cdots + a_t P_{n+t}$ and $G_2 = b_1 P_{n+1} + \cdots + b_t P_{n+t}$, where $a_i \leq b_i$ for all $i = 1, \ldots, t$ and $2g - 2 < \sum_{i=1}^{t} a_i < \sum_{i=1}^{t} b_i < n$. From construction, $\operatorname{supp} G_1 \cap \operatorname{supp} D = \emptyset$ and $\operatorname{supp} G_2 \cap \operatorname{supp} D = \emptyset$. Since $G_1 \leq G_2$, we have $\mathcal{L}(G_1) \subset \mathcal{L}(G_2)$; hence, $C_{\mathcal{L}}(D, G_1) \subset C_{\mathcal{L}}(D, G_2)$. From Theorem 4.5.1, the code $C_1 := C_{\mathcal{L}}(D, G_1)$ has parameters $[n, k_1, d_1]_q$, where $d_1 \geq n - \sum_{i=1}^{t} a_i$ and $k_1 = \sum_{i=1}^{t} a_i - g + 1$; the code $C_2 := C_{\mathcal{L}}(D, G_2)$ has parameters $[n, k_2, d_2]_q$, where $d_2 \geq n - \sum_{i=1}^{t} b_i$ and $k_2 = \sum_{i=1}^{t} b_i - g + 1$. On the other hand, from Theorems 4.5.2 and 4.5.3, the dual code $C_1^{\perp} = C_{\Omega}(D, G_1)$ of C_1 has parameters $[n, k_1^{\perp}, d_1^{\perp}]_q$, where $d_1^{\perp} \geq \sum_{i=1}^{t} a_i - (2g - 2)$ and $k_1^{\perp} = n + g - 1 - \sum_{i=1}^{t} a_i$; the dual code $C_2^{\perp} = C_{\Omega}(D, G_2)$ of C_2 has parameters $[n, k_2^{\perp}, d_2^{\perp}]_q$, with $d_2^{\perp} \geq \sum_{i=1}^{t} b_i - (2g - 2)$ and $k_2^{\perp} = n + g - 1 - \sum_{i=1}^{t} b_i$.

Applying the CSS construction to C_1 and C_2, we obtain an $[[n, k, d]]_q$ code, with $k = k_2 - k_1 = (\sum_{i=1}^{t} b_i - g + 1) - (\sum_{i=1}^{t} a_i - g + 1) = \sum_{i=1}^{t} b_i - \sum_{i=1}^{t} a_i$ and $d \geq \min\{d_2, d_1^{\perp}\}$, where $d_2 \geq n - \sum_{i=1}^{t} b_i$ and $d_1^{\perp} \geq \sum_{i=1}^{t} a_i - (2g - 2)$. The proof is complete. $\square$

Remark 5.4.1 In [27, 81], the authors utilized one-point AG codes to construct good/(asymptotically good) quantum codes. In [26], the author applied two-point AG codes to derive good/(asymptotically good) quantum codes. Note that, in this context, Theorem 5.4.1 is a natural generalization of the one-point as well as two-point AG code construction to the t-point ($t \geq 1$) AG code construction.

Corollary 5.4.1 (One-Point codes) *There exists a quantum code with parameters* $[[q(1 + (q - 1)m), b - a, d]]_{q^2}$, *where* $(q - 1)(m - 1) - 2 < a < b < q(1 + (q - 1)m)$, $m | (q + 1)$ *and* $d \geq \min\{q(1 + (q - 1)m) - b, a - (q - 1)(m - 1) + 2\}$.

Proof Let $F = \mathbb{F}_{q^2}(x, y)$, where $y^q + y = x^m$ and $m | (q + 1)$. It is known the genus of F is $g = (q - 1)(m - 1)/2$, and the number of places of degree one is equal to $N = 1 + q(1 + (q - 1)m)$ (see the Example 6.4.2. of [152]). Let $\{P_1, P_2, \ldots, P_n, P_{n+1}, \ldots, P_N\}$ be these pairwise distinct places. Without loss of generality, choose the $\mathbb{F}_{q^2}$-rational point P_N. Let $D = P_1 + \cdots + P_{N-1}$ be a divisor and let $G_1 = aP_N$ and $G_2 = bP_N$ be other two divisors such that $\operatorname{supp} G_1 \cap \operatorname{supp} D = \emptyset$ and $\operatorname{supp} G_2 \cap \operatorname{supp} D = \emptyset$, where $(q - 1)(m - 1) - 2 < a < b < q(1 + (q - 1)m)$. From Theorem 5.4.1, there exists a quantum code with parameters $[[q(1 + (q - 1)m), b - a, d]]_{q^2}$, where $d \geq \min\{q(1 + (q - 1)m) - b, a - (q - 1)(m - 1) + 2\}$. The proof is complete. $\qquad\square$

Remark 5.4.2 Note that the Hermitian curve defined by $y^q + y = x^{q+1}$, over $\mathbb{F}_{q^2}$, is a particular case of the curve $y^q + y = x^m$ considered in the proof of Corollary 5.4.1.

Corollary 5.4.2 (Two-Point codes) *There exists a quantum code with parameters* $[[q(1 + (q - 1)m) - 1, b_1 + b_2 - a_1 - a_2, d]]_{q^2}$, *where* $a_i \leq b_i$ *for* $i = 1, 2$, $(q - 1)(m - 1) - 2 < a_1 + a_2 < b_1 + b_2 < q[1 + (q - 1)m] - 1$, $m | (q + 1)$ *and* $d \geq \min\{q[1 + (q - 1)m] - b_1 - b_2 - 1, a_1 + a_2 - (q - 1)(m - 1) + 2\}$.

Proof Let $D = P_1 + \cdots + P_{N-2}$ be a divisor and let $G_1 = a_1 P_{N-2} + a_2 P_{N-1}$ and $G_2 = b_1 P_{N-2} + b_2 P_{N-1}$ be other two divisors with $\operatorname{supp} G_1 \cap \operatorname{supp} D = \emptyset$ and $\operatorname{supp} G_2 \cap \operatorname{supp} D = \emptyset$, where $(q - 1)(m - 1) - 2 < a_1 + a_2 < b_1 + b_2 < q(1 + (q - 1)m) - 1$. From Theorem 5.4.1, there exists an

$$[[q(1 + (q - 1)m) - 1, b_1 + b_2 - a_1 - a_2, d]]_{q^2}$$

code, where $d \geq \min\{q(1 + (q - 1)m) - 1 - b_1 - b_2, a_1 + a_2 - (q - 1)(m - 1) + 2\}$. This finishes the proof. $\qquad\square$

Corollary 5.4.3 (*t*-Point codes, $t \geq 2$) *There exists a quantum code with parameters* $[[q(1 + (q - 1)m) - t + 1, b_1 + \cdots + b_t - (a_1 + \cdots + a_t), d]]_{q^2}$, *where* $a_i \leq b_i$ *for* $i = 1, \ldots t$, $(q - 1)(m - 1) - 2 < a_1 + \cdots + a_t < b_1 + \cdots + b_t < q(1 + (q - 1)m) - t + 1$, $m | (q + 1)$ *and* $d \geq \min\{q(1 + (q - 1)m) - (b_1 + \cdots + b_t) - t + 1, a_1 + \cdots + a_t - (q - 1)(m - 1) + 2\}$.

Proof Similar to that of Corollary 5.4.2. $\qquad\square$

5.4.2.2 Construction II

We construct here quantum codes derived from AG codes whose divisors are multiples of a non-rational divisor G. The first result is given in the following.

Theorem 5.4.2 (General construction) *Let q be a prime power and let $F/\mathbb{F}_q$ be an algebraic function field of genus g, with n pairwise distinct rational places P_i, $i = 1, \ldots, n$. Assume that there exist pairwise distinct places $Q_1, \ldots, Q_t$ of $F/\mathbb{F}_q$, of degree $\alpha_i \geq 2$, respectively, $i = 1, \ldots, t$, where $t \geq 1$. Let $G_1 = \sum_{i=1}^{t} a_i Q_i$ and $G_2 = \sum_{i=1}^{t} b_i Q_i$, where $a_i \leq b_i$, for all $i = 1, \ldots, t$, and $2g - 2 < a_1\alpha_1 + \ldots + a_t\alpha_t < b_1\alpha_1 + \ldots + b_t\alpha_t < n$. Let $D = P_1 + \cdots + P_n$ be a divisor of $F/\mathbb{F}_q$, and consider that $\operatorname{supp} G_1 \cap \operatorname{supp} D = \emptyset$ and $\operatorname{supp} G_2 \cap \operatorname{supp} D = \emptyset$. Then there exists a quantum code with parameters $[[n, k, d]]_q$, where $k = (b_1 - a_1)\alpha_1 + \ldots + (b_t - a_t)\alpha_t$ and $d \geq \min\{n - (b_1\alpha_1 + \ldots + b_t\alpha_t), (a_1\alpha_1 + \ldots + a_t\alpha_t) - (2g - 2)\}$.*

Proof Similar to that of Theorem 5.4.1.   □

Corollary 5.4.4 *Let q be a prime power and let $F/\mathbb{F}_q$ be a hyperelliptic function field of genus $g \geq 2$, with n pairwise distinct rational places. Then there exists an $[[n, 2(t_2 - t_1), d]]_q$ code, where t_1, t_2 are positive integers satisfying $2g - 2 < t_1 < t_2 < n$ and $d \geq \min\{n - 2t_2, 2t_1 - 2g + 2\}$.*

Proof Since F is a hyperelliptic function field, there exists a place G of degree two (see Lemma 6.2.2.(a) of [152]). Let $D = P_1 + \cdots + P_n$ be a divisor, where P_i are all rational points of F. Let $G_2 = t_2 G$ and $G_1 = t_1 G$, where $2g - 2 < 2t_1 < 2t_2 < n$. We know that $\operatorname{supp} G_1 \cap \operatorname{supp} D = \emptyset$, $\operatorname{supp} G_2 \cap \operatorname{supp} D = \emptyset$ and $C_{\mathcal{L}}(D, G_1) \subset C_{\mathcal{L}}(D, G_2)$. From Theorem 4.5.1, the code $C_1 := C_{\mathcal{L}}(D, G_1)$ has parameters $[n, k_1, d_1]_q$, where $d_1 \geq n - 2t_1$ and $k_1 = 2t_1 - g + 1$. The code $C_2 := C_{\mathcal{L}}(D, G_2)$ has parameters $[n, k_2, d_2]_q$, where $d_2 \geq n - 2t_2$ and $k_2 = 2t_2 - g + 1$.

From Theorems 4.5.2 and 4.5.3, the dual code $C_1^{\perp} = C_{\Omega}(D, G_1)$ of C_1 has parameters $[n, k_1^{\perp}, d_1^{\perp}]_q$, where $d_1^{\perp} \geq 2t_1 - (2g - 2)$ and $k_1^{\perp} = n + g - 1 - 2t_1$. Analogously, the dual code $C_2^{\perp} = C_{\Omega}(D, G_2)$ of C_2 has parameters $[n, k_2^{\perp}, d_2^{\perp}]_q$, where $d_2^{\perp} \geq 2t_2 - (2g - 2)$ and $k_2^{\perp} = n + g - 1 - 2t_2$.

Applying the CSS construction to C_1 and C_2, we obtain an $[[n, k, d]]_q$ code, with $k = k_2 - k_1 = (2t_2 - g + 1) - (2t_1 - g + 1) = 2(t_2 - t_1)$ and $d \geq \min\{d_2, d_1^{\perp}\}$, where $d_2 \geq n - 2t_2$ and $d_1^{\perp} \geq 2t_1 - (2g - 2)$. The proof is complete.   □

Corollary 5.4.5 *There exists a quantum code with parameters $[[46, 2(t_2 - t_1), d]]_{25}$, where t_1, t_2 are positive integers such that $1 < t_1 < t_2 < 23$ and $d \geq \min\{46 - 2t_2, 2t_1 - 2\}$.*

Proof Let us consider the function field $F = \mathbb{F}_{q^2}(x, y)$ with $y^q + y = x^m$ and $m | (q + 1)$; take $m = 2$ and $q = 5$. As the genus of F is $g = 2$, then F is a hyperelliptic function field (see Lemma 6.2.2.(b) of [152]), and the result follows from Corollary 5.4.4.   □

5.4.2.3 Construction III

In this subsection, we propose constructions of sequences of asymptotically good quantum codes derived from AG codes.

Recall that a tower of function fields (see Definition 1.3 of [47]) over $\mathbb{F}_q$ is a sequence $\mathcal{T} = (F_1, F_2, \ldots)$ of function fields $F_i/\mathbb{F}_q$ with the following properties:

(1) $F_1 \subseteq F_2 \subseteq F_3 \cdots$.
(2) For each $n \geq 1$, the extension F_{n+1}/F_n is separable of degree $[F_{n+1} : F_n] > 1$.
(3) $g(F_j) > 1$, for some $j > 1$.

By the Hurwitz genus formula, the condition (3) implies that $g(F_n) \to \infty$ for $n \to \infty$. The tower is said to be *asymptotically good* if $\lambda(\mathcal{T}) = \lim \sup_{i \to \infty} N(F_i)/g(F_i) > 0$, where $N(F_i)$ and $g(F_i)$ denote the number of $\mathbb{F}_q$-rational points and the genus of F_i, respectively. In the case of tower of function fields one can replace $\lim \sup_{i \to \infty} N(F_i)/g(F_i)$ by $\lim_{i \to \infty} N(F_i)/g(F_i)$, because the sequence $(N(F_i)/g(F_i))_{i \geq 1}$ is convergent. We say that the tower $\mathcal{T}$ (over $\mathbb{F}_q$) attains the Drinfeld-Vladut bound if $\lambda(\mathcal{T}) = \lim \sup_{i \to \infty} N(F_i)/g(F_i) = \sqrt{q} - 1$. To simplify the notation we put $N(F_i) = N_i$ and $g(F_i) = g_i$.

Let $(\mathcal{Q}_i)_{i \geq 1}$ be a sequence of quantum codes over $\mathbb{F}_q$ with parameters $[[n_i, k_i, d_i]]_q$, respectively. We say that $(\mathcal{Q}_i)_{i \geq 1}$ is asymptotically good if $\lim \sup_{i \to \infty} k_i/n_i > 0$ and $\lim \sup_{i \to \infty} d_i/n_i > 0$. The next result shows how to construct asymptotically good quantum codes derived from classical two-point AG codes.

Theorem 5.4.3 (Two-point asymptotically good codes) *Assume that the tower $\mathcal{T} = (F_1, F_2, \ldots)$ of function fields over $\mathbb{F}_q$ attains the Drinfeld-Vladut bound. Then there exists a sequence $(\mathcal{Q}_i)_{i \geq 1}$ of asymptotically good quantum codes, over $\mathbb{F}_q$, derived from classical two-point AG codes.*

Proof For each F_i, let us consider the set of rational places $P_1(i), \ldots, P_{N_i-2}(i)$, $P_{N_i-1}(i), P_{N_i}(i)$ of F_i. We set the divisors $D(i) = P_1(i) + \cdots + P_{N_i-2}(i)$, $G_1(i) = a_1(i)P_{N_i-1}(i) + a_2(i)P_{N_i}(i)$ and $G_2(i) = b_1(i)P_{N_i-1}(i) + b_2(i)P_{N_i}(i)$, where $a_1(i) \leq b_1(i)$ and $a_2(i) \leq b_2(i)$, with $2g_i - 2 < a_1(i) + a_2(i) < b_1(i) + b_2(i) < N_i - 2$. Let $C_1(i) := C_{\mathcal{L}}(i)[D(i), G_1(i)]$ and $C_2(i) := C_{\mathcal{L}}(i)[D(i), G_2(i)]$ be the two-point AG codes, over $\mathbb{F}_q$, corresponding to $G_1(i)$ and $G_2(i)$, respectively; hence, $C_1(i) \subset C_2(i)$. The code $C_1(i)$ has parameters

$$[N_i - 2, a_1(i) + a_2(i) - g_i + 1, d_1(i)]_q,$$

where $d_1(i) \geq N_i - 2 - (a_1(i) + a_2(i))$, and $C_2(i)$ has parameters

$$[N_i - 2, b_1(i) + b_2(i) - g_i + 1, d_2(i)]_q,$$

where $d_2(i) \geq N_i - 2 - (b_1(i) + b_2(i))$. Therefore, the corresponding CSS code has parameters

$$[[N_i - 2, K_i = b_1(i) + b_2(i) - (a_1(i) + a_2(i)), D_i]]_q,$$

where $D_i \geq \min\{N_i - 2 - (b_1(i) + b_2(i)), a_1(i) + a_2(i) - (2g_i - 2)\}$.

We know that the K_i's assume all the values from 1 to $N_i - 2g_i - 2$, i.e., $0 < K_i \leq N_i - 2g_i - 2$. For any such K_i we set $b_1(i) + b_2(i) = \lfloor (N_i + 2g_i + K_i - 4)/2 \rfloor$; hence, it follows that $N_i - 2 - (b_1(i) + b_2(i)) \geq a_1(i) + a_2(i) - (2g_i - 2)$, where $a_1(i) + a_2(i) - (2g_i - 2) \geq (N_i - K_i - 2g_i - 1)/2$. The sequence of positive integers $(K_i)_{i \geq 1}$ satisfies

$$0 < \limsup_{i \to \infty} \frac{K_i}{N_i - 2} \leq$$

$$\leq \limsup_{i \to \infty} N_i/(N_i - 2)$$
$$- \limsup_{i \to \infty} 2g_i/(N_i - 2)$$
$$+ \limsup_{i \to \infty} -2/(N_i - 2)$$
$$= 1 - 2/(\sqrt{q} - 1),$$

where in the last equality we utilize the fact that $\limsup_{i \to \infty} N_i/g_i = \sqrt{q} - 1$. For each $0 < c < 1 - 2/(\sqrt{q} - 1)$, we can choose convenient values for K_i such that $\lim_{i \to \infty} K_i/N_i = c$. Thus, $\limsup_{i \to \infty} K_i/(N_i - 2) = c > 0$. Moreover, we have

$$\limsup_{i \to \infty} (N_i - K_i - 2g_i - 1)/2(N_i - 2) = 1/2 \left[1 - 2/(\sqrt{q} - 1) - c \right] > 0.$$

Therefore, there exists a sequence $(Q_i)_{i \geq 1}$ of asymptotically good quantum codes over $\mathbb{F}_q$. The proof is complete.   □

Remark 5.4.3 Since several works available in the literature already presented constructions of asymptotically good quantum codes derived from one-point AG codes (see [12, 26, 27]), we do not present such constructions here.

Theorem 5.4.4 (*t*-point asymptotically good codes) *Assume that the tower* $T = (F_1, F_2, \ldots)$ *of function fields over* $\mathbb{F}_q$ *attains the Drinfeld-Vladut bound. Then there exists a sequence* $(Q_i)_{i \geq 1}$ *of asymptotically good quantum codes, over* $\mathbb{F}_q$, *derived from classical t-point AG codes.*

Proof We adopt the same notation as in the proof of Theorem 5.4.3. For each F_i, let us consider the set of rational places $P_1(i), \ldots, P_{n_i}(i), P_{n_i+1}(i), \ldots, P_{n_i+t}(i)$ of F_i, where $N_i = n_i + t$. Set $D(i) = P_1(i) + \cdots + P_{n_i}(i)$, $G_1(i) = a_1(i)P_{n_i+1}(i) + \ldots + a_t(i)P_{n_i+t}(i)$ and $G_2(i) = b_1(i)P_{n_i+1}(i) + \ldots + b_t(i)P_{n_i+t}(i)$, where $a_j(i) \leq b_j(i)$ for all $j = 1, \ldots, t$, with $2g_i - 2 < \sum_{j=1}^{t} a_j(i) < \sum_{j=1}^{t} b_j(i) < N_i - t$. Let us consider the *t*-point AG codes $C_1(i) := C_{\mathcal{L}}(i)[D(i), G_1(i)]$ and $C_2(i) := C_{\mathcal{L}}(i)[D(i), G_2(i)]$. It follows that $C_1(i) \subset C_2(i)$, and $C_1(i)$ has parameters

$$\left[N_i - t, \sum_{j=1}^{t} a_j(i) - g_i + 1, d_1(i) \right]_q,$$

where $d_1(i) \geq N_i - t - \sum_{j=1}^{t} a_j(i)$. Moreover, $C_2(i)$ has parameters

$$\left[N_i - t, \sum_{j=1}^{t} b_j(i) - g_i + 1, d_2(i) \right]_q ,$$

where $d_2(i) \geq N_i - t - \sum_{j=1}^{t} b_j(i)$.

Setting $\sum_{j=1}^{t} b_j(i) = \lfloor (N_i + 2g_i + K_i - t - 2)/2 \rfloor$ and proceeding similarly as in the proof of Theorem 5.4.3, the result follows.   □

Let q be a prime power and let C be an $[n, k, d]_{q^m}$ code over $\mathbb{F}_{q^m}$. Let β be a basis of $\mathbb{F}_{q^m}$ over $\mathbb{F}_q$, and assume also that $\beta^\perp$ is a dual basis of β. Let $C^\perp$ be the Euclidean dual of C. Then we have $[\beta(C)]^\perp = \beta^\perp(C^\perp)$ (see for instance [56, 94]).

Theorem 5.4.5 *For any prime p, there exists a sequence $(\mathcal{Q}_i)_{i \geq 1}$ of asymptotically good quantum codes over $\mathbb{F}_p$.*

Proof Let $q^2 = p^{2r}$, p prime. Let us consider the tower of function fields $\mathcal{T} = (F_1, F_2, \ldots)$ over $\mathbb{F}_{q^2}$, shown in [47], defined by $F_t = \mathbb{F}_{q^2}(x_1, \ldots, x_t)$, where

$$x_{i+1}^q + x_{i+1} = x_i^q/(x_i^{q-1} + 1),$$

for $i = 1, \ldots t - 1$. This tower attains the Drinfeld-Vladut bound. We next expand the codes $C_1(i)$ and $C_2(i)$, shown in the proof of Theorem 5.4.3, with respect to some basis β of $\mathbb{F}_{q^2}$ over $\mathbb{F}_p$. Thus, we obtain codes $\beta(C_1(i))$ and $\beta(C_2(i))$, both over $\mathbb{F}_p$, with parameters

$$[2r(N_i - 2), 2r(a_1(i) + a_2(i) - g_i + 1), d_1^*(i)]_p,$$

where $d_1^*(i) \geq d_1(i) \geq N_i - 2 - (a_1(i) + a_2(i))$, and

$$[2r(N_i - 2), 2r(b_1(i) + b_2(i) - g_i + 1), d_2^*(i)]_p,$$

where $d_2^*(i) \geq d_2(i) \geq N_i - 2 - (b_1(i) + b_2(i))$, respectively. Since the inclusion $\beta(C_1(i)) \subset \beta(C_2(i))$ holds, we apply the CSS construction to these codes, obtaining therefore an

$$[[2r(N_i - 2), 2rK_i, D_i]]_p$$

quantum code, where $D_i \geq \min\{N_i - 2 - (b_1(i) + b_2(i)), a_1(i) + a_2(i) - (2g_i - 2)\}$ (note that since $[\beta(C_1(i))]^\perp = \beta^\perp(C_1(i))^\perp$, it follows that the minimum distance of $[\beta(C_1(i))]^{\perp}$ is at least $a_1(i) + a_2(i) - (2g_i - 2))$. Proceeding similarly as in the proof of Theorem 5.4.3, we get $N_i - 2 - (b_1(i) + b_2(i)) \geq a_1(i) + a_2(i) - (2g_i - 2) \geq (N_i - K_i - 2g_i + 1)/2$. Consequently, we have $\limsup_{i \to \infty} 2r K_i / 2r(N_i - 2) > 0$ and $\limsup_{i \to \infty}(N_i - K_i - 2g_i + 1)/4r(N_i - 2) = 1/4r$ $[1 - 2/(p^r - 1) - c] > 0$, as desired. The proof is complete. $\qquad\square$

Remark 5.4.4 Although the proofs of Theorems 5.4.3 and 5.4.5 are similar to the proofs of the corresponding results shown in [27, 81], we utilize t-point ($t \geq 2$) AG codes, whereas in such references, the authors utilized only one-point AG codes to perform their constructions. Another difference is that in such papers the authors utilized the technique of code concatenation to obtain quantum codes over prime fields; here, we utilize the technique of code expansion.

5.4.3 Examples

In Tables 5.15, 5.16 and 5.17, we exhibit some quantum codes derived from Corollaries 5.4.1, 5.4.2 and 5.4.5, respectively. In these tables, q is a prime power and $a, b, a_1, a_2, b_1, b_2, t_1, t_2, m$ are positive integers satisfying some conditions. More precisely: in Table 5.15 we consider that $(q - 1)(m - 1) - 2 < a < b$, $b < q(1 + (q - 1)m)$ and $m | (q + 1)$; in Table 5.16, we assume that $a_i \leq b_i$ $i = 1, 2, (q - 1)(m - 1) - 2 < a_1 + a_2 < b_1 + b_2$, $b_1 + b_2 < q[1 + (q - 1)m] - 1$ and $m | (q + 1)$; in Table 5.17, we suppose that $1 < t_1 < t_2 < 23$.

Recall that the parameters of an $Q := [[n, k, d]]_q$ quantum code satisfy the inequality $k + 2d \leq n + 2$ (quantum Singleton bound). The *Singleton defect* (SD_Q) of a code is defined as $SD_Q = n + 2 - k - 2d$. In this context, we measure the performance of a code by means of the Singleton defect. We adopt this method because, for large alphabets, it is difficult to find codes over them: "... for large q, it is difficult to find explicit known codes to compare with ours since there are no suitable tables for reference" (see p. 3 of [71]).

Our $[[26, 16, d \geq 3]]_9$ code is better than the $[[26, 14, 3]]_9$ code shown in Ref. [34], because the Singleton defect of $Q_1 := [[26, 16, d \geq 3]]_9$ is $SD_{Q_1} \leq 6$, whereas the Singleton defect of $Q_2 := [[26, 14, 3]]_9$ is $SD_{Q_2} = 8$. Our $[[26, 14, d \geq 4]]_9$ code with Singleton defect at most 8 is better than the $[[26, 4, 4]]_9$ code shown in Ref. [34], which has Singleton defect 16. Our codes of length 46 have Singleton defect at most 4. Moreover, the quantum codes of lengths $n = 26$ and $n = 27$, exhibited in Table 5.15, have Singleton defect at most 6.

Note that the $[[27, 3, d \geq 10]]_9$, $[[27, 5, d \geq 9]]_9$, $[[65, 9, d \geq 25]]_{25}$, $[[175, 31, d \geq 64]]_{49}$ codes have large minimum distances when compared to their code lengths.

The quantum codes shown in [71] were constructed over the field $\mathbb{F}_{q^2}$, where q is a power of 2, whereas we construct here quantum codes over $\mathbb{F}_q$ for all prime power q. Great part of the codes displayed in [71] were constructed over $\mathbb{F}_8$; this fact does not allow us to compare our codes with the ones shown in [71].

Table 5.15 Quantum codes

Codes from Corollary 5.4.1	q	m	a	b
$[[27, 17, d \geq 3]]_9$	3	4	7	24
$[[27, 15, d \geq 4]]_9$	3	4	8	23
$[[27, 13, d \geq 5]]_9$	3	4	9	22
$[[27, 11, d \geq 6]]_9$	3	4	10	21
$[[27, 9, d \geq 7]]_9$	3	4	11	20
$[[27, 7, d \geq 8]]_9$	3	4	12	19
$[[27, 5, d \geq 9]]_9$	3	4	13	18
$[[27, 3, d \geq 10]]_9$	3	4	14	17
$[[27, 1, d \geq 11]]_9$	3	4	15	16
$[[64, 48, d \geq 3]]_{16}$	4	5	13	61
$[[64, 46, d \geq 4]]_{16}$	4	5	14	60
$[[64, 44, d \geq 5]]_{16}$	4	5	15	59
$[[64, 24, d \geq 15]]_{16}$	4	5	25	49
$[[64, 4, d \geq 25]]_{16}$	4	5	35	39
$[[64, 2, d \geq 26]]_{16}$	4	5	36	38
$[[65, 53, d \geq 3]]_{25}$	5	3	9	62
$[[65, 51, d \geq 4]]_{25}$	5	3	10	61
$[[65, 49, d \geq 5]]_{25}$	5	3	11	60
$[[65, 9, d \geq 25]]_{25}$	5	3	31	40
$[[175, 153, d \geq 3]]_{49}$	7	4	19	172
$[[175, 151, d \geq 4]]_{49}$	7	4	20	171
$[[175, 149, d \geq 5]]_{49}$	7	4	21	170
$[[175, 109, d \geq 25]]_{49}$	7	4	41	150
$[[175, 31, d \geq 64]]_{49}$	7	4	80	111
$[[175, 1, d \geq 79]]_{49}$	7	4	95	96

Table 5.16 Quantum codes

Codes from Corollary 5.4.2	q	m	a_1	a_2	b_1	b_2
$[[26, 16, d \geq 3]]_9$	3	4	3	4	7	16
$[[26, 14, d \geq 4]]_9$	3	4	3	5	7	15
$[[26, 12, d \geq 5]]_9$	3	4	3	6	7	14
$[[26, 4, d \geq 9]]_9$	3	4	3	10	7	10
$[[26, 2, d \geq 10]]_9$	3	4	4	10	6	10

Table 5.17 Quantum codes

Codes from Corollary 5.4.5	q	m	t_1	t_2
$[[46, 36, d \geq 4]]_{25}$	5	2	3	21
$[[46, 32, d \geq 6]]_{25}$	5	2	4	20
$[[46, 28, d \geq 8]]_{25}$	5	2	5	19
$[[46, 4, d \geq 20]]_{25}$	5	2	11	13

The quantum codes exhibited in [119] were constructed over the fields $\mathbb{F}_2$, $\mathbb{F}_3$, $\mathbb{F}_4$, $\mathbb{F}_5$, $\mathbb{F}_8$, $\mathbb{F}_9$. In our case, we exhibit examples of quantum codes constructed over $\mathbb{F}_9$, $\mathbb{F}_{16}$, $\mathbb{F}_{25}$, $\mathbb{F}_{49}$. The codes over $\mathbb{F}_9$ constructed in [119] have parameters $[[15, 13, 2]]_9$, $[[15, 7, 4]]_9$, $[[15, 5, 5]]_9$, $[[15, 1, 7]]_9$, $[[243, 241, 2]]_9$, $[[243, 219, 6]]_9$, $[[243, 213, 9]]_9$. Our codes over $\mathbb{F}_9$, shown in Tables 7.2 and 5.16, have parameters $[[26, 16, d \geq 3]]_9$, $[[26, 14, d \geq 4]]_9$, $[[26, 12, d \geq 5]]_9$, $[[26, 4, d \geq 9]]_9$, $[[26, 2, d \geq 10]]_9$, $[[27, 17, d \geq 3]]_9$, $[[27, 15, d \geq 4]]_9$, $[[27, 13, d \geq 5]]_9$, $[[27, 11, d \geq 6]]_9$, $[[27, 9, d \geq 7]]_9$, $[[27, 7, d \geq 8]]_9$, $[[27, 5, d \geq 9]]_9$, $[[27, 3, d \geq 10]]_9$ and $[[27, 1, d \geq 11]]_9$. Since the parameters among these codes are different, we do not perform a fair comparison of such codes.

Summarizing the results obtained in this section: we have constructed several families of quantum codes with good as well as asymptotically good parameters. Great part of our quantum codes has large minimum distances when compared with their corresponding code lengths. Additionally, they have relatively small Singleton defects. We have also shown how to obtain sequences of asymptotically good quantum codes derived from t-point AG codes.

5.5 Quantum Synchronizable Codes

We provide here constructions of families of quantum synchronizable codes (QSCs) by applying the method discovered by Fujiwara [42] (see also [44]). More precisely, we show how to construct families of QSCs derived from the sum and intersection of cyclic codes, BCH codes and from the product of cyclic codes. All these results presented here can be found in [61]. For more details with respect to quantum synchronizable codes we refer the reader to [42].

Before presenting our constructions of QSCs, we will review this concept. The main goal of frame synchronization in communication systems is to ensure that information block boundaries can be correctly determined at the receiver. To achieve this goal, numerous synchronization techniques have been developed for classical communication systems. However, these techniques are not applicable to quantum communication systems since a qubit measurement typically destroys the quantum states and thus also the corresponding quantum information. To circumvent this problem, synchronization can be achieved using a classical system external to the quantum system, but such a solution does not take advantage of the benefits that quantum processing can provide.

In his seminal paper [42], Fujiwara provided a framework for quantum block synchronization. The approach is to employ QSCs, which allow the identification of codeword boundaries without destroying the quantum states. More precisely, an $(a_l, a_r) - [[n, k]]$ QSC is an binary $[[n, k]]$ code that encodes k logical qubits into a physical qubit, and corrects misalignments by up to al qubits to the left and up to ar qubits to the right. These quantum codes may correct more phase errors than bit errors. This is an advantage because, as shown by Ioffe and Mzard [70], in physical systems the noise is typically asymmetric in the sense that bit errors (can) occur less frequently than phase errors. In this light, QSCs can be considered as asymmetric quantum codes.

Several constructions of QSCs have been presented in the literature [44, 164, 167]. These constructions utilize BCH codes, cyclic codes related to finite geometries, punctured ReedMuller codes, quadratic residue codes and duadic codes. According to the authors in [44]: "One of the main hurdles in the theoretical study of quantum synchronizable codes is that it is quite difficult to find suitable classical error-correcting codes because the required algebraic constraints are very severe and difficult to analyze."

In order to proceed further we need a result due to Fujiwara.

Theorem 5.5.1 ([42, Theorem 1]) *Let C be a dual-containing $[n, k_1, d_1]$ cyclic code and let D be a C-containing cyclic code with parameters $[n, k_2, d_2]$, with $k_1 < k_2$. Then, for any pair of nonnegative integers (a_l, a_r) satisfying $a_l + a_r < k_2 - k_1$, there exists an $(a_l, a_r) - [[n + a_l + a_r, 2k_1 - n]]$ QSC that corrects up to at least $\lfloor \frac{d_1-1}{2} \rfloor$ phase errors and up to at least $\lfloor \frac{d_2-1}{2} \rfloor$ bit errors.*

5.5.1 Synchronizable Codes from Cyclic Codes

At this point we already have tools to develop the results. We start by presenting two constructions of QSCs from cyclic codes. The first result is based on the sum of cyclic codes.

Theorem 5.5.2 *Let $n \geq 3$ be an integer such that $\gcd(n, 2) = 1$ and assume that $m = \mathrm{ord}_n(2)$. Let C_1 be an $[n, k_1, d_1]$ dual-containing cyclic code and C_2 be an $[n, k_2, d_2]$ C_1-containing cyclic code. Let also C_3 be a cyclic code with parameters $[n, k_3, d_3]$ and C_4 be an $[n, k_4, d_4]$ C_3-containing cyclic code such that $\deg(\gcd(g_2(x), g_4(x))) < \deg(\gcd(g_1(x), g_3(x)))$, where $g_i(x)$ is the generator polynomial of C_i, $i = 1, 2, 3, 4$. Then, for any pair of nonnegative integers (a_l, a_r) satisfying $a_l + a_r < \deg(\gcd(g_1(x), g_3(x))) - \deg(\gcd(g_2(x), g_4(x)))$, there exists an $(a_l, a_r) - [[n + a_l + a_r, n - 2\deg(\gcd(g_1(x), g_3(x)))]]$ QSC that corrects up to at least $\lfloor \frac{d-1}{2} \rfloor$ phase errors and up to at least $\lfloor \frac{d^*-1}{2} \rfloor$ bit errors, where d is the minimum distance of the code $C_1 + C_3$ and d^* is the minimum distance of $C_2 + C_4$.*

Proof Since C_1 and C_3 are cyclic codes, the sum code $C_1 + C_3 = \{c_1 + c_3 | c_1 \in C_1 \text{ and } c_3 \in C_3\}$ is also cyclic. We know that $C_1 \subset C_1 + C_3$, so $(C_1 + C_3)^{\perp} \subset C_1^{\perp}$. As C_1 contains its dual code, it follows that $(C_1 + C_3)^{\perp} \subset C_1^{\perp} \subset C_1 \subset C_1 + C_3$. Thus the sum $C_1 + C_3$ is also a dual-containing cyclic code.

Let $g_o(x) = \gcd(g_2(x), g_4(x))$. Since $g_2(x) | g_1(x)$ and $g_4(x) | g_3(x)$, it follows that $g_o(x)$ divides $g(x)$, which implies that the inclusion $C_1 + C_3 \subset C_2 + C_4$ is true, where $C_2 + C_4$ is also a cyclic code. From the fact that $\deg(g_o(x)) < \deg(g(x))$, one has $C_1 + C_3 \subsetneq C_2 + C_4$. Applying Theorem 5.5.1 to the codes $C_1 + C_3$ and $C_2 + C_4$, we can construct a QSC with parameters $(a_l, a_r) - [[n + a_l + a_r, n - 2\deg(\gcd(g_1(x), g_3(x)))]]$, where $a_l + a_r < \deg(\gcd(g_1(x), g_3(x))) - \deg(\gcd(g_2(x), g_4(x)))$, and corrects up to at least $\lfloor \frac{d-1}{2} \rfloor$ phase errors and up to at least $\lfloor \frac{d^*-1}{2} \rfloor$ bit errors.   □

The second theorem is obtained by considering the intersection of cyclic codes.

Theorem 5.5.3 *Let $n \geq 3$ be an integer such that $\gcd(n, 2) = 1$ and consider that $m = \text{ord}_n(2)$. Let C_1 be an $[n, k_1, d_1]$ self-orthogonal cyclic code. Assume also that C_2 and C_3 are two cyclic codes with parameters $[n, k_2, d_2]$ and $[n, k_3, d_3]$, respectively, such that $\{0\} \subsetneq C_3^{\perp} \subsetneq C_1 \cap C_2$. Then for any pair of nonnegative integers (a_l, a_r) satisfying $a_l + a_r < n - \deg(g_3(x)) - \deg(\text{lcm}(g_1(x), g_2(x)))$, there exists an*

$$(a_l, a_r) - [[n + a_l + a_r, 2\deg(\text{lcm}(g_1(x), g_2(x))) - n]]$$

QSC that corrects up to at least $\lfloor \frac{d-1}{2} \rfloor$ phase errors and up to at least $\lfloor \frac{d_3-1}{2} \rfloor$ bit errors, where d is the minimum distance of $(C_1 \cap C_2)^{\perp}$, and $g_i(x)$ is the generator polynomial of C_i, $i = 1, 2, 3$.

Proof Since C_1 and C_2 are cyclic code, it follows that $C_1 \cap C_2$ is cyclic, which implies that its dual code $(C_1 \cap C_2)^{\perp}$ is also cyclic. As $C_1 \cap C_2 \subset C_1$, the inclusion $C_1^{\perp} \subset (C_1 \cap C_2)^{\perp}$ holds. Since C_1 is self-orthogonal, we have $C_1 \cap C_2 \subset C_1 \subset C_1^{\perp} \subset (C_1 \cap C_2)^{\perp}$, i.e., $C_1 \cap C_2$ is self-orthogonal, which implies that $(C_1 \cap C_2)^{\perp}$ is dual-containing cyclic code. As $C_3^{\perp} \subsetneq C_1 \cap C_2$, it follows that $(C_1 \cap C_2)^{\perp} \subsetneq C_3$. The dimension of the QSC is equal to $2\deg(\text{lcm}(g_1(x), g_2(x))) - n$ and $a_l + a_r < n - \deg(g_3(x)) - \deg(\text{lcm}(g_1(x), g_2(x)))$. Applying Theorem 5.5.1 to the codes $(C_1 \cap C_2)^{\perp}$ and C_3, for any pair of nonnegative integers (a_l, a_r) satisfying $a_l + a_r < n - \deg(g_3(x)) - \deg(\text{lcm}(g_1(x), g_2(x)))$, there exists an QSC with the specified parameters.   □

5.5.2 Synchronizable Codes from BCH Codes

The results presented here are concerned with constructions of QSCs derived from BCH codes.

Let $\gcd(n, q) = 1$. Recall that the q-coset, of s modulo n is defined as $\mathbb{C}_s = \{s, sq, \ldots, sq^{ms-1}\}$, where $sq^{ms} \equiv s \bmod n$. Let α be a primitive nth root of unity and $M_i(x)$ denotes the minimal polynomial of α^i.

Let us recall two results shown in [4].

Proposition 5.5.1 ([4, Theorems 3 and 10]) *Let n be a positive integer such that* $\gcd(n, 2) = 1$ *and let* $m = \mathrm{ord}_n(2)$. *If* $2 \le \delta \le \delta_{max} = \lfloor \kappa \rfloor$, *where* $\kappa = \frac{n}{2^m - 1}(2^{\lceil m/2 \rceil} - 1)$, *then the narrow-sense* $BCH(n, 2, \delta)$ *code contains its Euclidean dual* $BCH^{\perp}$ $(n, 2, \delta)$.

Lemma 5.5.1 ([4, Lemmas 8 and 9]) *Let $n \ge 1$ be an integer such that* $\gcd(n, 2) = 1$ *and* $2^{\lfloor m/2 \rfloor} < n \le 2^m - 1$, *where* $m = \mathrm{ord}_n(2)$. *Then the following hold:*
(i) The 2-coset $\mathbb{C}_x$ *has cardinality m for all x in the range* $1 \le x \le n2^{\lceil m/2 \rceil}/(2^m - 1)$.
(ii) If x and y are distinct integers in the range $1 \le x, y \le \min\{\lfloor n2^{\lceil m/2 \rceil}/(2^m - 1) - 1 \rfloor, n - 1\}$ *such that $x, y \not\equiv 0 \bmod 2$, then the 2-cosets of x and y modulo n are disjoint.*

We are now able to prove the first result of this subsection.

Theorem 5.5.4 *Let $n \ge 3$ be an integer such that* $\gcd(n, 2) = 1$ *and assume that* $2^{\lfloor m/2 \rfloor} < n \le 2^m - 1$, *where* $m = \mathrm{ord}_n(2)$. *Take integers a, b such that* $1 \le a < b < r = \min\{\lfloor n2^{\lceil m/2 \rceil}/(2^m - 1) - 1 \rfloor, n - 1, \lfloor \kappa \rfloor\}$, *where* $\kappa = \frac{n}{2^m - 1}(2^{\lceil m/2 \rceil} - 1)$ *and* $a, b \not\equiv 0 \bmod 2$. *Then, for any pair of nonnegative integers (a_l, a_r) satisfying* $a_l + a_r < m(t - u)$, *there exists an $(a_l, a_r) - [[n + a_l + a_r, n - 2m(t + 1)]]$ QSC that corrects up to at least $\lfloor \frac{d-1}{2} \rfloor$ phase errors and up to at least $\lfloor \frac{d^*-1}{2} \rfloor$ bit errors, where $d \ge b + 1$, $d^* \ge a + 1$, $t = (b - 1)/2$ and $u = (a - 1)/2$.*

Proof Let D be the binary BCH code of length n generated by

$$D = \langle M_1(x)M_3(x) \cdots M_a(x) \rangle,$$

where $a = 2u + 1$ and $u \ge 0$ is an integer. Furthermore, let C be the binary BCH code of length n generated by

$$C = \langle M_1(x)M_3(x) \cdots M_b(x) \rangle,$$

where $b = 2t + 1$ and $t \ge 1$ is an integer. From construction, we have $C \subset D$, and by Proposition 5.5.1, C is dual-containing. From Lemma 5.5.1 and from straightforward computation, the dimension of D is equal to $k_2 = n - m(u + 1)$ and the dimension of C is $k_1 = n - m(t + 1)$, which implies $k_2 - k_1 = m(t - u)$ and $2k_1 - n = n - 2m(t + 1)$. From the BCH bound, because the defining set of D contains a sequence of a consecutive integers, the minimum distance of D satisfies $d_2 \ge a + 1$. Similarly, as the defining set of C contains a sequence of b consecutive integers, the minimum distance of C satisfies $d_1 \ge b + 1$. The result follows from Theorem 5.5.1 and the proof is complete. $\square$

We next show how to construct QSCs from the sum of BCH codes.

Theorem 5.5.5 *Let $n \geq 3$ be an integer such that $\gcd(n, 2) = 1$ and assume that $2^{\lfloor m/2 \rfloor} < n \leq 2^m - 1$, where $m = \mathrm{ord}_n(2)$. Take integers a, b, e and f such that $2 \leq e < a < b < f < \min\{\lfloor n 2^{\lceil m/2 \rceil}/(2^m - 1) - 1 \rfloor, n - 1, \lfloor \kappa \rfloor\}$,   where $\kappa = \frac{n}{2^m - 1}(2^{\lceil m/2 \rceil} - 1)$ and $a, b, e, f \not\equiv 0 \bmod 2$. Then, for any pair of nonnegative integers (a_l, a_r) satisfying $a_l + a_r < m(t - w)$, there exists an $(a_l, a_r) - [[n + a_l + a_r, n - 2m(t + 1)]]$ QSC that corrects up to at least $\lfloor \frac{d-1}{2} \rfloor$ phase errors and up to at least $\lfloor \frac{d^*-1}{2} \rfloor$ bit errors, where $d \geq b + 1$, $d^* \geq e + 1$, $t = (b - 1)/2$ and $w = (e - 1)/2$.*

Proof Let C_1 be the binary BCH code of length n generated by

$$C_1 = \langle M_1(x) M_3(x) \cdots M_b(x) \rangle,$$

where $b = 2t + 1$ and t is a nonnegative integer. Let C_2 be the binary BCH code of length n generated by

$$C_2 = \langle M_1(x) M_3(x) \cdots M_a(x) \rangle,$$

where $a = 2u + 1$ and $u \geq 1$.

From construction, $C_1 \subset C_2$; by Proposition 5.5.1 it follows that C_1 contains its dual code. Further, consider the binary BCH codes of length n generated by

$$C_3 = \langle M_1(x) M_3(x) \cdots M_f(x) \rangle$$

and

$$C_4 = \langle M_1(x) M_3(x) \cdots M_e(x) \rangle,$$

where $f = 2v + 1$ with $v \geq 1$, and $e = 2w + 1$ with $w \geq 0$. From construction, $C_3 \subsetneqq C_4$. It then follows that $C_1 + C_3 \subsetneqq C_2 + C_4$. Since $e < a$ and $b < f$, from Lemma 5.5.1 and a straightforward computation, the codes $C_2 + C_4$ and $C_1 + C_3$ have dimensions $K_2 = n - m(w + 1)$ and $K_1 = n - m(t + 1)$, respectively, which implies that the dimension of the corresponding QSC is equal to $K = n - 2m(t + 1)$ and $K_2 - K_1 = m(t - w)$. Since C_1 is dual-containing, $C_1 + C_3$ is also dual-containing. From the BCH bound, the minimum distance d_{13} of $C_1 + C_3$ satisfies $d_{13} \geq b + 1$ and the minimum distance d_{24} of $C_2 + C_4$ satisfies $d_{24} \geq e + 1$. Applying Theorem 5.5.2 to $C_1 + C_3$ and $C_2 + C_4$ we get the result.   □

5.5.3 Synchronizable Codes from Product Codes

For the reader convenience, we recall some basic concepts on product codes. Let us consider C_1 and C_2 be two linear codes with parameters $[n_1, k_1, d_1]$ and $[n_2, k_2, d_2]$, respectively, both over $\mathbb{F}_q$. Suppose $G^{(1)}$ and $G^{(2)}$ are the generator matrices of C_1 and C_2, respectively. Then the product code $C_1 \otimes C_2$ is a linear code with parameters

Table 5.18 Some QSCs from Theorem 5.5.4

$[[n + a_l + a_r, n - 2m(t + 1)]]$	$d, d^*, a_l + a_r$
$(a_l, a_r) - [[63 + a_l + a_r, 27]]$	$d \geq 6, d^* \geq 2, 0 \leq a_i + a_r < 12$
$(a_l, a_r) - [[85 + a_l + a_r, 53]]$	$d \geq 4, d^* \geq 2, 0 \leq a_i + a_r < 8$
$(a_l, a_r) - [[127 + a_l + a_r, 29]]$	$d \geq 14, d^* \geq 2, 0 \leq a_i + a_r < 42$
$(a_l, a_r) - [[127 + a_l + a_r, 29]]$	$d \geq 14, d^* \geq 4, 0 \leq a_i + a_r < 35$
$(a_l, a_r) - [[127 + a_l + a_r, 29]]$	$d \geq 14, d^* \geq 6, 0 \leq a_i + a_r < 28$
$(a_l, a_r) - [[127 + a_l + a_r, 29]]$	$d \geq 14, d^* \geq 8, 0 \leq a_i + a_r < 21$
$(a_l, a_r) - [[127 + a_l + a_r, 29]]$	$d \geq 14, d^* \geq 10, 0 \leq a_i + a_r < 14$
$(a_l, a_r) - [[1365 + a_l + a_r, 1125]]$	$d \geq 20, d^* \geq 2, 0 \leq a_i + a_r < 108$
$(a_l, a_r) - [[1365 + a_l + a_r, 1125]]$	$d \geq 20, d^* \geq 4, 0 \leq a_i + a_r < 96$
$(a_l, a_r) - [[1365 + a_l + a_r, 1125]]$	$d \geq 20, d^* \geq 6, 0 \leq a_i + a_r < 84$
$(a_l, a_r) - [[1365 + a_l + a_r, 1125]]$	$d \geq 20, d^* \geq 8, 0 \leq a_i + a_r < 72$
$(a_l, a_r) - [[1365 + a_l + a_r, 1125]]$	$d \geq 20, d^* \geq 18, 0 \leq a_i + a_r < 12$
$(a_l, a_r) - [[1365 + a_l + a_r, 1245]]$	$d \geq 10, d^* \geq 2, 0 \leq a_i + a_r < 48$
$(a_l, a_r) - [[1365 + a_l + a_r, 1245]]$	$d \geq 10, d^* \geq 4, 0 \leq a_i + a_r < 36$
$(a_l, a_r) - [[1365 + a_l + a_r, 1245]]$	$d \geq 10, d^* \geq 6, 0 \leq a_i + a_r < 24$
$(a_l, a_r) - [[1365 + a_l + a_r, 1245]]$	$d \geq 10, d^* \geq 8, 0 \leq a_i + a_r < 12$

$[n_1 n_2, k_1 k_2, d_1 d_2]$ code over $\mathbb{F}_q$ generated by the Kronecker product matrix $G^{(1)} \otimes G^{(2)}$ (see Definition 1.7.14) defined as

$$G^{(1)} \otimes G^{(2)} = \begin{bmatrix} g_{11}^{(1)} G^{(2)} & g_{12}^{(1)} G^{(2)} & \cdots & g_{1n_1}^{(1)} G^{(2)} \\ g_{21}^{(1)} G^{(2)} & g_{22}^{(1)} G^{(2)} & \cdots & g_{2n_1}^{(1)} G^{(2)} \\ \vdots & \vdots & \vdots & \vdots \\ g_{k_1 1}^{(1)} G^{(2)} & g_{k_1 2}^{(1)} G^{(2)} & \cdots & g_{k_1 n_1}^{(1)} G^{(2)} \end{bmatrix}$$

We can now prove the following result.

Theorem 5.5.6 *Let n and n^* be two positive odd integers such that $\gcd(n, n^*) = 1$. Let C_1 be an $[n, k_1, d_1]$ self-orthogonal cyclic code and let C_2 be an $[n, k_2, d_2]$ cyclic code, both over $\mathbb{F}_2$. Assume also that C_3 and C_4 are two cyclic codes with parameters $[n^*, k_3, d_3]$ and $[n^*, k_4, d_4]$, respectively, both over $\mathbb{F}_2$, such that $(C_1 \otimes C_3)^{\perp} \subsetneq C_2 \otimes C_4$. Then for any pair of nonnegative integers (a_l, a_r) satisfying $a_l + a_r < k_1 k_3 + k_2 k_4 - nn^*$, there exists an $(a_l, a_r) - [[nn^* + a_l + a_r, nn^* - 2k_1 k_3]]$ QSC that corrects up to at least $\lfloor \frac{d-1}{2} \rfloor$ phase errors and up to at least $\lfloor \frac{d_2 d_4 - 1}{2} \rfloor$ bit errors, where d is the minimum distance of $(C_1 \otimes C_3)^{\perp}$, which satisfies $d \geq d_2 d_4$.*

Proof As $\gcd(n, n^*) = 1$, it follows that the product code $C_2 \otimes C_4$ (consequently $C_1 \otimes C_3$ and $(C_1 \otimes C_3)^{\perp}$), is also cyclic (see [114, Theorem 1, p. 570]). The elements of the code $C_1 \otimes C_3$ are linear combinations of vectors $v_i^{(1)} \otimes w_j^{(3)}$, where $v_i^{(1)} \in C_1$

and $w_j^{(3)} \in C_3$. In other words, each $c \in C_1 \otimes C_3$ can be written as $c = \sum_i v_i^{(1)} \otimes$ $w_j^{(3)}$. An (Euclidean) inner product on $C_1 \otimes C_3$ can be defined as

$$\left\langle v_i^{(1)} \otimes w_i^{(3)} | v_j^{(1)} \otimes w_j^{(3)} \right\rangle = \langle v_i^{(1)} | v_j^{(1)} \rangle \langle w_i^{(3)} | w_j^{(3)} \rangle, \tag{5.2}$$

and it is extended by linearity for all elements of $C_1 \otimes C_3$. Note that $\langle c_i^{(1)} | c_j^{(1)} \rangle$ and $\langle c_i^{(3)} | c_j^{(3)} \rangle$ are the Euclidean inner products on C_1 and C_3, respectively. Since C_1 is self-orthogonal, it follows from Eq. 5.2 that $C_1 \otimes C_3$ is also self-orthogonal, which implies that $(C_1 \otimes C_3)^\perp$ is dual-containing. The parameters of the codes $(C_1 \otimes C_3)^\perp$ and $C_2 \otimes C_4$ are $[nn^*, nn^* - k_1k_3, d]$ and $[nn^*, k_2k_2, d_2d_4]$, respectively. Since $(C_1 \otimes C_3)^\perp \subsetneqq C_2 \otimes C_4$, it follows that $d \geq d_2d_4$. Applying Theorem 5.5.1 to the cyclic codes $(C_1 \otimes C_3)^\perp$ and $C_2 \otimes C_4$ the proof is complete. $\square$

Chapter 6
Asymmetric Quantum Codes

To make reliable the transmission or storage of quantum information against noise caused by the environment, there exist many works available in the literature dealing with constructions of efficient quantum error-correcting codes (QECCs) over unbiased quantum channels [11, 12, 25, 55, 80, 89, 92, 133, 148].

In the past 10 years, these constructions have been extended to asymmetric quantum channels in a natural way [3, 35–37, 70, 92, 94, 95, 97, 101, 139, 140, 149, 157].

Asymmetric quantum error-correcting codes (AQECCs) are quantum codes defined over quantum channels where qudit-flip errors and phase-shift errors may have different probabilities. Steane [147] was the first author who introduced the notion of asymmetric quantum errors. As usual, the parameters of an AQECC is given by $[[n, k, d_z/d_x]]_q$, where n is the length, k means that the code has dimension q^k, d_z is the minimum distance corresponding to phase-shift errors and d_x is the minimum distance corresponding to qudit-flip errors. As an example of a quantum channel such that $d_z > d_x$ (i.e., a channel presenting asymmetry) is the combined amplitude damping and dephasing channel (specific to binary systems; see [139]).

To put the reader into context, we give here a brief summary of the papers available in the literature dealing with investigations of AQECCs. In [70], the authors utilized BCH codes to correct qubit-flip errors and LDPC codes to correct more frequently phase-shift errors. In [149], an investigation of AQECCs via code conversion was presented. In references [3, 92], several families of AQECCs derived from BCH codes were constructed. Asymmetric stabilizer codes obtained from LDPC codes were constructed in [139]; in [140], the same authors constructed several families of both binary and nonbinary AQECCs as well as they derived bounds such as the (quantum) Singleton and the linear programming bound to AQECCs. In [157], constructions of nonadditive AQECCs as well as constructions of asymptotically good AQECCs derived from algebraic geometry codes were presented. In [35], the CSS construction [25, 80, 121] was extended to include codes endowed with the Hermitian and also

© Springer Nature Switzerland AG 2020
G. G. La Guardia, *Quantum Error Correction*, Quantum Science and Technology,
https://doi.org/10.1007/978-3-030-48551-1_6

trace Hermitian inner product. In [37], asymmetric quantum MDS codes obtained from generalized Reed–Solomon (GRS) codes were constructed. In the papers [94, 95], we constructed families of AQECCs by expanding GRS codes and by applying product codes, respectively.

In this chapter, we construct families of asymmetric quantum codes derived from (classical) BCH codes (Sect. 6.3), Reed–Solomon and generalized Reed–Solomon codes (Sect. 6.5), tensor product codes (Sect. 6.6). In Sect. 6.7, we generalize to asymmetric quantum codes the known results which are valid to quantum codes. More precisely, we show how to construct AQECCs by applying to methods of puncturing, extending, expanding, direct sum and by the technique of $(\mathbf{u}|\mathbf{u} + \mathbf{v})$ construction. In particular, several families of AQECCs are obtained by employing these results to generalized Reed–Muller codes, character codes, BCH, quadratic residue and affine-invariant codes.

6.1 Preliminaries

As always, p is a prime, q is a prime power, $\mathbb{F}_q$ is the finite field with q elements and $\alpha \in \mathbb{F}_{q^m}$ is a primitive nth root of unity.

Recall that the trace map (see Definition 3.5.3) $\mathrm{tr}_{q^m/q} : \mathbb{F}_{q^m} \longrightarrow \mathbb{F}_q$ is defined as $\mathrm{tr}_{q^m/q}(a) := \sum_{i=0}^{m-1} a^{q^i}$. As it is usual in group theory, we write $H \leq G$ to denote that H is a subgroup of a group $(G, *)$. The center of G is denoted by $Z(G)$. If $S \leq G$, then we write $C_G(S)$ meaning the centralizer of S in G. Further, we write $SZ(G)$ to denote the subgroup generated by S and the center $Z(G)$ (these notations are essentially the same to that of [80]).

We write $\mathrm{wt}(C)$ to denote the minimum weight of a code C, and by $\mathrm{d}(C)$ its minimum distance. Sometimes we abuse the notation by writing $C = [n, k, d]_q$. If C is an $[n, k, d]_q$ linear code, recall that its Euclidean dual (see Definition 4.2.1) is defined as

$$C^{\perp} = \{\mathbf{y} \in \mathbb{F}_q^n \mid \mathbf{y} \cdot \mathbf{x} = 0, \forall\, \mathbf{x} \in C\}.$$

If C is an $[n, k, d]_{q^2}$ code over $\mathbb{F}_{q^2}$, recall that its Hermitian dual (see Definition 4.2.3) is defined by

$$C^{\perp_h} = \{\mathbf{y} \in \mathbb{F}_{q^2}^n \mid \mathbf{y}^q \cdot \mathbf{x} = 0, \forall\, \mathbf{x} \in C\},$$

where $\mathbf{y}^q = (y_1^q, \ldots, y_n^q)$ denotes the conjugate of the vector $\mathbf{y} = (y_1, \ldots, y_n)$.

If $C \leq \mathbb{F}_q^{2n}$ is an additive code, then we write $\mathrm{swt}(C)$ to denote the symplectic weight (see Definition 3.5.7) of C and $C^{\perp_s}$ to denote the trace-symplectic dual of C (see Definition 6.2.1).

Similarly, if $C \leq \mathbb{F}_{q^2}^n$ is an additive code, then $C^{\perp_a}$ denotes the trace-alternating dual of C, where the trace-alternating form of two vectors $\mathbf{v}, \mathbf{w} \in \mathbb{F}_{q^2}^n$ is defined as

$$\langle \mathbf{v}|\mathbf{w} \rangle_a = \mathrm{tr}_{q/p} \left(\frac{\mathbf{v} \cdot \mathbf{w}^q - \mathbf{v}^q \cdot \mathbf{w}}{\beta^{2q} - \beta^2} \right),$$

where (β, β^q) is a normal basis of $\mathbb{F}_q^2$ over $\mathbb{F}_q$.

6.2 Error Groups and Asymmetric Codes

To start, we need to define the error model adopted to asymmetric quantum noise channel.

Let $\mathcal{H}$ be the Hilbert space $\mathcal{H} = \mathbb{C}^{q^n} = \mathbb{C}^q \otimes \ldots \otimes \mathbb{C}^q$. Let $|x\rangle$ be the vectors of an orthonormal basis of $\mathbb{C}^q$, where the labels x are elements of $\mathbb{F}_q$. Consider $a, b \in \mathbb{F}_q$; the unitary operators $X(a)$ and $Z(b)$ on $\mathbb{C}^q$ are defined by $X(a)|x\rangle = |x + a\rangle$ and $Z(b)|x\rangle = w^{tr_{q/p}(bx)}|x\rangle$, respectively, where $w = \exp(2\pi i/p)$ is a pth root of unity.

Let $\mathbf{a} = (a_1, \ldots, a_n) \in \mathbb{F}_q^n$ and $\mathbf{b} = (b_1, \ldots, b_n) \in \mathbb{F}_q^n$. Let

$$X(\mathbf{a}) = X(a_1) \otimes \ldots \otimes X(a_n)$$

and

$$Z(\mathbf{b}) = Z(b_1) \otimes \ldots \otimes Z(b_n)$$

be the tensor products of n error operators. It is easy to see that

$$\mathbf{E}_n = \{X(\mathbf{a})Z(\mathbf{b}) \mid \mathbf{a}, \mathbf{b} \in \mathbb{F}_q^n\}$$

is an *error basis* on the complex vector space $\mathbb{C}^{q^n}$. The set

$$\mathbf{G}_n = \{w^c X(\mathbf{a})Z(\mathbf{b}) \mid \mathbf{a}, \mathbf{b} \in \mathbb{F}_q^n, c \in \mathbb{F}_p\}$$

is the *error group* associated with $\mathbf{E}_n$. For a quantum error $e = w^c X(\mathbf{a})Z(\mathbf{b}) \in \mathbf{G}_n$, the X-weight is defined as

$$\mathrm{wt}_X(e) = \#\{i : 1 \leq i \leq n | a_i \neq 0\},$$

whereas the Z-weight is defined by

$$\mathrm{wt}_Z(e) = \#\{i : 1 \leq i \leq n | b_i \neq 0\}.$$

Recall that the symplectic (or quantum) weight $\mathrm{wt}(e)$ is defined similarly as in the error model for quantum codes (see Definition 3.5.7):

$$\text{wt}(e) = \#\{i : 1 \le i \le n | (a_i, b_i) \neq (0, 0)\}.$$

Definition 6.2.1 The trace-symplectic form of two vectors $(\mathbf{a}|\mathbf{b})$, $(\mathbf{a}^*|\mathbf{b}^*) \in \mathbb{F}_q^{2n}$ is defined by

$$\langle (\mathbf{a}|\mathbf{b})|(\mathbf{a}^*|\mathbf{b}^*) \rangle_s = \text{tr}_{q/p}(\mathbf{b} \cdot \mathbf{a}^* - \mathbf{b}^* \cdot \mathbf{a}).$$

In the sequence, we define formally the concept of asymmetric quantum error-correcting codes.

Definition 6.2.2 An AQECC with parameters $((n, K, d_z/d_x))_q$ is an K-dimensional subspace of the Hilbert space $\mathbb{C}^{q^n}$. The code corrects all qudit-flip errors up to $\lfloor \frac{d_x-1}{2} \rfloor$ and all phase-shift errors up to $\lfloor \frac{d_z-1}{2} \rfloor$. An $((n, q^k, d_z/d_x))_q$ code is denoted by $[[n, k, d_z/d_x]]_q$.

Since the last two decades, many authors have focused the attention in the construction of good quantum codes [4, 11, 17, 25, 28, 30, 55, 56, 62, 80, 81, 89, 105, 108, 113, 138, 148, 153, 155, 156, 162]. On the other hand, some authors have also presented constructions of asymmetric quantum codes with good parameters [3, 62, 70, 139, 140, 149].

In the same manner that the CSS construction works as an important tool in the construction of quantum codes, it also works in the case of asymmetric errors.

Lemma 6.2.1 ([25, 80, 121]) (CSS construction) *Let C_1 and C_2 denote two classical linear codes with parameters $[n, k_1, d_1]_q$ and $[n, k_2, d_2]_q$, respectively. Assume that $C_2 \subset C_1$. Then there exists an AQECC with parameters $[[n, K = k_1 - k_2, d_z/d_x]]_q$, where $d_x = wt(C_2^\perp \setminus C_1^\perp)\}$ and $d_z = \text{wt}(C_1 \setminus C_2)$. The resulting code is said pure if, in the above construction, $d_x = d(C_2^\perp)$ and $d_z = d(C_1)$.*

6.3 Asymmetric BCH Codes

In this section, we deal with constructions of families of nonbinary asymmetric quantum BCH codes. These quantum codes have good parameters and they can be applied in quantum systems where the asymmetry between qudit-flip and phase-shift errors is large.

More precisely, we construct several families of asymmetric q-ary (q is an odd prime power) quantum BCH codes by means of the CSS construction (Lemma 6.2.1) applied to two distinct q-ary classical BCH codes. To do this, we construct subclasses of (classical) BCH codes with great dimension, also computing lower bounds to the corresponding minimum distances d_z and d_x by applying the well-known BCH bound. The proposed families have parameters better than the ones available in the literature.

Although in [105] families of quantum codes were constructed by applying similar technique, the parameters (dimension and minimum distances) of the proposed families are quite different from the ones shown in [105]. Additionally, the lower bounds for the minimum distances d_z and d_x of codes displayed in [105] are the same, whereas we here construct AQECCs where the lower bound for d_z is greater than the lower bound for d_x. In other words, in the proposed codes one has d_z more large than d_x. This fact allows us to generate quantum codes capable of correcting quantum errors with great asymmetry.

To compare the parameters of the asymmetric quantum BCH codes constructed here with the ones shown in the literature, we utilize the usual criterion: for fixed values of the code length n, and for fixed values of the lower bounds for d_z and d_x, our codes achieve greater values of the number of qudits than the ones available in the literature.

The asymmetric quantum BCH codes have parameters given by

(i) $[[n, n - m(2c - l - 4) - 2, d_z \geq c/d_x \geq (c - l)]]_q$, $2 \leq c \leq q$ and $0 \leq l \leq c - 2$;

(ii) $[[n, n - m(2c - l - 6) - 2, d_z \geq c/d_x \geq (c - l)]]_q$, $q + 2 < c \leq 2q$ and $0 \leq l \leq c - q - 3$;

(iii) $[[n, n - m(4q - l - 5) - 1, d_z \geq (2q + 1)/d_x \geq (2q - l)]]_q$, $0 \leq l \leq q - 2$;

(iv) $[[n, n - m(4q - l - 5) - 2, d_z \geq (2q + 2)/d_x \geq (2q - l)]]_q$, $0 \leq l \leq q - 2$,

where q is an odd prime power and $n = q^m - 1$.

Notation. We always assume that q is an odd prime power, $n = q^m - 1$ is the code length, $\mathbb{F}_q$ denotes the finite field with q elements, α denotes a primitive element of $\mathbb{F}_{q^m}$, $M^{(J)}(x)$ denotes the minimal polynomial of $\alpha^j \in \mathbb{F}_{q^m}$, the congruence $\equiv$ is considered modulo n (mod n), $\text{CSS}(C_1, C_2)$ denotes the asymmetric CSS code derived from two distinct classical linear codes C_1 and C_2, $C^{\perp}$ denotes the Euclidean dual code of a code C and $\mathbb{C}_{[a]}$ denotes the cyclotomic coset containing a, where a is not necessarily the smallest number in the coset $\mathbb{C}_{[a]}$.

The following lemma will be applied in the proposed construction:

Lemma 6.3.1 ([4, Lemmas 8 and 9]) *Let $n \geq 1$ be an integer and q be a power of a prime such that $\gcd(n, q) = 1$ and $q^{\lfloor m/2 \rfloor} < n \leq q^m - 1$, where $m = ord_n(q)$ denotes the* multiplicative order *of q modulo n. Then the coset $\mathbb{C}_x = \{xq^j \bmod n \mid 0 \leq j < m\}$ has cardinality m for all x in the range $1 \leq x \leq nq^{\lceil m/2 \rceil}/(q^m - 1)$. Moreover, if x and y are distinct integers in the range $1 \leq x, y \leq \min\{\lfloor nq^{\lceil m/2 \rceil}/(q^m - 1) - 1 \rfloor, n - 1\}$ such that the congruence $x, y \equiv 0 \bmod q$ does not hold, then the q-ary cyclotomic cosets of x and y modulo n are distinct.*

6.3.1 Code Constructions

The main results presented in this subsection are Theorems 6.3.1 and 6.3.2 and Corollary 6.3.1. They provide several families of asymmetric quantum codes derived from BCH codes. In order to obtain good codes, the main idea subjacent Theorem 6.3.1 is as follows: the smaller the cardinality of the defining set is, the greater is its dimension. According to this idea, we need to show the existence of distinct and specific singleton cyclotomic cosets contained in the defining sets of codes C_1 and C, where C is the code equivalent to the code $C_2^\perp$. Additionally, we need to find the cardinality of their defining sets as well as to show that they are mutually disjoint. These results are presented in Lemma 6.3.2 to Lemma 6.3.7. They enable us to compute the exact dimension of the corresponding AQECC, which is a hard task, since the dimension of BCH codes is not known in general. In our construction, we use the code C_1 to correct phase-shift errors and $C_2^\perp$ to correct qudit-flip errors.

We will utilize the following lemmas shown in [105].

Lemma 6.3.2 *Let $n = q^m - 1$, where $q \geq 3$ is an odd prime power and $m \geq 3$ is an integer. Then*

(i) *the cyclotomic coset $\mathbb{C}_{\lceil \frac{q^m-1}{2} \rceil}$ contains only one element;*

(ii) *the coset $\mathbb{C}_{\lceil \frac{q^m-1}{2} - 1 \rceil}$ contains the element $\frac{q^m-1}{2} - q$;*

(iii) *the coset $\mathbb{C}_{\lceil \frac{q^m-1}{2} + 1 \rceil}$ contains the element $\frac{q^m-1}{2} + q$.*

Proof See [105, Lemma 3.1] for a detailed proof.

Lemma 6.3.3 *If $n = q^m - 1$, where $q \geq 3$ is an odd prime power and $m \geq 3$ is an integer, then the q-cosets $\mathbb{C}_1, \mathbb{C}_2, \ldots, \mathbb{C}_{q-1}, \mathbb{C}_{q+1}, \ldots, \mathbb{C}_{2q-1}$ (modulo n) are mutually disjoint and each of them has m elements.*

Proof See [105, Lemma 3.2].   □

Lemma 6.3.4 *If $n = q^m - 1$, where $q \geq 3$ is an odd prime power and $m \geq 3$ is an integer (if $q = 3$, $m \geq 4$), then the q-cosets*

$$\mathbb{C}_0, \mathbb{C}_1, \mathbb{C}_2, \ldots, \mathbb{C}_{q-1}, \mathbb{C}_{q+1}, \ldots, \mathbb{C}_{2q-1}$$

are disjoint from the q-cosets $\mathbb{C}_{\lceil \frac{q^m-1}{2} + k \rceil}$, where $k = 0, 1, \ldots, q - 1$.

Proof See [105, Lemma 3.3].   □

Lemma 6.3.5 *If $n = q^m - 1$, where $q \geq 3$ is an odd prime power and $m \geq 3$ is an integer (if $q = 3$, $m \geq 4$), then the q-cosets*

$$\mathbb{C}_0, \mathbb{C}_1, \mathbb{C}_2, \ldots, \mathbb{C}_{q-1}, \mathbb{C}_{q+1}, \ldots, \mathbb{C}_{2q-1}$$

are disjoint from the q-cosets $\mathbb{C}_{\lceil \frac{q^m-1}{2} - k \rceil}$, where $k = 1, \ldots, q - 1$.

Proof See [105, Lemma 3.4].   □

Lemma 6.3.6 *Let $n = q^m - 1$, where $q \geq 3$ is an odd prime power and $m \geq 3$ is an integer. Then we have*

(i) *the q-cosets* $\mathbb{C}_{[\frac{q^m-1}{2}+k]}$ *are mutually disjoint, where* $k = 1, \ldots, q - 1$;

(ii) *the q-cosets* $\mathbb{C}_{[\frac{q^m-1}{2}-k]}$ *are mutually disjoint, where* $k = 1, \ldots, q - 1$;

(iii) *the q-cosets* $\mathbb{C}_{[\frac{q^m-1}{2}+i]}$ *are disjoint from the cosets* $\mathbb{C}_{[\frac{q^m-1}{2}-j]}$, *where* $1 \leq i, j \leq$ $q - 1$.

Proof See [105, Lemma 3.5]. □

Lemma 6.3.7 *Let* $n = q^m - 1$, *where* $q \geq 3$ *is an odd prime power and* $m \geq 3$ *is an integer (if* $q = 3$, $m \geq 4$). *Then each of the q-cosets* $\mathbb{C}_{[\frac{q^m-1}{2}+i]}$ *and* $\mathbb{C}_{[\frac{q^m-1}{2}-j]}$, *where* $1 \leq i, j \leq q - 1$, *has m elements.*

Proof See [105, Lemma 3.6]. □

Based on these previous results, we are now able to show how to construct asymmetric quantum codes.

Theorem 6.3.1 *Let* $n = q^m - 1$, *where* q *is an odd prime power and* $m \geq 3$ *is an integer (if* $q = 3$, $m \geq 4$). *Then there exists an*

$$[[n, n - m(4q - 5) - 2, d_z \geq (2q + 2)/d_x \geq 2q]]_q$$

AQECC.

Proof Let $C_1 = [n, k_1, d_1]_q$ be the BCH code generated by the product of the minimal polynomials

$$g_1(x) = M^{(0)}(x)M^{(1)}(x) \ldots M^{(q-1)}(x)M^{(q+1)}(x) \ldots M^{(2q-1)}(x),$$

and $C_2 = [n, k_2, d_2]_q$ be the cyclic code generated by the product of the minimal polynomials

$$g_2(x) = \prod_i M^{(i)}(x),$$

where each $M^{(i)}(x)$ is the minimal polynomial of α^i such that

$$i \notin \{a - q + 2, \ldots, a - 1, a, a + 1, \ldots, a + q - 1\},$$

$a = \frac{q^m-1}{2}$ and i runs through the coset representatives modulo n, where $n = q^m - 1$.

We next construct asymmetric quantum BCH codes derived from codes C_1 and C_2 by applying the CSS construction. From the BCH bound one has $d_1 \geq 2q + 2$, because the defining set of C_1 contains the sequence of $2q + 1$ consecutive integers given by $0, 1, \ldots, 2q$. Similarly, the defining set of the code C generated by the polynomial $h_2(x) = (x^n - 1)/g_2(x)$ contains the sequence of $2q - 1$ consecutive integers given by $a - q + 2, \ldots, a - 1, a, a + 1, \ldots, a + q$, (note that, from Lemma 6.3.2, the q-coset $\mathbb{C}_{[\frac{q^m-1}{2}+1]}$ contains the element $\frac{q^m-1}{2} + q$). Thus, from the

BCH bound, C has minimum distance greater than or equal to $2q$. Since C is equivalent to $C_2^\perp$, it follows that $C_2^\perp$ also has minimum distance greater than or equal to $2q$. Therefore, the resulting asymmetric quantum code has minimum distances satisfying $d_z \geq 2q + 2$ and $d_x \geq 2q$. Furthermore, from Lemmas 6.3.4 and 6.3.5 and by construction, one has $C_2 \subsetneq C_1$.

From Lemma 6.3.3, the $(2q - 2)$ q-cosets

$$\mathbb{C}_1, \mathbb{C}_2, \ldots, \mathbb{C}_{q-1}, \mathbb{C}_{q+1}, \ldots, \mathbb{C}_{2q-1}$$

are disjoint and each of them has m elements. Since $\mathbb{C}_0$ has only one element, the defining set of C_1 has $2m(q - 1) + 1$ elements.

Thus C_1 has dimension $k_1 = n - 2m(q - 1) - 1$. From Lemma 6.3.2, the coset $\mathbb{C}_{[\frac{q^m-1}{2}]}$ contains only one element. From Lemmas 6.3.6 and 6.3.7, the $(2q - 2)$ q-cosets $\mathbb{C}_{[\frac{q^m-1}{2}+j]}$ and $\mathbb{C}_{[\frac{q^m-1}{2}-i]}$, where $1 \leq i, j \leq q - 1$, are mutually disjoint and each of them has m elements. Since $\mathbb{C}_{[\frac{q^m-1}{2}]}$ has only one element and each of the q-cosets $\mathbb{C}_{[\frac{q^m-1}{2}+j]}$ and $\mathbb{C}_{[\frac{q^m-1}{2}-i]}$, $1 \leq i, j \leq q - 1$, has m elements, $m \geq 3$ ($m \geq 4$ if $q = 3$), it follows that the q-coset $\mathbb{C}_{[\frac{q^m-1}{2}]}$ is disjoint of the q-cosets $\mathbb{C}_{[\frac{q^m-1}{2}+j]}$ and $\mathbb{C}_{[\frac{q^m-1}{2}-i]}$, $1 \leq i, j \leq q - 1$. Therefore, C_2 has dimension $k_2 = m(2q - 3) + 1$; hence, the dimension of the AQECC equals $k_1 - k_2 = n - m(4q - 5) - 2$, where $n = q^m - 1$.

Applying the CSS construction to the codes C_1 and C_2 one obtains an

$$[[n, n - m(4q - 5) - 2, d_z \geq (2q + 2)/d_x \geq 2q]]_q,$$

AQECC, as desired. $\square$

Theorem 6.3.2 is a generalization of Theorem 6.3.1. It is one of the main results of this subsection.

Theorem 6.3.2 *Let $n = q^m - 1$, where q is an odd prime power and $m \geq 3$ is an integer (if $q = 3$, $m \geq 4$). Then there exists an*

$$[[n, n - m(4q - c - 5) - 2, d_z \geq (2q + 2)/d_x \geq (2q - c)]]_q$$

AQECC, where $0 \leq c \leq q - 2$.

Proof Let $C_1 = [n, k_1, d_1]_q$ be the BCH code generated by

$$g_1(x) = M^{(0)}(x)M^{(1)}(x)\ldots M^{(q-1)}(x)M^{(q+1)}(x)\ldots M^{(2q-1)}(x),$$

and let $C_2 = [n, k_2, d_2]_q$ be the cyclic code generated by

$$g_2(x) = \prod_i M^{(i)}(x),$$

where each $M^{(i)}(x)$ is the minimal polynomial of α^i such that

$$i \notin \{a - q + 2 + c, \ldots, a, a + 1, \ldots, a + q - 1\},$$

$a = \frac{q^m - 1}{2}$, i runs through the coset representatives mod n and $0 \le c \le q - 2$.

From the BCH bound, it follows that the minimum distances of the quantum code are lower bounded by $d_z \ge 2q + 2$ and $d_x \ge 2q - c$. Applying the same method shown in the proof of Theorem 6.3.1, we know that $k_1 = n - 2m(q - 1) - 1$ and $k_2 = m(2q - c - 3) + 1$; so $k_1 - k_2 = n - m(4q - c - 5) - 2$.

Therefore, one can get an

$$[[n, n - m(4q - c - 5) - 2, d_z \ge (2q + 2)/d_x \ge (2q - c)]]_q$$

AQECC. □

Corollary 6.3.1 *Let $n = q^m - 1$, q is an odd prime power and $m \ge 3$ is an integer. Then we have*

(i) there exists an

$$[[n, n - m(2c - l - 4) - 2, d_z \ge c/d_x \ge (c - l)]]_q$$

AQECC, where $2 \le c \le q$ and $0 \le l \le c - 2$;
(ii) there exists an

$$[[n, n - m(2c - l - 6) - 2, d_z \ge c/d_x \ge (c - l)]]_q$$

AQECC, where $q + 2 < c \le 2q$ and $0 \le l \le c - q - 3$;
(iii) there exists an

$$[[n, n - m(4q - l - 5) - 1, d_z \ge (2q + 1)/d_x \ge (2q - l)]]_q$$

AQECC, where $0 \le l \le q - 2$.

Proof (i) It suffices to consider C_1 as the BCH code generated by

$$g_1(x) = M^{(0)}(x)M^{(1)}(x) \ldots M^{(c-2)}(x),$$

and C_2 as the cyclic code generated by

$$g_2(x) = \prod_i M^{(i)}(x),$$

where each $M^{(i)}(x)$ is the minimal polynomial of α^i such that $i \notin \{a, \ldots, a + c - 2 - l\}$, $a = \frac{q^m - 1}{2}$ and i runs through the coset representatives mod $(q^m - 1)$. Proceeding similarly as in the proof of Theorem 6.3.1 the result follows.

(ii) Let C_1 be the BCH code generated by

$$g_1(x) = M^{(0)}(x)M^{(1)}(x)\ldots M^{(q-1)}(x)M^{(q+1)}(x)\ldots M^{(c-2)}(x),$$

and C_2 generated by polynomials

$$g_2(x) = \prod_i M^{(i)}(x),$$

where each $M^{(i)}(x)$ is the minimal polynomial of α^i such that $i \notin \{a - r + l, \ldots, a - 1, a, a + 1, \ldots, a + q - 1\}$, $a = \frac{q^m - 1}{2}$, r is an integer such that $r = c - 2 - q$, $0 \le l \le c - q - 3$ and i runs through the coset representatives mod $(q^m - 1)$. Proceeding similarly as in the proof of Theorem 6.3.1 the result follows.

(iii) Let C_1 be the BCH code generated by

$$g_1(x) = M^{(1)}(x)\ldots M^{(q-1)}(x)M^{(q+1)}(x)\ldots M^{(2q-1)}(x),$$

and C_2 generated by

$$g_2(x) = \prod_i M^{(i)}(x),$$

where each $M^{(i)}(x)$ is the minimal polynomial of α^i such that $i \notin \{a - q + 2 + l, \ldots, a, a + 1, \ldots, a + q - 1\}$, $a = \frac{q^m - 1}{2}$ and i runs through the coset representatives mod $(q^m - 1)$. Applying the CSS construction to C_1 and C_2 and proceeding similarly as in the proof of Theorem 6.3.1 the result follows. $\square$

6.3.1.1   The Case $m = 3$ and $q = 3$

We now investigate the case $m = 3$ and $q = 3$. For $q = 3$ and $n = 3^3 - 1 = 26$ the cosets are given by $\mathbb{C}_0 = \{0\}$, $\mathbb{C}_1 = \{1, 3, 9\}$, $\mathbb{C}_2 = \{2, 6, 18\}$, $\mathbb{C}_4 = \{4, 12, 10\}$, $\mathbb{C}_5 = \{5, 15, 19\}$, $\mathbb{C}_7 = \{7, 21, 11\}$, $\mathbb{C}_8 = \{8, 24, 20\}$, $\mathbb{C}_{13} = \{13\}$, $\mathbb{C}_{14} = \{14, 16, 22\}$, $\mathbb{C}_{17} = \{17, 25, 23\}$.

Corollary 6.3.2 *There exist quantum codes with parameters*

$$[[26, 13, d_z \ge 5/d_x \ge 4]]_3,$$

$$[[26, 15, d_z \ge 5/d_x \ge 3]]_3,$$

$$[[26, 16, d_z \geq 4/d_x \geq 3]]_3.$$

Proof Let $C_1 = [n, k_1, d_1]_3$ be the BCH code generated by

$$C_1 = \langle g_1(x) \rangle = \langle M^{(0)}(x) M^{(1)}(x) M^{(2)}(x) \rangle$$

and let $C_2 = [n, k_2, d_2]_3$ be the cyclic code generated by

$$\prod_i M^{(i)}(x),$$

where each $M^{(i)}(x)$ is the minimal polynomial of α^i such that $i \notin \{5, 14\}$, and i runs through the coset representatives mod 26.

The sequence $0, 1, 2, 3$ belongs to the defining set of C_1; from the BCH bound, it follows that $d_1 \geq 5$. Moreover, it follows that $k_1 = 19$. The sequence $14, 15, 16$ belongs to the defining set of code C which is generated by the polynomial $h_2(x) = (x^n - 1)/g_2(x)$. Since C is equivalent to $C_2^\perp$, from the BCH bound, it follows that $C_2^\perp$ has minimum distance $d_2^\perp \geq 4$. Moreover, C_2 has dimension $k_2 = 6$. Therefore, an $[[26, 13, d_z \geq 5/d_x \geq 4]]_3$ AQECC can be constructed.

Analogously, let us consider C_1 generated by

$$\langle g_1(x) \rangle = \langle M^{(0)}(x) M^{(1)}(x) M^{(2)}(x) \rangle$$

and C_2 generated by $\prod_i M^{(i)}(x)$, where each $M^{(i)}(x)$ is the minimal polynomial of α^i such that $i \notin \{13, 14\}$, and i runs through the coset representatives mod 26. Then one has an $[[26, 15, d_z \geq 5/d_x \geq 3]]_3$ AQECC.

Furthermore, if $C_1 = \langle g_1(x) \rangle = \langle M^{(1)}(x) M^{(2)}(x) \rangle$ and if C_2 is generated by $\prod_i M^{(i)}(x)$, where each $M^{(i)}(x)$ is the minimal polynomial of α^i such that $i \notin \{13, 14\}$, and i runs through the coset representatives mod 26, then an $[[26, 16, d_z \geq 4/d_x \geq 3]]_3$ asymmetric quantum BCH code can be constructed. $\square$

6.4 Examples

In this section, we present illustrative examples to show how the proposed construction works.

Example 6.4.1 Let C_1 be the BCH code and C_2 be the cyclic code both of length 80 over $\mathbb{F}_3$, generated, respectively, by the polynomials

$$g_1 = M^{(0)}(x) M^{(1)}(x) M^{(2)}(x) M^{(4)}(x) M^{(5)}(x),$$

and

$$g_2(x) = \prod_i M^{(i)}(x),$$

where each $M^{(i)}(x)$ is the minimal polynomial of α^i such that $i \notin \{14, 40, 41\}$, and i runs through the coset representatives mod 80.

The sequence $0, 1, 2, 3, 4, 5, 6$ belongs to the defining set of C_1; from the BCH bound one has $d_1 \geq 8$. Analogously, the sequence $40, 41, 42, 43$ belongs to the defining set of code C which is generated by the polynomial $h_2(x) = (x^n - 1)/g_2(x)$. Since C is equivalent to $C_2^{\perp}$, from the BCH bound, $C_2^{\perp}$ has minimum distance greater than or equal to 5. The cosets corresponding to the code C_1 are $\mathbb{C}_0 = \{0\}$, $\mathbb{C}_1 = \{1, 3, 9, 27\}$, $\mathbb{C}_2 = \{2, 6, 18, 54\}$, $\mathbb{C}_4 = \{4, 12, 36, 28\}$, $\mathbb{C}_5 = \{5, 15, 45, 55\}$. The cosets corresponding to C_2 are all cosets except the cosets $\mathbb{C}_{14} = \{14, 42, 46, 58\}$, $\mathbb{C}_{40} = \{40\}$, $\mathbb{C}_{41} = \{41, 43, 49, 67\}$. Hence, C_1 has dimension $k_1 = 63$ and C_2 has dimension $k_2 = 9$, which means that the dimension of the quantum code is equal to $k_1 - k_2 = 54$. Therefore, an $[[80, 54, d_z \geq 8/d_x \geq 5]]_3$ AQECC can be constructed. Similarly, an $[[80, 58, d_z \geq 6/d_x \geq 5]]_3$ AQECC can be also constructed.

Example 6.4.2 Let us now consider C_1 and C_2 codes of length 124 over $\mathbb{F}_5$, generated, respectively, by the polynomials

$$g_1(x) = M^{(0)}(x) M^{(1)}(x) M^{(2)}(x) M^{(3)}(x),$$

and

$$g_2(x) = \prod_i M^{(i)}(x),$$

where each $M^{(i)}(x)$ is the minimal polynomial of α^i such that $i \notin \{62, 63, 64\}$, and i runs through the coset representatives mod $n = 124$. Proceeding similarly as above, an $[[124, 107, d_z \geq 5/d_x \geq 4]]_5$ asymmetric quantum code can be obtained. Analogously, an $[[124, 110, d_z \geq 5/d_x \geq 3]]_5$ quantum code can be constructed and so on.

6.4.1   Code Comparison

In this section, we compare the parameters of the asymmetric quantum BCH codes constructed here with the best asymmetric CSS codes available in [3]. In order to do this, let us recall a result shown in [3].

Theorem 6.4.1 ([3, Theorem 8]) *Let q be a prime power and $\gcd(q, n) = 1$, with $ord_n(q) = m$. Let C_1 and C_2 be two narrow-sense BCH codes of length $q^{\lfloor m/2 \rfloor} < n \leq q^m - 1$ over $\mathbb{F}_q$ with designed distances δ_1 and δ_2 in the range $2 \leq \delta_1, \delta_2 \leq \delta_{max} = \min\{\lfloor nq^{\lceil m/2 \rceil}/(q^m - 1)\rfloor, n\}$ and $\delta_1 < \delta_2^{\perp} \leq \delta_2 < \delta_1^{\perp}$. Assume*

$S_1 \cup \ldots \cup S_{\delta_1-1} \neq S_1 \cup \ldots \cup S_{\delta_2-1}$, *then there exists an asymmetric quantum error control code with parameters*

$$[[n, n - m\lceil(\delta_1 - 1)(1 - 1/q)\rceil - m\lceil(\delta_2 - 1)(1 - 1/q)\rceil, d_z^*/d_x^*]]_q,$$

where $d_z^* = wt(C_2\backslash C_1^{\perp}) \geq \delta_2 > d_x^* = wt(C_1\backslash C_2^{\perp}) \geq \delta_1$.

In Table 6.1, the parameters of the asymmetric quantum BCH codes shown in [3] are given by $[[n, k^*, d_z^*/d_x^*]]_q =$

$$= [[n, n - m\lceil(\delta_1 - 1)(1 - 1/q)\rceil - m\lceil(\delta_2 - 1)(1 - 1/q)\rceil, d_z^*/d_x^*]]_q,$$

where $d_z^* = wt(C_2\backslash C_1^{\perp}) \geq \delta_2 > d_x^* = wt(C_1\backslash C_2^{\perp}) \geq \delta_1$. Here, $n = q^m - 1$ is the code length, (q is an odd prime power), q^{k^*} is the code dimension and d_z^*/d_x^* are the corresponding minimum distances with respect to phase-shift and qudit-flip errors, respectively.

Our code parameters are denoted by $[[n, k, d_z \geq d/d_x \geq (d - c)]]_q$ and they are given in the following:

- $[[n, n - m(2c - l - 4) - 2, d_z \geq c/d_x \geq (c - l)]]_q$, where $2 \leq c \leq q$ and $0 \leq l \leq c - 2$;
- $[[n, n - m(2c - l - 6) - 2, d_z \geq c/d_x \geq (c - l)]]_q$, where $q + 2 < c \leq 2q$ and $0 \leq l \leq c - q - 3$;
- $[[n, n - m(4q - l - 5) - 1, d_z \geq (2q + 1)/d_x \geq (2q - l)]]_q$, where $0 \leq l \leq q - 2$;
- $[[n, n - m(4q - l - 5) - 2, d_z \geq (2q + 2)/d_x \geq (2q - l)]]_q$, where $0 \leq l \leq q - 2$,

where $n = q^m - 1$ is the code length, q is an odd prime power, q^k is the code dimension and d_z/d_x are the corresponding minimum distances with respect to phase-shift and qudit-flip errors, respectively.

6.5   Reed–Solomon and GRS Codes

The aim here is to construct asymmetric quantum error-correcting codes derived from classical Reed–Solomon (RS) and generalized Reed–Solomon (GRS) codes by applying the CSS construction (see Lemma 6.2.1). The results presented in this subsection are obtained from Ref. [94].

6.5.1   Review of RS and GRS Codes

We review basics concepts on Reed–Solomon (RS) and generalized Reed–Solomon (GRS) codes. The reader can consult [67, 114] for more details concerning RS and GRS codes. Let us recall the definition of such codes.

Let q be a prime power and n be a positive integer with $\gcd(q, n) = 1$. Recall that a cyclic code C (see Definition 4.4.4) of length n over $\mathbb{F}_q$ is a BCH code with designed distance δ if

Table 6.1 Quantum code comparison

New asymmetric codes	Asymmetric codes shown in [3]
$[[n, k, d_z \geq d/d_x \geq (d - c)]]_q$	$[[n, k^*, d_{z^*}/d_{x^*}]]_q$
$[[26, 16, d_z \geq 4/d_x \geq 3]]_3$	$[[26, 14, d_{z^*} \geq 4/d_{x^*} \geq 3]]_3$
$[[26, 15, d_z \geq 5/d_x \geq 3]]_3$	$[[26, 11, d_{z^*} \geq 5/d_{x^*} \geq 3]]_3$
$[[26, 13, d_z \geq 5/d_x \geq 4]]_3$	$[[26, 11, d_{z^*} \geq 5/d_{x^*} \geq 4]]_3$
$[[80, 58, d_z \geq 6/d_x \geq 5]]_3$	$[[80, 52, d_{z^*} \geq 6/d_{x^*} \geq 5]]_3$
$[[80, 54, d_z \geq 8/d_x \geq 5]]_3$	$[[80, 48, d_{z^*} \geq 8/d_{x^*} \geq 5]]_3$
$[[242, 210, d_z \geq 8/d_x \geq 5]]_3$	$[[242, 202, d_{z^*} \geq 8/d_{x^*} \geq 5]]_3$
$[[242, 205, d_z \geq 8/d_x \geq 6]]_3$	$[[242, 197, d_{z^*} \geq 8/d_{x^*} \geq 6]]_3$
$[[728, 690, d_z \geq 8/d_x \geq 5]]_3$	$[[728, 680, d_{z^*} \geq 8/d_{x^*} \geq 5]]_3$
$[[728, 684, d_z \geq 8/d_x \geq 6]]_3$	$[[728, 674, d_{z^*} \geq 8/d_{x^*} \geq 6]]_3$
$[[728, 691, d_z \geq 7/d_x \geq 5]]_3$	$[[728, 686, d_{z^*} \geq 7/d_{x^*} \geq 5]]_3$
$[[124, 110, d_z \geq 5/d_x \geq 3]]_5$	$[[124, 106, d_{z^*} \geq 5/d_{x^*} \geq 3]]_5$
$[[124, 107, d_z \geq 5/d_x \geq 4]]_5$	$[[124, 103, d_{z^*} \geq 5/d_{x^*} \geq 4]]_5$
$[[124, 86, d_z \geq 10/d_x \geq 8]]_5$	$[[124, 82, d_{z^*} \geq 10/d_{x^*} \geq 8]]_5$
$[[124, 87, d_z \geq 11/d_x \geq 7]]_5$	$[[124, 85, d_{z^*} \geq 11/d_{x^*} \geq 7]]_5$
$[[124, 86, d_z \geq 12/d_x \geq 7]]_5$	$[[124, 82, d_{z^*} \geq 12/d_{x^*} \geq 7]]_5$
$[[124, 83, d_z \geq 12/d_x \geq 8]]_5$	$[[124, 79, d_{z^*} \geq 12/d_{x^*} \geq 8]]_5$
$[[124, 80, d_z \geq 12/d_x \geq 9]]_5$	$[[124, 76, d_{z^*} \geq 12/d_{x^*} \geq 9]]_5$
$[[124, 77, d_z \geq 12/d_x \geq 10]]_5$	$[[124, 73, d_{z^*} \geq 12/d_{x^*} \geq 10]]_5$
$[[624, 606, d_z \geq 5/d_x \geq 3]]_5$	$[[624, 600, d_{z^*} \geq 5/d_{x^*} \geq 3]]_5$
$[[624, 602, d_z \geq 5/d_x \geq 4]]_5$	$[[624, 596, d_{z^*} \geq 5/d_{x^*} \geq 4]]_5$
$[[624, 574, d_z \geq 10/d_x \geq 8]]_5$	$[[624, 568, d_{z^*} \geq 10/d_{x^*} \geq 8]]_5$
$[[624, 575, d_z \geq 11/d_x \geq 7]]_5$	$[[624, 572, d_{z^*} \geq 11/d_{x^*} \geq 7]]_5$
$[[624, 574, d_z \geq 12/d_x \geq 7]]_5$	$[[624, 568, d_{z^*} \geq 12/d_{x^*} \geq 7]]_5$
$[[624, 562, d_z \geq 12/d_x \geq 10]]_5$	$[[624, 556, d_{z^*} \geq 12/d_{x^*} \geq 10]]_5$
$[[342, 322, d_z \geq 7/d_x \geq 3]]_7$	$[[342, 318, d_{z^*} \geq 7/d_{x^*} \geq 3]]_7$
$[[342, 316, d_z \geq 7/d_x \geq 5]]_7$	$[[342, 312, d_{z^*} \geq 7/d_{x^*} \geq 5]]_7$
$[[342, 292, d_z \geq 12/d_x \geq 10]]_7$	$[[342, 288, d_{z^*} \geq 12/d_{x^*} \geq 10]]_7$
$[[342, 284, d_z \geq 15/d_x \geq 10]]_7$	$[[342, 282, d_{z^*} \geq 15/d_{x^*} \geq 10]]_7$
$[[342, 286, d_z \geq 16/d_x \geq 9]]_7$	$[[342, 282, d_{z^*} \geq 16/d_{x^*} \geq 9]]_7$

$$g(x) = \mathrm{lcm}\{M^{(b)}(x), M^{(b+1)}(x), \ldots, M^{(b+\delta-2)}(x)\},$$

i.e., $g(x)$ is the monic polynomial of smallest degree over $\mathbb{F}_q$ having α^b, α^{b+1}, $\ldots, \alpha^{b+\delta-2}$ as zeros.

Definition 6.5.1 A *Reed–Solomon* (RS) code over $\mathbb{F}_q$ is a BCH code of length $n = q - 1$.

The dimension of a RS code is $k = n - \deg(g(x)) = n - \delta + 1$, and its minimum distance coincides with its designed distance, i.e., $d = n - k + 1$. Thus, a RS code is a maximum-distance-separable code. Let us next define the concept of generalized RS code.

Definition 6.5.2 Choose n distinct elements a_i of $\mathbb{F}_{q^m}$ and form the vector $\mathbf{a} = (a_1, a_2, \ldots, a_n)$. After this, choose n nonzero elements $v_i \in \mathbb{F}_{q^m}$ and form the vector $\mathbf{v} = (v_1, v_2, \ldots, v_n)$. Then the generalized Reed–Solomon (GRS) code $\mathrm{GRS}_{n,k}(\mathbf{a}, \mathbf{v})$ consists of all vectors $(v_1 f(a_1), v_2 f(a_2), \ldots, v_n f(a_n))$, where $f(x)$ ranges over all polynomials of degree at most $k - 1$ with coefficients from $\mathbb{F}_{q^m}$. More precisely,

$$\mathrm{GRS}_{n,k}(\mathbf{a}, \mathbf{v}) = \{(v_1 f(a_1), v_2 f(a_2), \ldots, v_n f(a_n))|$$
$$f(x) \in \mathbb{F}_q[x], \deg(f(x)) \leq k - 1\},$$

where $1 \leq k \leq n$ and $2 < n \leq q^m$.

The code $\mathrm{GRS}_{n,k}(\mathbf{a}, \mathbf{v})$ is linear over $\mathbb{F}_{q^m}$ and it is a MDS code with parameters $[n, k, n - k + 1]_{q^m}$.

Theorem 6.5.1 *The Euclidean dual* $\mathrm{GRS}_{n,k}^{\perp}(\mathbf{a}, \mathbf{v})$ *of an GRS code is also the GRS code* $\mathrm{GRS}_{n,k}(\mathbf{a}, \mathbf{v})$ *is the GRS code* $\mathrm{GRS}_{n,n-k}(\mathbf{a}, \mathbf{u})$, *where* $\mathbf{u} = (u_1, u_2, \ldots, u_n)$, $(u_i \in \mathbb{F}_{q^m}, i = 1, 2, \ldots, n)$ *and* u_i *is given by* $u_i^{-1} = v_i \prod_{j \neq i}(a_i - a_j)$ *for all* $i = 1, 2, \ldots, n$.

6.5.2 Construction I

In this subsection, we utilize classical RS codes to obtain AQECCs. We first consider the case $q = p$, where p is a prime. Assume that $d \geq 3$ is a fixed integer. The construction of p-ary quantum codes starts by finding the smallest positive integer m such that the following inequality $p^m \geq 2d - c - 1$ is satisfied, where $d > c + 1$ and $c \geq 1$. This inequality provides the field of smallest cardinality such that each of the RS codes (of length $p^m - 1$) $C_1 = [n, k_1, d_1]_{p^m}$, $C_2 = [n, k_2, d_2]_{p^m}$, $C_2^{\perp} = [n, k_2^{\perp}, d_2^{\perp}]_{p^m}$ may be constructed such that $d_1 = d$ and $d_2^{\perp} = d - c$, respectively, and, at the same time, providing the smallest p-ary expansion of C_1 and C_2 with respect to the basis β of $\mathbb{F}_{p^m}$ over $\mathbb{F}_p$.

For convenience, we only construct AQECCs for $c = 1$, since the other cases are analogous to this one. In other words, we construct an RS code C_1 with minimum distance d and an RS code $C_2^{\perp}$ with minimum distance $d - 1$, so the inequality $p^m \geq 2d - 2$ must be satisfied.

Before proceeding further, we need the following result shown in [12, 56].

Lemma 6.5.1 (i) *Let C be an $[N, K, D_1]_{p^m}$ linear code and β be a basis of $\mathbb{F}_{p^m}$ over $\mathbb{F}_p$. Then the p-ary expansion $\beta(C)$ of C with respect to β is an $[mN, mK, D_2 \geq D_1]_p$ linear code.*

(ii) *Let C be an $[N, K, D]_{p^m}$ linear code with Euclidean dual $C^{\perp}$. Then the dual code of the p-ary expansion $\beta(C)$ of C with respect to β is the p-ary expansion $\beta^{\perp}(C^{\perp})$ of the dual code $C^{\perp}$ with respect to $\beta^{\perp}$.*

We next construct AQECCs derived from RS codes.

Theorem 6.5.2 *Let p be a prime number. Then there exists an*

$$[[N = m(p^m - 1), K = m(p^m - 2d + 2), d_z \geq d/d_x \geq (d - 1)]]_p$$

asymmetric quantum error-correcting code, where $m \geq 1$ is an integer.

Proof Let $C_1 = [n, k_1, d_1]_{p^m}$ be the RS code generated by the polynomial

$$g_1(x) = (x - 1)(x - \alpha)(x - \alpha^2) \cdots (x - \alpha^{(d-2)}),$$

where α is a primitive element in $\mathbb{F}_{p^m}$ and $d \geq 3$ is a fixed integer. We know that C_1 is an $[p^m - 1, p^m - d, d]_{p^m}$ code.

Let $U_{\mathbb{F}_{p^m}}$ be the set of units of $\mathbb{F}_{p^m}$ ordered in increasing order of the powers of α:

$$U_{\mathbb{F}_{p^m}} = \{\alpha^0 = 1, \alpha, \alpha^2, \ldots, \alpha^{(p^m - 2)}\}.$$

The set of units of $\mathbb{F}_{p^m}$, except the roots of $g_1(x)$, is denoted by

$$NR(g_1(x)) = U_{\mathbb{F}_{p^m}} - R(g_1(x)) =$$
$$= \{\alpha^{d-1}, \alpha^d, \alpha^{d+1}, \ldots, \alpha^{(p^m - 2)}\},$$

where $NR(g_1(x))$ is also ordered in increasing order of the powers of α.

Let us next consider A be the set consisting of the last $d - 2$ elements of $NR(g_1(x))$, i.e.,

$$A = \{\alpha^{[p^m - (d-1)]}, \alpha^{[p^m - (d-2)]}, \ldots, \alpha^{(p^m - 2)}\}.$$

Let C_2 be the code generated by

$$g_2(x) = \prod_{j=0}^{p^m - d}(x - \alpha^j).$$

From construction, it follows that $C_2 \subset C_1$. We know that $C_2^\perp$ is equivalent to the code C_3 generated by

$$g_3(x) = (x^{(p^m-1)} - 1)/g_2(x) = \prod_{\alpha^i \in A} (x - \alpha^i).$$

The minimum distance of $C_2^\perp$ is equal to $d - 1$, because C_3 is an RS code with minimum distance $d - 1$.

Let β be a basis of $\mathbb{F}_{p^m}$ over $\mathbb{F}_p$ and let $\beta^\perp$ its dual basis. Let us consider the codes C_1, C_2 and $C_2^\perp$ constructed above. Since $C_2 \subset C_1$, then $\beta(C_2) \subset \beta(C_1)$. Because C_1 has minimum distance d, it follows from Item (i) of Lemma 6.5.1 that $\beta(C_1)$ has minimum distance greater than or equal to d. The code $C_2^\perp$ has minimum distance $d - 1$. From Item (ii) of Lemma 6.5.1 (applied to $C_2^\perp$) we have $[\beta(C_2)]^\perp = \beta^\perp(C_2^\perp)$. Since $\beta^\perp$ is a basis of $\mathbb{F}_{p^m}$ over $\mathbb{F}_p$, applying again Item (i) of Lemma 6.5.1, to the code $\beta^\perp(C_2^\perp)$, it follows that this code has minimum distance greater than or equal to $d - 1$; hence, $[\beta(C_2)]^\perp$ also has minimum distance greater than or equal to $d - 1$. Again, from Item (i) of Lemma 6.5.1, the codes $\beta(C_1)$, $\beta(C_2)$ and $[\beta(C_2)]^\perp$ are linear. The code $\beta(C_1)$ has dimension $K_1 = m(p^m - d)$ and $K_2 = m(d - 2)$. Applying the CSS construction we have an AQECC with the required parameters. The proof is complete.   □

Let us recall Lemma 4.3.1 of Sect. 4.3

Lemma 6.5.2 *Let $C = [n, k, d]_{q^m}$ be a linear code over $\mathbb{F}_{q^m}$, where q is a prime power. Let $C^\perp$ be the dual of the code C. Then the dual code of the q-ary expansion $\beta(C)$ of the code C with respect to the basis β is the q-ary expansion $\beta^\perp(C^\perp)$ of the dual code $C^\perp$ with respect to $\beta^\perp$.*

Hence, by applying the previous result one can extend in a natural way Theorem 6.5.2.

Theorem 6.5.3 *Let q be a prime power. Then there exists an*

$$[[N = m(q^m - 1), K = m(q^m - 2d + c + 1), d_z \geq d/d_x \geq (d - c)]]_q$$

AQECC, where $d > c + 1$, and $c \geq 1$, $m \geq 1$ are integers.

Proof For $c = 1$ the proof is that of Theorem 6.5.2. The cases when $c > 1$ is similar to such proof and we omit it.   □

Example 6.5.1 We now construct an 7-ary AQECC with $d_z \geq 10$ and $d_x \geq 9$ ($c = 1$). The codes C_1, C_2 and $C_2^\perp$ have length 48 because $2d - 2 = 18 \leq 7^2$. Let C_1, C_2 and $C_2^\perp$ be the codes generated, respectively, by

$$g_1(x) = (x - 1)(x - \alpha) \cdots (x - \alpha^8),$$
$$g_2(x) = (x - 1)(x - \alpha) \cdots (x - \alpha^{39}),$$
$$g_2^\perp(x) = (x - \alpha^{40})(x - \alpha^{41}) \cdots (x - \alpha^{47}).$$

Then an $[[96, 62, d_z \geq 10/d_x \geq 9]]_7$ AQECC is constructed. Analogously, we can constructed AQECCs with parameters $[[96, 64, d_z \geq 10/d_x \geq 8]]_7$, $[[96, 70, d_z \geq 10/d_x \geq 5]]_7$, $[[96, 82, d_z \geq 5/d_x \geq 4]]_7$, $[[96, 84, d_z \geq 5/d_x \geq 3]]_7$, $[[372, 111, d_z \geq 45/d_x \geq 44]]_5$.

6.5.3 Construction II

In this subsection, we utilize GRS codes to obtain more asymmetric quantum codes. The method of construction is similar to that of Theorem 6.5.2.

Theorem 6.5.4 *Let q be a prime power. Then there exists an*

$$[[N = mn, K = m(2k - n + c), d_z \geq d/d_x \geq (d - c)]]_q$$

AQECC, where $1 < k < n < 2k + c \leq q^m$, $k = n - d + 1$, $d > c + 1$, $c \geq 1$ and $m \geq 1$.

Proof Choose a vector $\mathbf{a} = (a_1, a_2, \ldots, a_n)$, where the a_i's are distinct elements of $\mathbb{F}_{q^m}$ and a vector $\mathbf{v} = (v_1, \ldots, v_n)$, where the v_i's are nonzero elements of F_q^m. Let C_1 be the GRS code $\mathrm{GRS}_{n,k}(\mathbf{a}, \mathbf{v})$. We know that C_1 is an $[n, k = n - d + 1, d]_{q^m}$ code. Assume that the inequality $n < 2k + c$ is true. Let C_2 be the GRS code $\mathrm{GRS}_{n,(n-k-c)}(\mathbf{a}, \mathbf{v})$. Then C_2 has parameters $[n, n - k - c, k + c + 1]_{q^m}$. From construction, $C_2 \subset C_1$. Further, the dual code $C_2^\perp$ is the GRS code $\mathrm{GRS}_{n,(k+c)}(\mathbf{a}, \mathbf{u})$, where $u = (u_1, u_2, \ldots, u_n)$ and each u_i is given by $u_i^{-1} = v_i \prod_{j \neq i}(a_i - a_j)$ for all $i = 1, 2, \ldots, n$. Note that $C_2^\perp$ has minimum distance $d - c$.

Let β be a basis of $\mathbb{F}_{q^m}$ over $\mathbb{F}_q$ and $\beta^\perp$ its dual basis. Since $C_2 \subset C_1$, it follows that $\beta(C_2) \subset \beta(C_1)$. We know that the minimum distance of $\beta(C_1)$ is greater than or equal to d. Moreover, the minimum distance of $[\beta(C_2)]^\perp$ is greater than or equal to $d - c$. The codes $\beta(C_1)$, $\beta(C_2)$ and $[\beta(C_2)]^\perp$ are linear. Applying the CSS construction to $\beta(C_1)$ and $\beta(C_2)$, the code follows. $\square$

Definition 6.5.3 Given an $[[n, k, d_z/d_x]]_q$ AQECC, the asymmetric quantum Singleton bound (see Lemma [140, Lemma 3.2]) asserts that $k \leq n - d_z - d_x + 2$. If the parameters of the AQECC satisfies the equality $k = n - d_z - d_x + 2$, then the code is called maximum-distance-separable (MDS) or optimal.

Remark 6.5.1 According with Definition 6.5.3, if we consider $m = 1$ in Theorem 6.5.2, we have asymmetric quantum MDS codes.

6.5.4 Code Comparison

In this subsection, we compare the parameters of our codes with the ones available in the literature. The criterion adopted is the usual one: for fixed values of the code

Table 6.2 Quantum MDS

Our AQECCs	Codes in [3, Theorem 8]
$[[N, K, d_z \geq d/d_x \geq (d-c)]]_q$	$[[N', K', d_z/d_x]]_q$
$[[14, 4, d_z \geq 4/d_x \geq 3]]_3$	$[[14, 2, d_z \geq 2/d_x \geq 2]]_5$
$[[16, 2, d_z \geq 5/d_x \geq 4]]_3$	$[[16, 0, d_z \geq 5/d_x \geq 2]]_5$
$[[16, 4, d_z \geq 5/d_x \geq 3]]_3$	$[[16, 0, d_z \geq 5/d_x \geq 2]]_5$
$[[18, 2, d_z \geq 7/d_x \geq 3]]_5$	$[[18, 0, d_z \geq 3/d_x \geq 2]]_5$
$[[22, 10, d_z \geq 5/d_x \geq 3]]_5$	$[[22, 2, d_z \geq 4/d_x \geq 2]]_5$
$[[22, 8, d_z \geq 5/d_x \geq 4]]_5$	$[[22, 2, d_z \geq 4/d_x \geq 2]]_5$
$[[22, 8, d_z \geq 6/d_x \geq 3]]_5$	$[[22, 2, d_z \geq 4/d_x \geq 2]]_5$
$[[24, 12, d_z \geq 5/d_x \geq 3]]_5$	$[[24, 12, d_z \geq 5/d_x \geq 3]]_5$
$[[48, 18, d_z \geq 12/d_x \geq 5]]_5$	$[[48, 0, d_z \geq 12/d_x \geq 4]]_5$
$[[124, 88, d_z \geq 8/d_x \geq 3]]_5$	$[[124, 100, d_z \geq 8/d_x \geq 3]]_5$
$[[124, 64, d_z \geq 12/d_x \geq 5]]_5$	$[[124, 85, d_z \geq 12/d_x \geq 5]]_5$
$[[78, 66, d_z \geq 3/d_x \geq 3]]_5$	$[[78, 62, d_z \geq 3/d_x \geq 3]]_5$
$[[156, 128, d_z \geq 6/d_x \geq 3]]_5$	$[[156, 132, d_z \geq 6/d_x \geq 3]]_5$
$[[312, 252, d_z \geq 12/d_x \geq 5]]_5$	$[[312, 260, d_z \geq 12/d_x \geq 5]]_5$
$[[372, 318, d_z \geq 15/d_x \geq 5]]_5$	$[[372, 276, d_z \geq 15/d_x \geq 5]]_5$
$[[372, 282, d_z \geq 25/d_x \geq 7]]_5$	$[[372, 222, d_z \geq 25/d_x \geq 7]]_5$
$[[30, 18, d_z \geq 5/d_x \geq 3]]_7$	$[[30, 6, d_z \geq 5/d_x \geq 3]]_7$
$[[30, 8, d_z \geq 10/d_x \geq 3]]_7$	$[[30, 2, d_z \geq 6/d_x \geq 3]]_7$
$[[96, 70, d_z \geq 10/d_x \geq 5]]_7$	$[[96, 48, d_z \geq 10/d_x \geq 5]]_7$
$[[96, 48, d_z \geq 21/d_x \geq 5]]_7$	$[[96, 8, d_z \geq 21/d_x \geq 5]]_7$

length n, and for fixed values of d_z and d_x, the better code is the one which achieves greater values of the number of qudits than the other.

In Table 6.2, $[[N, K, d_z \geq d/d_x \geq (d-c)]]_q$ denotes the parameters of our codes, where $N = mn$, $K = m(2k - n + c)$ and $k = n - d + 1$. On the other hand, we denote $[[N', K', d_z/d_x]]_q$ meaning the parameters of the codes shown in [3, Theorem 8].

As we can see in Table 6.2, our AQECCs have parameters comparable with the ones available in [3, Theorem 8]. Since our construction and that construction showed in such paper are quite different, there exist cases where our codes are better than the ones shown in the cited article and reciprocally.

6.6   Tensor Product Codes

This section is devoted to construct AQECCs derived from product codes. Although the parameters of our codes are not so good, these codes have great asymmetry.

Let us recall the following result on tensor product of cyclic codes.

Theorem 6.6.1 ([19, Chap. 10.2]) *Let $C_1 = [n_1, k_1, d_1]_q$ and $C_2 = [n_2, k_2, d_2]_q$ be cyclic codes with generator polynomial $g_1(x)$ and $g_2(y)$, respectively. Then the code $C_1 \otimes C_2$ is a bicyclic code generate by $g_1(x)g_2(y)$. The codewords of $C_1 \otimes C_2$ correspond to all polynomials of two variables of the form $c(x, y) = l(x, y)g_1(x)g_2(y)$ modulo de ideal generated by $X^{n_1} - 1$ and $Y^{n_2} - 1$, where $l(x, y) \in \mathbb{F}_q[x, y]$. The Euclidean dual $(C_1 \otimes C_2)^{\perp}$ of $C_1 \otimes C_2$ consists of all polynomials that are multiple of $h_1(x)$ or $h_2(y)$, where $h_1(x)$ and $h_2(y)$ are the generator polynomials of $C_1^{\perp}$ and $C_2^{\perp}$, respectively.*

The next result can be found in [57] (see Theorem 8).

Theorem 6.6.2 *The product code $C_1 \otimes C_2$ of two Reed–Solomon codes $C_1 = [q - 1, q - \delta_1, \delta_1]_q$ and $C_2 = [q - 1, q - \delta_2, \delta_2]_q$ over $\mathbb{F}_q$ is an $[(q - 1)^2, (q - \delta_1)(q - \delta_2), \delta_1\delta_2]_q$ code. The Euclidean dual $(C_1 \otimes C_2)^{\perp}$ of $C_1 \otimes C_2$ has parameters $[(q - 1)^2, k^{\perp}, d^{\perp}]_q$, where $k^{\perp} = q(\delta_1 + \delta_2 - 2) - \delta_1\delta_2 + 1$, and $d^{\perp} = \min\{q - \delta_1, q - \delta_2\}$, where δ_1 and δ_2 are the minimum distances of C_1 and C_2, respectively.*

Lemmas 6.6.1 establishes conditions under which product codes are nested.

Lemma 6.6.1 *Let C_1, C_2, C_3 and C_4 be four cyclic codes over $\mathbb{F}_q$ of same length such that the inclusions $C_2 \subset C_1$ and $C_4 \subset C_3$ hold. Then the product code $C_1 \otimes C_3$ contains the product code $C_2 \otimes C_4$.*

Proof Let C_1, C_2, C_3 and C_4 be cyclic code generated by the polynomials $g_1(x)$, $g_2(x)$, $g_3(y)$ and $g_4(y)$, respectively. Since $C_2 \subset C_1$, then there exists a polynomial $a(x)$ such that C_2 is generated by $a(x)g_1(x)$, i.e., $g_2(x) = a(x)g_1(x)$. Analogously, as $C_4 \subset C_3$, t follows that $g_4(y) = b(y)g_3(y)$. From Theorem 6.6.1, the code $C_2 \otimes C_4$ is bicyclic code generated by the polynomial $p(x, y) = g_2(x)g_4(y)$, i.e., $p(x, y) = a(x)g_1(x)b(y)g_3(x)$. Furthermore, applying again Theorem 6.6.1 we can construct a bicyclic code $C_1 \otimes C_3$ generated by the polynomial $q(x, y) = g_1(x)g_3(x)$. Because $q(x, y)|p(x, y)$, we have $C_2 \otimes C_4 \subset C_1 \otimes C_3$. The proof is complete.   □

As an immediate consequence of the previous lemma, we can conclude the following.

Corollary 6.6.1 *Let C_1 and C_2 be cyclic codes over $\mathbb{F}_q$ of same length such that $C_2 \subset C_1$. Then the inclusion $C_2 \otimes C_2 \subset C_1 \otimes C_1$ holds.*

In Theorem 6.6.3 we construct asymmetric quantum product codes.

Theorem 6.6.3 *There exists an*

$$[[(q - 1)^2, (q - d_1)(q - d_3) - (q - d_2)(q - d_4), d_z/d_x]]_q$$

asymmetric quantum code, where d_1, d_2, d_3 and d_4 are integers satisfying the inequalities $2 \leq d_1 \leq d_2 < q - 1$, $2 \leq d_3 \leq d_4 < q - 1$ and $d_z \geq \max\{d_1d_3, \min\{q - d_2, q - d_4\}\}$, and $d_x \geq \min\{d_1d_3, \min\{q - d_2, q - d_4\}\}$.

Proof Let α be a primitive element of $\mathbb{F}_q$. Let C_1 be the Reed–Solomon (RS) code over $\mathbb{F}_q$ generated by the polynomial

$$g_1(x) = \prod_{i=0}^{d_1-2}(x - \alpha^i),$$

where $d_1 \geq 2$.

Assume that C_2 is the RS code over $\mathbb{F}_q$ generated by

$$g_2(x) = \prod_{i=0}^{d_2-2}(x - \alpha^i),$$

where $d_1 \leq d_2 < q - 1$. We know that C_1 is an $[q - 1, q - d_1, d_1]_q$ code, and C_2 is an $[q - 1, q - d_2, d_2]_q$ code. From construction, it follows that $C_2 \subset C_1$.

Let C_3 be the RS code over $\mathbb{F}_q$ generated by

$$g_3(x) = \prod_{i=0}^{d_3-2}(x - \alpha^i),$$

where $d_3 \geq 2$, and suppose C_4 is the RS over $\mathbb{F}_q$ generated by

$$g_4(x) = \prod_{i=0}^{d_4-2}(x - \alpha^i),$$

where $d_3 \leq d_4 < q - 1$. Similarly, C_3 is an $[q - 1, q - d_3, d_3]_q$ code, C_4 is an $[q - 1, q - d_4, d_4]_q$ code and $C_4 \subset C_3$. We know that the product code $C_1 \otimes C_3$ is an

$$[(q - 1)^2, (q - d_1)(q - d_3), d_1 d_3]_q$$

code, and $C_2 \otimes C_4$ is an

$$[(q - 1)^2, (q - d_2)(q - d_4), d_2 d_4]_q$$

code. Applying Lemma 6.6.1, we have $C_2 \otimes C_4 \subset C_1 \otimes C_3$. From Theorem 6.6.2, the Euclidean dual $(C_2 \otimes C_4)^\perp$ of $C_2 \otimes C_4$ is an

$$[(q - 1)^2, (q - 1)^2 - (q - d_2)(q - d_4), d^\perp]_q$$

code, where $d^\perp = \min\{q - d_2, q - d_4\}$. Applying the CSS construction to the codes $C_1 \otimes C_3$, $C_2 \otimes C_4$ and $(C_2 \otimes C_4)^\perp$ we obtain the desired code. The proof is complete. □

Corollary 6.6.2 *There exists an*

$$[[(q-1)^2, (q-d_1)^2 - (q-d_2)^2, d_z/d_x]]_q$$

asymmetric quantum code, where $d_z \geq \max\{(d_1)^2, q - d_2\}$, $d_x \geq \min\{(d_1)^2, q - d_2\}$, *and* d_1, d_2 *satisfy* $2 \leq d_1 \leq d_2 < q - 1$.

Exercise 6.6.1 Show Corollary 6.6.2.

Example 6.6.1 In this example we construct an $[[36, 11, d_z/d_x]]_7$ AQECC, where $d_z \geq 8$ and $d_x \geq 2$. To do this, it suffices to consider in Theorem 6.6.3, $q = 7, d_1 = 4$, $d_3 = 2, d_2 = d_4 = 5$. Similarly, we can construct an $[[36, 11, d_z/d_x]]_7$ AQECC, with $d_z \geq 6$ and $d_x \geq 3$ and an $[[36, 6, d_z/d_x]]_7$ AQECC, where $d_z \geq 10$ and $d_x \geq 2$. From Corollary 6.6.2, we obtain an $[[64, 7, d_z/d_x]]_9$ AQECC, where $d_z \geq 25$ and $d_x \geq 3$.

6.6.1 Code Comparison

Here, we compare the parameters of our codes with the ones displayed in the literature. To make this comparison, we proceed as follows: fixing the code length n and the lower bound for the minimum distances d_z and d_x, the AQECC with greater dimension is better than the other. This comparison is usual in the literature.

For example, our $[[100, 55, d_z/d_x]]_{11}$ code, where $d_z \geq 9$ and $d_x \geq 3$ is better that the $[[100, 0, d_z^*/d_x^*]]_{11}$ code shown in [3], where $d_z^* \geq 9$ and $d_x^* \geq 3$. On the other hand, our $[[36, 0, d_z/d_x]]_7$ code, where $d_z \geq 16$ and $d_x \geq 3$ is not as good as the $[[36, 0, d_z'/d_x']]_7$ code, where $d_z' \geq 16$ and $d_x' \geq 8$.

We point out that this construction does not provide AQECC with good parameters due to the fact that the code length increases quickly in taking tensor products. However, maybe the ideas presented here can motivate the reader to improve our construction.

As can be seen, in the cases shown in Table 6.3, the new quantum codes have parameters better than the ones shown in Theorem 8 in [3]. On the other hand, the parameters of the new codes are not as good as the parameters of the quantum codes given by the referee. Concerning Refs. [90, 94], it seems that the new code parameters are not as good as those shown in such papers, but we do not perform the comparison properly, since the corresponding parameters are quite distinct. Hence, the AQECCs constructed in this subsection are comparable with AQECCs discussed in the literature. In other words, there exist cases where our codes are better than ones available in the literature and there exist other cases where our codes are not as good as the ones available in the literature.

Table 6.3 Quantum MDS codes

Our AQECCs	Codes in [3]	Codes in the literature
$[[n, k, d_z/d_x]]_q$	$[[n, k^*, d_z^*/d_x^*]]_q$	$[[n, k', d_z'/d_x']]_q$
$[[16, 5, 4/2]]_5$	$[[16, 0, 4/2]]_5$	
$[[16, 0, 4/3]]_5$		$[[16, 4, 4/3]]_5$
$[[16, 0, 9/2]]_5$		$[[16, 0, 9/3]]_5$
$[[36, 21, 4/2]]_7$	$[[36, 12, 4/2]]_7$	
$[[36, 16, 4/3]]_7$	$[[36, 6, 4/3]]_7$	
$[[36, 9, 4/4]]_7$	$[[36, 0, 4/4]]_7$	
$[[36, 11, 8/2]]_7$		
$[[36, 11, 6/3]]_7$		
$[[36, 12, 9/2]]_7$		
$[[36, 0, 16/3]]_7$		$[[36, 0, 16/8]]_7$
$[[36, 0, 25/2]]_7$		$[[36, 0, 25/4]]_7$
$[[49, 27, 4/3]]_8$	$[[49, 14, 4/3]]_8$	
$[[64, 27, 9/3]]_9$	$[[64, 0, 7/3]]_9$	
$[[64, 20, 9/4]]_9$	$[[64, 0, 7/3]]_9$	
$[[64, 0, 16/5]]_9$		$[[64, 0, 16/13]]_9$
$[[64, 7, 25/3]]_9$		$[[64, 19, 26/4]]_9$
$[[100, 55, 9/3]]_{11}$	$[[100, 0, 9/3]]_{11}$	
$[[676, 259, 25/15]]_{27}$	$[[676, 4, 25/5]]_{27}$	

6.7 New Codes from Old

In this subsection, we extend to asymmetric quantum error-correcting codes the methods which are valid to quantum error-correcting codes: puncturing, code extension, code expansion, direct sum and the $(\mathbf{u}|\mathbf{u} + \mathbf{v})$ code constructions. By applying these methods, we can construct several families of asymmetric quantum codes. As an example of application of quantum code expansion developed here, we construct families of AQECCs derived from generalized Reed–Muller (GRM) codes, quadratic residue (QR), Bose–Chaudhuri–Hocquenghem (BCH), character codes and affine-invariant codes.

Remark 6.7.1 Because the Euclidean dual $C^{\perp}$ of a linear code C and its Hermitian dual $C^{\perp_H}$ (in the case of fields with cardinality q^2, of course) are isomorphic under Galois conjugation which preserves Hamming metric, a similar result can be derived when considering in Lemma 6.2.1 the Hermitian inner product instead of considering the Euclidean one (we call CSS-type construction in the Hermitian case).

Remark 6.7.2 It is interesting to note that the CSS construction was extended to include additive codes in the paper by Ezerman et al. [35, Theorem 4.5].

The following result will be utilized to perform our construction techniques.

Theorem 6.7.1 ([80, Theorem 13]) *An* $((n, K, d))_q$ *stabilizer code exists if and only if there exists an additive code* $C \le \mathbb{F}_q^{2n}$ *of size* $|C| = q^n/K$ *such that* $C \le C^{\perp_s}$ *and* $swt(C^{\perp_s} \backslash C) = d$ *if* $K > 1$ *(and* $swt(C^{\perp_s}) = d$ *if* $K = 1$*).*

6.7.1 Code Expansion

Let us recall the concept of dual basis (see Definition 4.3.5). Let $\beta = \{b_1, b_2, \ldots, b_m\}$ be a basis of $\mathbb{F}_{q^m}$ over $\mathbb{F}_q$. A *dual basis* of β is defined as $\beta^{\perp} = \{b_1^*, b_2^*, \ldots, b_m^*\}$, where $tr_{q^m/q}(b_i b_j^*) = \delta_{ij}$, for all $i, j \in \{1, \ldots, m\}$. A self-dual basis β is a basis satisfying $\beta = \beta^{\perp}$. If C is an $[n, k, d_1]_{q^m}$ code and $\beta = \{b_1, b_2, \ldots, b_m\}$ is a basis of $\mathbb{F}_{q^m}$ over $\mathbb{F}_q$, then the q-ary expansion $\beta(C)$ of C with respect to β is an $[mn, mk, d_2 \ge d_1]_q$ code given by $\beta(C) := \{(c_{ij})_{i,j} \in \mathbb{F}_q^{mn} \mid \mathbf{c} = (\sum_j c_{ij} b_j)_i \in C\}$.

Let us recall the following result concerning the dual of a code obtaining by means of code expansion.

Lemma 6.7.1 ([12, 56, 94]) *Let* $C = [n, k, d]_{q^m}$ *be a linear code over* $\mathbb{F}_{q^m}$, *where* q *is a prime power. Let* $C^{\perp}$ *be the dual of the code* C. *Then the dual code of the* q-ary expansion $\beta(C)$ *of code* C *with respect to the basis* β *is the* q-ary expansion $\beta^{\perp}(C^{\perp})$ *of the dual code* $C^{\perp}$ *with respect to* $\beta^{\perp}$.

Theorem 6.7.2 establishes a method to construct AQECCs by expanding linear codes.

Theorem 6.7.2 *Let* q *be a prime power. Assume there exists an* $[[n, k, d_z/d_x]]_{q^m}$ *AQECC derived from linear codes* $C_1 = [n, k_1, d_1]_{q^m}$ *and* $C_2 = [n, k_2, d_2]_{q^m}$. *Then there exists an AQECC with parameters* $[[mn, mk, d_z^*/d_x^*]]_q$, *where* $k = k_1 - k_2$, $d_z^* \ge d_1$ *and* $d_x^* \ge d_2^{\perp}$, *where* $d_2^{\perp}$ *is the minimum distance of the dual code* $C_2^{\perp}$.

Proof The proof presented here utilizes the same idea and generalizes the proof of [94, Theorem 1] to all linear codes. Note first that $[\beta(C)]^{\perp} = \beta^{\perp}(C^{\perp})$. Let $C_1 = [n, k_1, d_1]_{q^m}$ and $C_2 = [n, k_2, d_2]_{q^m}$ be two linear nested codes $C_2 \subset C_1$. Let β be any basis of $\mathbb{F}_{q^m}$ over $\mathbb{F}_q$ and let $\beta^{\perp}$ be its dual basis. Perform the expansions $\beta(C_1)$ of the codes C_1 and $\beta(C_2)$ of C_2 with respect to β. Hence, the inclusion $\beta(C_2) \subset \beta(C_1)$ holds. The codes $\beta(C_1)$, $\beta(C_2)$ and $[\beta(C_2)]^{\perp}$ are linear. Furthermore, it follows that $\beta(C_1) = [mn, mk_1, D_1 \ge d_1]_q$ and $\beta(C_2) = [mn, mk_2, D_2 \ge d_2]_q$. Since $C_2^{\perp}$ has minimum distance $d_2^{\perp}$, then $\beta^{\perp}(C_2^{\perp})$ has minimum distance greater than or equal to $d_2^{\perp}$ because $\beta^{\perp}$ is a basis of $\mathbb{F}_{q^m}$ over $\mathbb{F}_q$. From Lemma 6.7.1, the equality $[\beta(C_2)]^{\perp} = \beta^{\perp}(C_2^{\perp})$ is true; hence, $[\beta(C_2)]^{\perp}$ has also minimum distance greater than or equal to $d_2^{\perp}$. Applying the CSS construction to the codes $\beta(C_1)$, $\beta(C_2)$ and $[\beta(C_2)]^{\perp}$, we obtain an $[[mn, m(k_1 - k_2), d_z^*/d_x^*]]_q$ asymmetric quantum error-correcting code, where $d_z^* \ge d_1$ and $d_x^* \ge d_2^{\perp}$. The proof is complete. $\square$

The following result is more general than Theorem 6.7.2.

Theorem 6.7.3 *Let $q = p^t$ be a prime power. If there exists an $((n, K, d_z/d_x))_{q^m}$ stabilizer code, then there exists an $((nm, K, d_z^*/d_x^*))_q$ stabilizer code, where $d_z^* \geq d_z$ and $d_x^* \geq d_x$.*

Proof If $\mathbf{a}$ is an element of $\mathbb{F}_{q^m}$, we can expand $\mathbf{a}$ with respect to a given basis $B = \{\beta_1, \ldots, \beta_m\}$ of $\mathbb{F}_{q^m}$ over $\mathbb{F}_q$ and put the coordinates of $\mathbf{a}$ in the vector form $c_B(\mathbf{a}) = (a_1, \ldots, a_m) \in \mathbb{F}_q^m$. Let us consider the non-degenerate symmetric form $\text{tr}_{q^m/q}(ab)$ on the vector space $\mathbb{F}_{q^m}$ (over $\mathbb{F}_q$). Assume that φ_B is the $\mathbb{F}_p$-vector space isomorphism from $\mathbb{F}_{q^m}^{2n}$ to $\mathbb{F}_q^{2nm}$ given (in the proof of [80, Lemma 76]) by $\varphi_B((\mathbf{u}|\mathbf{v})) = ((c_B(u_1), \ldots, c_B(u_n))|(Mc_B(v_1), \ldots, Mc_B(v_n)))$, where $\mathbf{u}, \mathbf{v} \in \mathbb{F}_{q^m}^n$ are given by $\mathbf{u} = (u_1, \ldots, u_n)$ and $\mathbf{v} = (v_1, \ldots, v_n)$, $M = (\text{tr}_{q^m/q}(\beta_i \beta_j))_{1 \leq i,j \leq m}$ is the Gram matrix and $\text{tr}_{q^m/q}(ab) = c_B(a)^t M c_B(b)$ for all $a, b \in \mathbb{F}_{q^m}$. Note that the inner product considered here is the usual Euclidean inner product of $\mathbb{F}_q$.

Assume that an $((n, K, d_z/d_x))_{q^m}$ stabilizer code exists. From [80, Theorem 13], there exists an additive code $C \leq \mathbb{F}_{q^m}^{2n}$ of size $|C| = q^{mn}/K$ such that $C \leq C^{\perp_s}$, $\text{wt}_X(C^{\perp_s} \backslash C) = d_x$ if $K > 1$ (and $\text{wt}_X(C^{\perp_s}) = d_x$ if $K = 1$) and $\text{wt}_Z(C^{\perp_s} \backslash C) = d_z$ if $K > 1$ (and $\text{wt}_Z(C^{\perp_s}) = d_z$ if $K = 1$). We know that φ_B preserves trace-symplectic orthogonality, that is, the code $\varphi_B(C)$ satisfies $\varphi_B(C) \leq [\varphi_B(C)]^{\perp_s}$. Let $(\mathbf{u}|\mathbf{v}) \in \mathbb{F}_{q^m}^{2n}$ and $u_i \neq 0$ (resp. $v_j \neq 0$) for some $i \in \{1, \ldots, n\}$ (resp. $j \in \{1, \ldots, n\}$). Hence, at least one coordinate of the corresponding vector $c_B(u_i)$ (resp. $Mc_B(v_j)$) is nonzero. Thus $\text{wt}_X([\varphi_B(C)]^{\perp_s} \backslash \varphi_B(C)) \geq d_x$ if $K > 1$ (and $\text{wt}_X([\varphi_B(C)]^{\perp_s}) \geq d_x$ if $K = 1$) and $\text{wt}_Z([\varphi_B(C)]^{\perp_s} \backslash \varphi_B(C)) \geq d_z$ if $K > 1$ (and $\text{wt}_Z([\varphi_B(C)]^{\perp_s}) \geq d_z$ if $K = 1$). Because the alphabet considered now is $\mathbb{F}_q$, then there exists an $((nm, K, d_z^*/d_x^*))_q$ stabilizer code, where $d_z^* \geq d_z$ and $d_x^* \geq d_x$. The proof is complete. $\square$

6.7.2 Direct Sum Codes

Let us recall the direct sum of codes (see Definition 4.3.6). Let $C_1 = [n_1, k_1, d_1]_q$ and $C_2 = [n_2, k_2, d_2]_q$ be two linear codes. Then the direct sum code $C_1 \oplus C_2$ is the linear code given by

$$C_1 \oplus C_2 = \{(\mathbf{c}_1, \mathbf{c}_2) | \mathbf{c}_1 \in C_1, \mathbf{c}_2 \in C_2\},$$

with parameters $[n_1 + n_2, k_1 + k_2, \min\{d_1, d_2\}]_q$.

In the sequence, we show how to obtain new AQECCs by applying direct sum of codes.

Theorem 6.7.4 *Let q be a prime power. Assume there exists an $[[n, k, d_z/d_x]]_q$ AQECC derived from two linear nested codes $C_1 = [n, k_1, d_1]_q$ and $C_2 = [n, k_2, d_2]_q$ with $C_2 \subset C_1$. Suppose also the existence of an $[[n^*, k^*, d_z^*/d_x^*]]_q$ AQECC derived from linear nested codes $C_3 = [n^*, k_3, d_3]_q$ and $C_4 = [n^*, k_4, d_4]_q$ with $C_4 \subset C_3$. Then there exists an $[[n + n^*, k + k^*, d_z^\diamond/d_x^\diamond]]_q = [[n + n^*, (k_1 + k_3) - (k_2 + k_4), d_z^\diamond/d_x^\diamond]]_q$ AQECC, where $d_z^\diamond \geq \min\{d_1, d_3\}$, $d_x^\diamond \geq \min\{d_2^\perp, d_4^\perp\}$ and $d_2^\perp, d_4^\perp$ are the minimum distances of the dual codes $C_2^\perp$ and $C_4^\perp$, respectively.*

Proof Let us consider the direct sum codes $C_1 \oplus C_3 = [n + n^*, k_1 + k_3, \min\{d_1,$ $d_3\}]_q$ and $C_2 \oplus C_4 = [n + n^*, k_2 + k_4, \min\{d_2, d_4\}]_q$. Since the inclusions $C_2 \subset C_1$ and $C_4 \subset C_3$ hold, it follows that the inclusion $C_2 \oplus C_4 \subset C_1 \oplus C_3$ also holds. We know that a parity check matrix of the code $(C_2 \oplus C_4)^{\perp}$ is given by

$$G_2 \oplus G_4 = \begin{bmatrix} G_2 & 0 \\ 0 & G_4 \end{bmatrix}.$$

Thus, the minimum distance of $(C_2 \oplus C_4)^{\perp}$ equals $\min\{d_2^{\perp}, d_4^{\perp}\}$. Therefore, applying the CSS construction to the codes $C_1 \oplus C_3$ and $C_2 \oplus C_4$ and $(C_2 \oplus C_4)^{\perp}$, we obtain an $[[n + n^*, (k_1 + k_3) - (k_2 + k_4), d_z^{\diamond}/d_x^{\diamond}]]_q$ AQECC, where $d_z^{\diamond} \geq \min\{d_1, d_3\}$ and $d_x^{\diamond} \geq \min\{d_2^{\perp}, d_4^{\perp}\}$. We are done. $\qquad\square$

Theorem 6.7.4 also holds in a more general setting, as states the following result.

Theorem 6.7.5 *Assume that there exist two stabilizer codes with parameters $((n_1, K_1, d_z^{(1)}/d_x^{(1)}))_q$ and $((n_2, K_2, d_z^{(2)}/d_x^{(2)}))_q$. Then there exists an $((n_1 + n_2, K_1 K_2, d_z^*/d_x^*))_q$ code, where $d_z^* = \min\{d_z^{(1)}, d_z^{(2)}\}$ and $d_x^* = \min\{d_x^{(1)}, d_x^{(2)}\}$.*

Proof The proof follows the same line of [80, Lemma 73]. We only show the result in the case of X-weight because the proof for Z-weight is analogous. Note that if $((n_1, K_1, d_z^{(1)}/d_x^{(1)}))_q$ and $((n_2, K_2, d_z^{(2)}/d_x^{(2)}))_q$ are stabilizer codes with orthogonal projectors P_1 and P_2, respectively, and stabilizer S_1 and S_2, respectively, then $P_1 \otimes P_2$ is an orthogonal projector onto a $K_1 K_2$-dimensional subspace $Q^{\oplus}$ of $\mathbb{C}^{q^{(n_1+n_2)}}$. The stabilizer of $Q^{\oplus}$ is given by

$$S^{\oplus} = \{E_1 \otimes E_2 | E_1 \in S_1, E_2 \in S_2\}.$$

Assume that an error $F_1 \otimes F_2 \in \mathbf{G}_{n_1} \otimes \mathbf{G}_{n_2}$ is not detectable. Hence, it follows that $F_1 \in C_{\mathbf{G}_{n_1}}(S_1)$ and $F_2 \in C_{\mathbf{G}_{n_2}}(S_2)$. Moreover, either $F_1 \notin S_1 Z(\mathbf{G}_{n_1})$ or $F_2 \notin S_2 Z(\mathbf{G}_{n_2})$, otherwise $F_1 \otimes F_2$ would be detectable. Thus, from [80, Lemma 11], either F_1 or F_2 is not detectable, so $\mathrm{wt}_X(F_1 \otimes F_2)$ is, at least, $\min\{d_x^{(1)}, d_x^{(2)}\}$. Therefore, the result follows. $\qquad\square$

6.7.3 Puncturing Codes

The technique of puncturing codes is well known in the literature as in the classical [67, 114] as well as in the quantum case [25, 80, 133]. In this subsection, we show how to construct AQECCs by puncturing classical codes.

Let C be an $[n, k, d]_q$ code. Then we write C^{P_i} to denote the punctured code in the coordinate i. Recall that the dual of a punctured code is a shortened code (see [67]).

We are now ready to show how to construct new stabilizer codes by applying puncturing techniques in the corresponding classical linear codes.

Theorem 6.7.6 *Assume there exists an $[[n, k, d_z/d_x]]_q$ stabilizer code derived from two linear codes $C_1 = [n, k_1, d_1]_q$ and $C_2 = [n, k_2, d_2]_q$, with $C_2 \subset C_1$, $n \geq 2$, $k = k_1 - k_2$, $d_z \geq d_1$ and $d_x \geq d_2^{\perp}$, where $d_2^{\perp}$ is the minimum distance of the dual code $C_2^{\perp}$. Suppose also that $d_1 \geq 2$, $d_2^{\perp} \geq 2$, and assume that $C_2^{\perp}$ contains at least a nonzero codeword with ith coordinate zero. Then the following hold:*

(i) *If C_1 has a minimum weight codeword with a nonzero ith coordinate, then there exists an $[[n - 1, k, d_z^{P_i}/d_x^{P_i}]]_q$ AQECC, where $k = k_1 - k_2$, $d_z^{P_i} \geq d_1 - 1$ and $d_x^{P_i} \geq d_2^{\perp}$;*

(ii) *If C_1 has no minimum weight codeword with a nonzero ith coordinate, then there exists an $[[n - 1, k, d_z^{P_i}/d_x^{P_i}]]_q$ AQECC, where $k = k_1 - k_2$, $d_z^{P_i} \geq d_1$ and $d_x^{P_i} \geq d_2^{\perp} \geq 2$.*

Proof We only prove Item (ii) since the proof of Item (i) is similar to this one. Consider the punctured codes $C_1^{P_i}$ and $C_2^{P_i}$. Since the inclusion $C_2 \subset C_1$ holds, it follows that $C_2^{P_i} \subset C_1^{P_i}$. Since from hypothesis one has $d_1 > 1$ then it follows that $d_2 > 1$ because $C_2 \subset C_1$; again from the hypothesis C_1 has no minimum weight codeword with a nonzero ith coordinate. Thus, by Theorem [67, Theorem 1.5.1], the punctured codes $C_1^{P_i}$ and $C_2^{P_i}$ have parameters $[n - 1, k_1, d_1]_q$ and $[n - 1, k_2, d_2^i]_q$, respectively, where $d_2^i = d_2$ or $d_2^i = d_2 - 1$.

We need to compute the minimum distance of $[C_2^{P_i}]^{\perp}$ in order to apply the CSS construction. To do this, let us consider the code $[C_2^{P_i}]^{\perp}$. Since $C_2^{\perp}$ contains at least a nonzero codeword whose ith coordinate is equal to zero, it follows that $C_2^{\perp}$ has a subcode $C_2^{\perp}(\{i\}) \neq \{0\}$ and, consequently, the minimum distance $d_{(C_2^{\perp})_i}$ of $C_2^{\perp}(\{i\})$ satisfies $d_{(C_2^{\perp})_i} \geq d_2^{\perp}$, where $d_2^{\perp} > 1$. Since $d_{(C_2^{\perp})_i} > 1$, and from definition, the code $C_2^{\perp}(\{i\})$ has no minimum weight codeword with a nonzero ith coordinate. Applying again Theorem [67, Theorem 1.5.1], we conclude that the shortened code $[C_2^{\perp}]_{S_i}$ has minimum distance $d_{(C_2^{\perp})_i}$. From [67, Theorem 1.5.7], we know that $[C_2^{P_i}]^{\perp} = [C_2^{\perp}]_{S_i}$; hence, the code $[C_2^{P_i}]^{\perp}$ has minimum distance $d_{(C_2^{\perp})_i}$, where $d_{(C_2^{\perp})_i} \geq d_2^{\perp}$. Therefore, applying the CSS construction to the codes $C_1^{P_i}$, $C_2^{P_i}$ and $[C_2^{P_i}]^{\perp}$, one can derive an $[[n - 1, k, d_z^{P_i}/d_x^{P_i}]]_q$ AQECC, where $k = k_1 - k_2$, $d_z^{P_i} \geq d_1$ and $d_x^{P_i} \geq d_{(C_2^{\perp})_i} \geq d_2^{\perp} \geq 2$. The proof is complete. $\square$

Following the lines adopted in [80], we can show a more general result.

Theorem 6.7.7 *Assume that a pure $[[n, k, d_z/d_x]]_q$ stabilizer code exists, where $n \geq 2$ and $d_x, d_z \geq 2$. Then there exists a pure $[[n - 1, k, d_z^*/d_x^*]]_q$ stabilizer code, where $d_z^* \geq d_z - 1$ and $d_x^* \geq d_x - 1$.*

Proof Assume that a pure $[[n, k, d_z/d_x]]_q$ stabilizer code with minimum distance d exists. From [80, Corollary 72], there exists a pure $[[n - 1, k, d^* \geq d - 1]]_q$ stabilizer code derived from an additive self-orthogonal (with respect to the trace-alternating form) code $D^{\perp_a} \leq \mathbb{F}_{q^2}^{n-1}$ with $\mathrm{wt}(D^{\perp_a}) \geq d - 1$. Let us consider the vectors $\mathbf{v}, \mathbf{w} \in \mathbb{F}_q^{2(n-1)}$. Let (β, β^q) be a normal basis of $\mathbb{F}_{q^2}$ over $\mathbb{F}_q$. We know that the bijective map $\phi((\mathbf{v}|\mathbf{w})) = \beta\mathbf{v} + \beta^q\mathbf{w}$ from $\mathbb{F}_q^{2(n-1)}$ onto $\mathbb{F}_{q^2}^{n-1}$ is an isometry (sym-plectic/Hamming weights, resp.) (see also [80, Lemma 14]). Considering the inverse map ϕ^{-1} and the corresponding additive code $\phi^{-1}(D^{\perp_a}) \le \mathbb{F}_q^{2(n-1)}$, it follows that $\phi^{-1}(D^{\perp_a})$ has minimum X-weight d_x^* satisfying $d_x^* \ge d_x - 1$ and the minimum Z-weight d_z^* is, at least, $d_z^* \ge d_z - 1$. The proof is complete. $\qquad\square$

Remark 6.7.3 Notice that the procedure adopted in Theorems 6.7.6 and 6.7.7 can be generalized by puncturing codes on two or more coordinates.

6.7.4 Code Extension

The technique of code extension was derived also in the quantum case [25, 80]. Here, we extend to AQECCs this technique.

Let C be an $[n, k, d]_q$ linear code over $\mathbb{F}_q$. Recall that the extended code C^e (see Definition 4.3.2) is the linear code given by

$$C^e = \{(x_1, \ldots, x_n, x_{n+1}) \in \mathbb{F}_q^{n+1} | (x_1, \ldots, x_n) \in C, x_1 + \cdots + x_n + x_{n+1} = 0\}.$$

The code C^e is linear and has parameters $[n + 1, k, d^e]_q$, where $d^e = d$ or $d^e = d + 1$. Recall that a vector $\mathbf{v} = (v_1, \ldots, v_n) \in \mathbb{F}_q^n$ is called *even-like* if it satisfies the equality $\sum_{i=1}^{n} v_i = 0$, and *odd-like* otherwise. For an $[n, k, d]_q$ code C, the minimum weight of the even-like codewords of C are called *minimum even-like weight* and denoted by d_{even} (or $(d)_{even}$). Similarly, the minimum weight of the odd-like code-words of C are called *minimum odd-like weight* and denoted by d_{odd} (or $(d)_{odd}$).

Theorem 6.7.8 *Assume that there exists an $[[n, k, d_z/d_x]]_q$ AQECC derived from two nested linear codes $C_1 = [n, k_1, d_1]_q$ and $C_2 = [n, k_2, d_2]_q$, with $C_2 C_1$. Then the following are true:*

(a) *If $(d_1)_{even} \le (d_1)_{odd}$, then there exists an $[[n + 1, k, d_z^e/d_x^e]]_q$ AQECC, where $d_z^e \ge d_1$ and $d_x^e \ge (d_2^e)^\perp$, where $(d_2^e)^\perp$ is the minimum distance of the dual $(C_2^e)^\perp$ of the extended code C_2^e;*

(b) *If $(d_1)_{odd} < (d_1)_{even}$, then there exists an $[[n + 1, k, d_z^e/d_x^e]]_q$ AQECC, where $d_z^e \ge d_1 + 1$ and $d_x^e \ge (d_2^e)^\perp$.*

Proof We only show Item (b), since Item (a) is similar. We note first that the inclusion $C_2^e \subset C_1^e$ holds. The parameters of the extended codes C_1^e and C_2^e are $[n + 1, k_1, d_1^e]_q$ and $[n + 1, k_2, d_2^e]_q$, respectively, where $d_1^e = d_1$ or $d_1^e = d_1 + 1$. Since $(d_1)_{odd} < (d_1)_{even}$, it follows from the remark shown in [67, pg. 15] that $d_1^e = d_1 + 1$. From hypothesis, we know that $k = k_1 - k_2$; so, the corresponding CSS code has dimension k. Applying the CSS construction to the codes C_1^e, C_2^e and $(C_2^e)^\perp$, we obtain an $[[n + 1, k, d_z^e/d_x^e]]_q$ AQECC, where $d_z^e \ge d_1 + 1$ and $d_x^e \ge (d_2^e)^\perp$. This finishes the proof. $\qquad\square$

6.7.5   The (u|u + v) Construction

The $(\mathbf{u}|\mathbf{u} + \mathbf{v})$ construction is an interesting method for constructing new (classical) linear codes from old ones. Our intention in this subsection is to apply this technique (valid for classical linear codes as well as to quantum codes) in order to generate a similar construction method for asymmetric quantum codes.

Let us recall how to perform this construction for classical codes (see Definition 4.3.8). Let C_1 and C_2 be two linear codes of same length both over $\mathbb{F}_q$ with parameters $[n, k_1, d_1]_q$ and $[n, k_2, d_2]_q$, respectively. From the $(\mathbf{u}|\mathbf{u} + \mathbf{v})$ construction, a new code $C = \{(\mathbf{u}, \mathbf{u} + \mathbf{v})|\mathbf{u} \in C_1, \mathbf{v} \in C_2\}$ with parameters $[2n, k_1 + k_2, \min\{2d_1, d_2\}]_q$ can be obtained. To simplify the notation, we denote by $(C_1|C_1 + C_2)$ the code derived from the $(\mathbf{u}|\mathbf{u} + \mathbf{v})$ technique applied to codes C_1 and C_2.

We have

Theorem 6.7.9 *Assume that there exist two asymmetric stabilizer codes $[[n, k^*, d_z^*/d_x^*]]_q$, derived from linear codes $C_1 = [n, k_1, d_1]_q$ and $C_2 = [n, k_2, d_2]_q$, with $C_2 \subset C_1$, and $[[n, k^\circ, d_z^\circ/d_x^\circ]]_q$, derived from codes $C_3 = [n, k_3, d_3]_q$ and $C_4 = [n, k_4, d_4]_q$, where $C_4 \subset C_3$. Then there exists an $[[2n, k^* + k^\circ, d_z/d_x]]_q$ AQECC, where $d_z \geq \min\{2d_1, d_3\}$, $d_x \geq \min\{2d_4^\perp, d_2^\perp\}$, with $d_z^* \geq d_1$, $d_x^* \geq d_2^\perp$, $d_z^\circ \geq d_3$ and $d_x^\circ \geq d_4^\perp$, where $d_2^\perp$ and $d_4^\perp$ are the minimum distances of the dual codes $C_2^\perp$ and $C_4^\perp$, respectively.*

Proof Since the inclusions $C_2 \subset C_1$ and $C_4 \subset C_3$ hold it follows that the inclusion $(C_2|C_2 + C_4) \subset (C_1|C_1 + C_3)$ also holds. We know that the codes $(C_2|C_2 + C_4)$ and $(C_1|C_1 + C_3)$ have parameters $[2n, k_2 + k_4, \min\{2d_2, d_4\}]_q$ and $[2n, k_1 + k_3, \min\{2d_1, d_3\}]_q$, respectively. Let us compute the minimum distance of the dual code $[(C_2|C_2 + C_4)]^\perp$. We know that a generator matrix of $[(C_2|C_2 + C_4)]^\perp$ is the matrix

$$\begin{bmatrix} H_2 & 0 \\ -H_4 & H_4 \end{bmatrix},$$

where H_2 and H_4 are the parity check matrices of C_2 and C_4, respectively. The codewords of $[(C_2|C_2 + C_4)]^\perp$ are of the form $\{(\mathbf{u} - \mathbf{v}, \mathbf{v})|\mathbf{u} \in C_2^\perp, \mathbf{v} \in C_4^\perp\}$. Let us consider the codeword $\mathbf{w} = (\mathbf{u} - \mathbf{v}, \mathbf{v})$. If $\mathbf{u} = 0$, then $\mathbf{w} = (-\mathbf{v}, \mathbf{v})$; hence, the minimum weight of $[(C_2|C_2 + C_4)]^\perp$ is equal to $2d_4^\perp$. On the other hand, if $\mathbf{u} \neq 0$, then it follows that

$$\text{wt}(\mathbf{w}) =$$
$$= \text{wt}(\mathbf{u} - \mathbf{v}) + \text{wt}(\mathbf{v})$$
$$= d(\mathbf{u}, \mathbf{v}) + d(\mathbf{v}, \mathbf{0}) \geq d(\mathbf{u}, \mathbf{0})$$
$$= \text{wt}(\mathbf{u}).$$

Thus, the minimum weight is given by $d_2^\perp$ and, consequently, the minimum distance of $[(C_2|C_2 + C_4)]^\perp$ is $\min\{2d_4^\perp, d_2^\perp\}$. Applying the CSS construction to $(C_2|C_2 +$

C_4), $(C_1|C_1 + C_3)$ and $[(C_2|C_2 + C_4)]^\perp$, one obtains an $[[2n, (k_1 + k_3) - (k_2 + k_4), d_z/d_x]]_q = [[2n, k^* + k^\circ, d_z/d_x]]_q$ asymmetric stabilizer code, where $d_z \geq \min \{2d_1, d_3\}$ and $d_x \geq \min\{2d_4^\perp, d_2^\perp\}$, as required.

As an alternative proof, we can also write the codewords of $[(C_2|C_2 + C_4)]^\perp$ in the form $\{(\mathbf{u} + \mathbf{v}, -\mathbf{v}) | \mathbf{u} \in C_2^\perp, \mathbf{v} \in C_4^\perp\}$, and because the Hamming weights of $\mathbf{v}$ and $-\mathbf{v}$ are the same, the latter code is equivalent to $\{(\mathbf{u} + \mathbf{v}, \mathbf{v}) | \mathbf{u} \in C_2^\perp, \mathbf{v} \in C_4^\perp\}$, and the result follows. $\square$

6.7.6 Code Constructions

In this subsection, we only apply the asymmetric quantum code expansion shown in Sect. 6.7.1 in order to generate families of AQECCs, although it is clear that all construction techniques proposed in Sect. 6.7.1–6.7.5 can be also applied. In the sequence, we construct AQECCs derived from generalized Reed–Muller (GRM), character codes, BCH, quadratic residue (QR) and affine-invariant codes, respectively.

Remark 6.7.4 It is important to observe that in all results presented in the following, we expand the codes defined over $\mathbb{F}_q$ (where $q = p^t, t \geq 1$ and p prime) with respect to the prime field $\mathbb{F}_p$. However, it is clear that the method also holds if the expansion is performed over any subfield of the field $\mathbb{F}_q$.

6.7.6.1   Construction I—Generalized RM Codes

The first family of AQECCs derived from binary Reed–Muller (RM) codes were constructed in [140, Lemma 4.1]. In this subsection, we present a construction of AQECCs derived from generalized Reed–Muller (GRM) codes [114, 129].

The GRM code $\mathcal{R}_q(\alpha, m)$ over $\mathbb{F}_q$ of order $\alpha, 0 \leq \alpha < q(m - 1)$, has parameters $[q^m, k(\alpha), d(\alpha)]_q$, where

$$k(\alpha) = \sum_{i=0}^{m} (-1)^i \binom{m}{i} \binom{m + \alpha - iq}{\alpha - iq} \tag{6.1}$$

and

$$d(\alpha) = (t + 1)q^u, \tag{6.2}$$

where $m(q - 1) - \alpha = (q - 1)u + t$ and $0 \leq t < q - 1$. The dual of a GRM code $\mathcal{R}_q(\alpha, m)$ is also a GRM code given by

$$[\mathcal{R}_q(\alpha, m)]^\perp = \mathcal{R}_q(\alpha^\perp, m),$$

where $\alpha^\perp = m(q-1) - 1 - \alpha$.

We utilize the properties of GRM codes in order to obtain more asymmetric quantum error-correcting codes:

Theorem 6.7.10 *Let $0 \le \alpha_1 \le \alpha_2 < m(q-1)$ and assume that $q = p^t$ is a prime power, where $t \ge 1$. Then there exists an p-ary asymmetric quantum GRM code with parameters $[[tq^m, t[k(\alpha_2) - k(\alpha_1)], d_z/d_x]]_p$, where $d_z \ge d(\alpha_2), d_x \ge d(\alpha_1^\perp), k(\alpha_2)$ and $k(\alpha_1)$ are given in Eq. (6.1), $d(\alpha_2)$ is shown in Eq. (6.2) and $d(\alpha_1^\perp) = (a+1)q^b$, where $\alpha_1 + 1 = (q-1)b + a$ and $0 \le a \le q - 1$.*

Proof We first note that since the inequality $\alpha_1 \le \alpha_2$ holds, then the inclusion $\mathcal{R}_q(\alpha_1, m) \subset \mathcal{R}_q(\alpha_2, m)$ also holds. The codes $\beta(\mathcal{R}_q(\alpha_1, m))$ and $\beta(\mathcal{R}_q(\alpha_2, m))$ have parameters $[tq^m, tk(\alpha_1), d(\alpha_1)]_p$ and $[tq^m, tk(\alpha_2), d(\alpha_2)]_p$, respectively, where $k(\alpha_1)$ and $k(\alpha_2)$ are computed according to Eq. (6.1) and $d(\alpha_1), d(\alpha_2)$ are computed by applying Eq. (6.2). We know that the parameter $\alpha_1^\perp$ of the dual code $[\mathcal{R}_q(\alpha_1, m)]^\perp = \mathcal{R}_q(\alpha_1^\perp, m)$ is equal to $\alpha_1^\perp = m(q-1) - 1 - \alpha_1$; thus, the minimum distance of $[\mathcal{R}_q(\alpha_1, m)]^\perp$ is $d(\alpha_1^\perp) = (a+1)q^b$, where $\alpha_1 + 1 = (q-1)b + a$ and $0 \le a \le q - 1$. Hence, the code $[\beta(\mathcal{R}_q(\alpha_1, m))]^\perp$ has minimum distance greater than or equal to $d(\alpha_1^\perp)$. Applying Theorem 6.7.3, we obtain an $[[tq^m, t[k(\alpha_2) - k(\alpha_1)], d_z/d_x]]_p$ asymmetric stabilizer code, where $d_z \ge d(\alpha_2)$ and $d_x \ge d(\alpha_1^\perp)$. The proof is complete. □

6.7.6.2 Construction II—Character Codes

The class of (classical) character codes was introduced by Ding et al. [33].

Before proceeding further, we need to define some concepts on character codes. Let us consider the commutative additive group $G = \mathbb{Z}_2^m$, where $m \ge 1$, and a finite field $\mathbb{F}_q$ of odd characteristic. Recall that the code $C_q(r, m) = C_X$, where $X \subset \mathbb{Z}_2^m$ consists of all elements with Hamming weight greater than r, has parameters $[2^m,$

$$s_m(r), 2^{m-r}]_q \text{ (see [33, Theorem 6]), where } s_m(r) = \sum_{i=0}^{r} \binom{m}{i}. \text{ The Euclidean dual}$$

code $[C_q(r, m)]^\perp$ of $C_q(r, m)$ is equivalent to $C_q(m - r - 1, m)$ (see [33, Theorem 8]) and, consequently, it has parameters $[2^m, s_m(m - r - 1), 2^{r+1}]_q$.

In the following, we utilize the code expansion applied to character codes to generate AQECCs, as establishes the next result.

Theorem 6.7.11 *Assume that $0 \le r_1 < r_2 \le m$ and let $q = p^t$ be a power of an odd prime p, where $t \ge 1$. Then there exists an $[[t2^m, t[k(r_2) - k(r_1)], d_z/d_x]]_p$ AQECC,*

$$\text{where } k(r) = \sum_{i=0}^{r} \binom{m}{i}, d_z \ge 2^{m-r_2} \text{ and } d_x \ge 2^{r_1+1}.$$

Proof It is easy to see that the inclusion $C_q(r_1, m) \subset C_q(r_2, m)$ holds. The dual code $[C_q(r_1, m)]^\perp$ is equivalent to the code $C_q(m - r_1 - 1, m)$. Applying Theorem 6.7.3, we get an $[[t2^m, t(k(r_2) - k(r_1)), d_z/d_x]]_p$ AQECC, where $t, k(r_1), k(r_2), d_x$ and d_z are specified in the hypotheses. □

6.7.6.3 Construction III—BCH Codes

In this subsection, we construct more families of asymmetric stabilizer codes derived from BCH codes. The first families of AQECCs derived from BCH codes were constructed by Aly [3, Theorem 8].

Recall that a cyclic code of length n over $\mathbb{F}_q$ is a BCH code with designed distance δ if, for some integer $b \geq 0$, one has

$$g(x) = \mathrm{lcm}\{M^{(b)}(x), M^{(b+1)}(x), \ldots, M^{(b+\delta-2)}(x)\},$$

i.e., $g(x)$ is the monic polynomial of smallest degree over $\mathbb{F}_q$ having α^b, α^{b+1}, $\ldots, \alpha^{b+\delta-2}$ as zeros.

The next result shows how to construct more AQECCs by expanding BCH codes.

Theorem 6.7.12 *Suppose that* $n = q^m - 1$, *where* $q = p^t$ *is a power of an odd prime* p, $t \geq 1$ *and* $m \geq 3$ *are integers (if* $q = 3$, $m \geq 4$). *Then there exist AQECCs with parameters*

- $[[tn, t(n - m(4q - 5) - 2), d_z \geq (2q + 2)/d_x \geq 2q]]_p$;
- $[[tn, t(n - m(4q - c - 5) - 2), d_z \geq (2q + 2)/d_x \geq (2q - c)]]_p$, *where* $0 \leq c \leq q - 2$;
- $[[tn, t(n - m(2c - l - 4) - 2), d_z \geq c/d_x \geq (c - l)]]_p$, *where* $2 \leq c \leq q$ *and* $0 \leq l \leq c - 2$;
- $[[tn, t(n - m(2c - l - 6) - 2), d_z \geq c/d_x \geq (c - l)]]_p$, *where* $q + 2 < c \leq 2q$ *and* $0 \leq l \leq c - q - 3$;
- $[[tn, t(n - m(4q - l - 5) - 1), d_z \geq (2q + 1)/d_x \geq (2q - l)]]_p$, *where* $0 \leq l \leq q - 2$.

Proof Consider the codes constructed in [92, Theorems 4 and 5 and Corollary 1]. These codes are derived from two distinct nested cyclic codes $C_2 \subset C_1$. Thus, applying Theorem 6.7.2 the result holds.   □

Theorem 6.7.13 *Let* $q = p^t$ *be a power of a prime* p, $t \geq 1$, $\gcd(q, n) = 1$ *and* $\mathrm{ord}_n(q) = m$. *Let* C_1 *and* C_2 *be two narrow-sense BCH codes of length* $q^{\lfloor m/2 \rfloor} < n \leq q^m - 1$ *over* $\mathbb{F}_q$ *with designed distances* δ_1 *and* δ_2 *in the range* $2 \leq \delta_1, \delta_2 \leq \delta_{max} = \min\{\lfloor nq^{\lceil m/2 \rceil}/(q^m - 1)\rfloor, n\}$ *and* $\delta_1 < \delta_2^\perp \leq \delta_2 < \delta_1^\perp$. *Assume also that* $S_1 \cup \ldots \cup S_{\delta_1-1} \neq S_1 \cup \ldots \cup S_{\delta_2-1}$, *where* S_i *denotes a* q-*coset. Then there exists an* $[[tn, t(n - m\lceil(\delta_1 - 1)(1 - 1/q)\rceil - m\lceil(\delta_2 - 1)(1 - 1/q)\rceil), d_z^*/d_x^*]]_p$ *AQECC, where* $d_z^* = wt(C_2\backslash C_1^\perp) \geq \delta_2$ *and* $d_x^* = wt(C_1\backslash C_2^\perp) \geq \delta_1$.

Proof It suffices to apply Theorem 6.7.2 in those codes shown in [3, Theorem 8]. □

Remark 6.7.5 Note that one can obtain more families of AQECCs by applying Theorem 6.7.3 in the existing families shown in [89]. Moreover, by expanding generalized Reed–Solomon (GRS) codes we have [94, Theorem 7.1] as a particular case of Theorem 6.7.2.

6.7.6.4   Construction IV—QR Codes

Here, we construct families of AQECCs derived from quadratic residue (QR) codes
[67, 114]. A family of quantum codes derived from classical QR codes was constructed in [80, Theorems 40 and 41].

Definition 6.7.1 Let p be an odd prime not dividing q, where q is a prime power
that is a square modulo p. Let Q be the set of nonzero squares modulo p and let C
consisting of non-squares modulo p. The quadratic residue codes $Q, Q^\diamond, C$ and $C^\diamond$ are
cyclic codes with generator polynomials $q(x)$, $(x-1)q(x)$, $c(x)$, $(x-1)c(x)$,
respectively, where

$$q(x) = \prod_{r \in Q}(x - \alpha^r), \quad c(x) = \prod_{s \in C}(x - \alpha^s)$$

have coefficients from $\mathbb{F}_q$, and α is a primitive pth root of unity belonging to some
extension field of $\mathbb{F}_q$.

The codes Q and C have the same parameters $[p, (p+1)/2, d_1]_q$, where $(d_1)^2 \geq$
p. Analogously, the codes $Q^\diamond$ and $C^\diamond$ also have the same parameters $[p, (p-1)/2, d_2]_q$,
where $(d_2)^2 \geq p$.

In the next result, we construct families of AQECCs by expanding quadratic
residue codes.

Theorem 6.7.14 *Let p be a prime number of the form $p \equiv 1 \mod 4$, and let $q =$
p_*^t $(t \geq 1)$ be a power of a prime that is not divisible by p. If q is a quadratic
residue modulo p, then there exists an $[[tp, t, d_z/d_x]]_{p_*}$ asymmetric quantum error-
correcting code, where d_z and d_x satisfy $d_z \geq \sqrt{p}$ and $d_x \geq \sqrt{p}$.*

Proof Let us consider the codes $Q, Q^\diamond$ and C given above. Since $p = 4k + 1$, we have
$Q^\diamond = C^\perp$, hence, $C^\perp \subset Q$. The codes Q and $C^\perp$ have parameters, respectively, given
by $[p, (p+1)/2, d_1]_q$, with $(d_1)^2 \geq p$ and $[p, (p-1)/2, d_2]_q$, where $(d_2)^2 \geq p$.
Proceeding similarly as in the proof of Theorem 6.7.2 we have an $[[tp, t, d_z/d_x]]_{p_*}$
AQECC, where d_z and d_x satisfy $d_z \geq \sqrt{p}$ and $d_x \geq \sqrt{p}$. $\qquad\square$

Theorem 6.7.15 *Let p be a prime of the form $p \equiv 3 \mod 4$, and let $q = p_*^t (t \geq 1)$
be a power of a prime that is not divisible by p. If q is a quadratic residue modulo p,
then there exists an $[[tp, t, d_z/d_x]]_{p_*}$ AQECC, where $d_z \geq d$, $d_x \geq d$ and d satisfies
$d^2 - d + 1 \geq p$.*

Proof Since $p = 4k - 1$, the dual code $Q^\perp$ of the code Q is equal to $Q^\perp =$
$Q^\diamond$, so $Q^\perp \subset Q$. The codes Q and $Q^\perp$ have parameters $[p, (p+1)/2, d]_q$ and
$[p, (p-1)/2, d^\diamond \geq d]_q$, respectively, and the minimum distance is bounded by
$d^2 - d + 1 \geq p$ (see, for instance, the proof of Theorem 40 in [80]). Applying Theorem 6.7.3, we get an $[[tp, t, d_z/d_x]]_{p_*}$ code, where $d_z \geq d$, $d_x \geq d$ and
$d^2 - d + 1 \geq p$. $\qquad\square$

Table 6.4 Families of AQECCs

Code Family / $[[n, k, d_z/d_x]]_q$	Range of Parameters	Ref.
BCH		
$[[n, n - m\lceil(\delta_1 - 1)(1 - 1/q)\rceil - m\lceil(\delta_2 - 1)(1 - 1/q)\rceil, \; d_z^*/d_x^*]]_q$	$\gcd(q, n) = 1$, $ord_n(q) = m$,	[3]
	$q^{\lfloor m/2 \rfloor} < n \leq q^m - 1$,	
	$2 \leq \delta_1, \delta_2 \leq \delta_{max} =$ $\min\{\lfloor nq^{\lceil m/2 \rceil}/(q^m - 1)\rfloor, n\}$,	
	$\delta_1 < \delta_2^{\frac{1}{2}} \leq \delta_2 < \delta_1^{\frac{1}{2}}$,	
	$d_z^* = wt(C_2 \backslash C_1^{\perp}) \geq \delta_2$,	
	$d_x^* = wt(C_1 \backslash C_2^{\perp}) \geq \delta_1$	
$[[2^m - 1, m(\delta_2 - \delta_1)/2, d_x/d_z]]_q$	$m \geq 2, 2 \leq \delta_1 < \delta_2 < \delta_{max} =$ $2^{\lceil m/2 \rceil} - 1$,	[140]
	$\delta_i \equiv 1 \bmod 2 \; d_x \geq \delta_1$, $d_z \geq \delta_{max} + 1$	
$[[n, k, d_z/d_x]]_q$	$n = q^m - 1, m \geq 3$ (if $q = 3$, $m \geq 4$):	[91]
$[[n, n - m(4q - 5) - 2, d_z \geq (2q + 2)/d_x \geq 2q]]_q$		
$[[n, n - m(4q - c - 5) - 2, d_z \geq (2q + 2)/d_x \geq (2q - c)]]_q$	$0 \leq c \leq q - 2$	
$[[n, n - m(2c - l - 4) - 2, d_z \geq c/d_x \geq (c - l)]]_q$	$2 \leq c \leq q$ and $0 \leq l \leq c - 2$	
$[[n, n - m(2c - l - 6) - 2, d_z \geq c/d_x \geq (c - l)]]_q$	$q + 2 < c \leq 2q$ and $0 \leq l \leq c - q - 3$	
$[[n, n - m(4q - l - 5) - 1, d_z \geq (2q + 1)/d_x \geq (2q - l)]]_q$	$0 \leq l \leq q - 2$	
Expanded BCH		
$[[tn, t[n - m\lceil(\delta_1 - 1)(1 - 1/q)\rceil - m\lceil(\delta_2 - 1)(1 - 1/q)\rceil], \; d_z^*/d_x^*]]_q$	$\gcd(q, n) = 1$, $ord_n(q) = m$,	[3]
	$t \geq 1, q^{\lfloor m/2 \rfloor} < n \leq q^m - 1$,	
	$2 \leq \delta_1, \delta_2 \leq \delta_{max} =$ $\min\{\lfloor nq^{\lceil m/2 \rceil}/(q^m - 1)\rfloor, n\}$,	
	$\delta_1 < \delta_2^{\frac{1}{2}} \leq \delta_2 < \delta_1^{\frac{1}{2}}$,	
	$d_z^* = wt(C_2 \backslash C_1^{\perp}) \geq \delta_2$,	
	$d_x^* = wt(C_1 \backslash C_2^{\perp}) \geq \delta_1$	
$[[tn, tk, d_z/d_x]]_q$	$n = q^m - 1, q = p^t$, p odd prime, $t \geq 1$,	
	$m \geq 3$ (if $q = 3, m \geq 4$):	
$[[tn, t(n - m(4q - 5) - 2), d_z \geq (2q + 2)/d_x \geq 2q]]_p$		
$[[tn, t(n - m(4q - c - 5) - 2), d_z \geq (2q + 2)/d_x \geq (2q - c)]]_p$	$0 \leq c \leq q - 2$	
$[[tn, t(n - m(2c - l - 4) - 2), d_z \geq c/d_x \geq (c - l)]]_p$	$2 \leq c \leq q, 0 \leq l \leq c - 2$	
$[[tn, t(n - m(2c - l - 6) - 2), d_z \geq c/d_x \geq (c - l)]]_p$	$q + 2 < c \leq 2q$, $0 \leq l \leq c - q - 3$	
$[[tn, t(n - m(4q - l - 5) - 1), d_z \geq (2q + 1)/d_x \geq (2q - l)]]_p$	$0 \leq l \leq q - 2$	
BCH-LDPC		
$[[p^{ms} - 1, k_x + k_z - p^{ms} + 1, d_z/d_x]]_p$	$\delta \leq \delta_0 = p^{\mu s} - 1$	[140]
	$k_x = \dim \mathrm{BCH}(\delta) \subseteq \mathbb{F}_p^n$,	
	$k_z = \dim C_{EG,c}^{(1)}(m, \mu, 0, s, p)$,	
	$d_x \geq \delta$, $d_z \geq A_{EG}(m, \mu, \mu - 1, s, p)$	
$[[2^{2s} - 1, 2^{2s} - 3^s - s(\delta - 1), \delta/2^s + 1]]_2$	$\delta = 2t + 1 \leq 2^s - 1$	[140]

(continued)

Table 6.4 (continued)

Code Family / $[[n, k, d_z/d_x]]_q$	Range of Parameters	Ref.
$[[n, k_x + k_z - n, d_z/d_x]]_p$	$n = (p^{(m+1)s} - 1)/(p^s - 1)$	[140]
	$\delta \le \delta_0 =$ $(p^{(\mu+1)s} - 1)/(p^s - 1)$, $k_x = \dim \text{BCH}_p(\delta, n)$, $k_z = \dim C_{PG}^{(1)}(m, \mu, 0, s, p)$, $d_x \ge \delta$,	
	$d_z \ge A_{EG}(m, \mu, \mu - 1, s, p)$	
$[[n, n - 3^s - 3s\lceil(\delta - 1)/2\rceil - 1, \delta/(2^s + 2)]]_2$	$n = 2^{2s} + 2^s + 1, \delta \le 2^{s/2} + 1$	[140]
LDPC-LDPC		
$[[p^{ms}, k_x + k_z - p^{ms}, d_z/d_x]]_p$	p prime, $q = p^s, s \ge 1, m \ge 2$,	[140]
	$1 < \mu_z < m$, $m - \mu_z + 1 \le \mu_x < m$, $k_x = \dim C_{EG}^{(1)}(m, \mu_x, 0, s, p)$, $k_z = \dim C_{EG}^{(1)}(m, \mu_z, 0, s, p)$,	
	$d_x \ge$ $A_{EG}(m, \mu_x, \mu_x - 1, s, p) + 1$,	
	$d_z \ge$ $A_{EG}(m, \mu_z, \mu_z - 1, s, p) + 1$	
concatenated RS		
$[[2mq, mk - 1, (\ge 2(q - k + 1))/2]]_4$	$n = 4^m, 1 \le k \le q$	[36]
GRS		
$[[mn, m(2k - n + c), d_z \ge d/d_x \ge (d - c)]]_q$	$1 < k < n < 2k + c \le q^m$,	[94]
	$k = n - d + 1, d > c + 1$, $c, m \ge 1$	

Remark 6.7.6 It is interesting to note that a refined statement can be made if we consider the code $\mathcal{Q}^\circ$ instead of considering the code $\mathcal{Q}$, because $d_{\mathcal{Q}^\circ} = d_{\mathcal{Q}} + 1$ (see [114, Chap. 16, Problem (2), p. 494]).

6.7.6.5 Construction V—Affine-Invariant Codes

We assume that the reader is familiar with the class of affine-invariant codes. The structure and results on this class of codes can be found in [67].

Quantum affine-invariant codes were investigated in the literature [60].

Lemma 6.7.2 ([60, Lemma 22]) *Let C^e be an extended maximal affine-invariant code $[p^m, p^m - 1 - m/t, d]_{p^t}$, then if $p > 3$ or $m > 2$ or $t \ne 1$, we have $(C^e)^\perp \subset C^e$.*

Applying Lemma 6.7.2, it is possible to construct a family of AQECCs derived from affine-invariant codes, as states the next result.

Theorem 6.7.16 *Assume that $q = p^t$, m is a positive integer and $n = p^m - 1$. If $p > 3$ or $m > 2$ or $t \neq 1$, then there exists an $[[tp^m, t(p^m - 2 - 2\frac{m}{t}), d_z/d_x]]_p$ AQECC, where $d_z \geq d_a$, $d_x \geq d_a$, and d_a is the minimum distance of an extended maximal affine-invariant code.*

Proof Consider the dual-containing extended maximal affine-invariant code C^e with parameters $[p^m, p^m - 1 - m/t, d]$ given in Lemma 6.7.2, where $p > 3$ (or $m > 2$ or $t \neq 1$). Applying Theorem 6.7.3 we have an $[[tp^m, t(p^m - 2 - 2\frac{m}{t}), d_z/d_x]]_p$ AQECC, where $d_z \geq d_a$, $d_x \geq d_a$ and d_a is the minimum distance of C^e. $\qquad\square$

Table 6.5 Families of AQECCs

Code Family / $[[n, k, d_z/d_x]]_q$	Range of Parameters	Ref.
RM		
$[[2^m, k, 2^{m-r_2} \geq 2^{r_1+1}]]_2$	$0 \leq r_1 < r_2 < m,$ $k = \sum_{j=r_1+1}^{r_2} \binom{m}{j}$	[140]
Expanded GRM		
$[[lq^m, l[k(\alpha_2) - k(\alpha_1)], d_z/d_x]]_p$	$0 \leq \alpha_1 \leq \alpha_2 < m(q-1), q = p^l,$ p prime, $l \geq 1,$	
	$k(\alpha) =$ $\sum_{i=0}^{m} (-1)^i \binom{m}{i} \binom{m+\alpha-iq}{\alpha-iq},$	
	$d_z \geq d(\alpha_2), d_x \geq d(\alpha_1^\perp),$ $d(\alpha_2) = (t+1)q^u,$	
	$m(q-1) - \alpha_2 = (q-1)u + t,$ $0 \leq t < q - 1,$	
	$d(\alpha_1^\perp) = (a+1)q^b,$ $\alpha_1 + 1 = (q-1)b + a,$ $0 \leq a \leq q - 1$	
MDS		
$[[n, n - d_1 - d_2 + 2, d_z/d_x]]_q$	$n = q - 1, d_x = d_1 < d_z = d_2$	[3]
$[[n, n - 2, 2/2]]_q$	q prime power, $n \geq 3$	[157]
$[[n, k - 1, (n - k + 1)/2]]_q$	$q \geq n > 3, 1 < k \leq n - 2$	[157]
$[[2^m + 2, 2, 2^m/2]]_{2^m}$	$m > 0$ integer	[157]
$[[2^m + 2, 2^m - 2, 4/2]]_{2^m}$	$m > 0, m \neq 2$ integer	[157]
$[[n, j, d_z/d_x]]_q$	$n, k, j \in \mathbb{Z}, q \geq 5, n \leq q,$ $2 \leq k \leq n - 3,$	[157]
	$j \leq n - k - 2,$ $\{d_z, d_x\} = \{n - k - j + 1, k + 1\}$	
$[[q + 1, 2j, d_z/d_x]]_q$	$n, k, j \in \mathbb{Z}, q \geq 5, k \geq 2,$ $k + 2j \leq q - 1,$	[157]
	$\{d_z, d_x\} = \{q - k - 2j + 2, k + 1\}$	
$[[q + 1, q - 1 - 2s, (2s + 1)/3]]_q$	$q = 2^m \geq 4, s \leq q/2 - 1$	[157]
$[[2^m + 2, 2^m - 4, 4/4]]_{2^m}$	$2^m \geq 4$	[157]
$[[n, 2k - n + c, d_z \geq d/d_x \geq (d - c)]]_q$	$1 < k < n < 2k + c \leq q,$ $k = n - d + 1, d > c + 1, c \geq 1$	[94]

(continued)

Table 6.5 (continued)

Code Family / $[[n, k, d_z/d_x]]_q$	Range of Parameters	Ref.
Expanded Character		
$[[t2^m, t[k(r_2) - k(r_1)], d_z/d_x]]_p$	$q = p^t$, p odd prime, $t \geq 1$,	
	$k(r) = \sum_{i=0}^{r} \binom{m}{i}$, $d_z \geq 2^{m-r_2}$,	
	$d_x \geq 2^{r_1+1}$	
QR		
$[[p, 1, d_z/d_x]]_q$	p prime, $p \equiv 1 \mod 4$,	[80]
	$q = p_1^t$, $p \nmid p_1$, q is a quadratic residue mod p,	
	$d_z \geq \sqrt{p}$, $d_x \geq \sqrt{p}$	
$[[p, 1, d_z/d_x]]_q$	p prime, $p \equiv 3 \mod 4$,	[80]
	$q = p_1^t$, $p \nmid p_1$, q is a quadratic residue mod p,	
	$d_z \geq d$, $d_x \geq d$, $d^2 - d + 1 \geq p$	
Expanded QR		
$[[tp, t, d_z/d_x]]_{p*}$	p prime, $p \equiv 1 \mod 4$, $q = p_*^t$,	
	$t \geq 1$, $p \nmid p_*$, q is a quadratic residue mod p,	
	$d_z \geq \sqrt{p}$, $d_x \geq \sqrt{p}$	
$[[tp, t, d_z/d_x]]_{p*}$	p prime, $p \equiv 3 \mod 4$, $q = p_*^t$,	
	$t \geq 1$, $p \nmid p_*$, q is a quadratic residue mod p,	
	$d_z \geq d$, $d_x \geq d$, $d^2 - d + 1 \geq p$	
Affine-Invariant		
$[[tp^m, t(p^m - 2 - 2\frac{m}{t}), d_z/d_x]]_p$	$q = p^t$,	
	$p > 3$, $m > 2$, $d_z \geq d_a$, $d_x \geq d_a$,	
	d_a is given in Theorem 6.7.16	
Product code		
$[[(q-1)^2, (q-d_1)(q-d_3) - (q-d_2)(q-d_4), d_z/d_x]]_q$	$2 \leq d_1 \leq d_2 < q - 1$, $2 \leq d_3 \leq d_4 < q - 1$,	[95]
	$d_z \geq \max\{d_1 d_3, \min\{q - d_2, q - d_4\}\}$,	
	$d_x \geq \min\{d_1 d_3, \min\{q - d_2, q - d_4\}\}$	

6.7.6.6 Code Tables

Here, we show Tables 6.4 and 6.5 containing families of AQECCs available in the literature as well as the code families constructed in this chapter. In the first column, we exhibit the class and the parameters $[[n, k, d_z/d_x]]_q$ of an AQECC; in the second column the parameter's range and in the third column, the corresponding references are shown.

Chapter 7
Constructions of QCCs

This chapter is devoted to the construction of quantum convolutional codes with good or even optimal parameters (in the sense that the code parameters attain the generalized quantum Singleton bound). We begin by presenting the necessary background for the constructions, i.e., the class of classical convolutional codes.

In Sects. 7.1 and 7.2, we give the necessary background to our code constructions. Families of maximum-distance-separable (optimal) quantum convolutional BCH codes are presented in Sect. 7.3. In Sect. 7.4, we construct more families of quantum convolutional BCH codes. In Sect. 7.5, we construct families of quantum convolutional codes derived from negacyclic codes. More families of quantum convolutional codes derived from algebraic geometry are presented in Sect. 7.6. Finally, in Sect. 7.7, we introduce the first class of asymmetric quantum convolutional codes displayed in the literature and we show how to construct such codes.

7.1 Convolutional Codes

The class of classical convolutional codes is extensively investigated in the literature [29, 40, 75, 107, 130, 135, 145]. Forney [40] was one of the pioneers in the investigations concerning convolutional codes. Constructions of convolutional codes with good parameters or even maximum-distance-separable (MDS) convolutional codes (optimal in the sense that they attain the generalized Singleton bound [135]) have also been presented in the literature [40, 51, 98, 99, 101, 107, 130, 135, 145]. Rosenthal et al. introduced the generalized Singleton bound for convolutional codes in 1999 [135] (see also [145]).

A convolutional code of length n and rank k can be understood as a free module (see Definition 1.5.4) over some finite field $\mathbb{F}_q$ given by a direct summand $\mathbb{F}_q[D]^n$. Since free modules have basis, we can codify it by means of a polynomial matrix with suitable properties.

© Springer Nature Switzerland AG 2020
G. G. La Guardia, *Quantum Error Correction*, Quantum Science and Technology,
https://doi.org/10.1007/978-3-030-48551-1_7

Definition 7.1.1 A polynomial encoder matrix $G(D) \in \mathbb{F}_q[D]^{k \times n}$ is called *basic* if $G(D)$ has a polynomial right inverse. A basic generator matrix is called *reduced* (or minimal), if the overall constraint length $\gamma = \sum_{i=1}^{k} \gamma_i$, where $\gamma_i = \max_{1 \leq j \leq n} \{\deg g_{ij}\}$, has the smallest value among all basic generator matrices. In this case, we say that γ is the *degree* of the code.

Definition 7.1.1 gives us the ingredients to define formally what we know by a convolutional code.

Definition 7.1.2 A rate k/n convolutional code C with parameters $(n, k, \gamma; \mu, d_f)_q$ is a submodule of $\mathbb{F}_q[D]^n$ generated by a reduced basic matrix $G(D) = (g_{ij}) \in \mathbb{F}_q[D]^{k \times n}$, i.e., $C = \{\mathbf{u}(D)G(D) | \mathbf{u}(D) \in \mathbb{F}_q[D]^k\}$, where n is the length, k is the dimension, $\gamma = \sum_{i=1}^{k} \gamma_i$ is the degree, $\mu = \max_{1 \leq i \leq k} \{\gamma_i\}$ is the memory and $d_f = \mathrm{wt}(C) = \min\{wt(\mathbf{v}(D)) \mid \mathbf{v}(D) \in C, \mathbf{v}(D) \neq 0\}$ is the free distance of the code.

In Definition 7.1.2, the *weight* of an element $\mathbf{v}(D) \in \mathbb{F}_q[D]^n$ is defined as

$$\mathrm{wt}(\mathbf{v}(D)) = \sum_{i=1}^{n} \mathrm{wt}(v_i(D)),$$

where $\mathrm{wt}(v_i(D))$ is the number of nonzero coefficients of $v_i(D)$. In the field of Laurent series $\mathbb{F}_q((D))$, whose elements are given by $\mathbf{u}(D) = \sum_i u_i D^i$, where $u_i \in \mathbb{F}_q$ and $u_i = 0$ for $i \leq r$, for some $r \in \mathbb{Z}$, we define the weight of $\mathbf{u}(D)$ as

$$\mathrm{wt}(\mathbf{u}(D)) = \sum_{\mathbb{Z}} \mathrm{wt}(u_i).$$

Definition 7.1.3 A generator matrix $G(D)$ is called *catastrophic* if there exists a $\mathbf{u}(D)^k \in \mathbb{F}_q((D))^k$ of infinite Hamming weight such that $\mathbf{u}(D)^k G(D)$ has finite Hamming weight.

We need to define the Euclidean dual of a convolutional code. To do this, we need to define what we mean by Euclidean inner products of polynomials.

Definition 7.1.4 The Euclidean inner product of two n-tuples $\mathbf{u}(D) = \sum_i \mathbf{u}_i D^i$ and $\mathbf{v}(D) = \sum_j \mathbf{u}_j D^j$ in $\mathbb{F}_q[D]^n$ is defined as $\langle \mathbf{u}(D) \mid \mathbf{v}(D) \rangle = \sum_i \mathbf{u}_i \cdot \mathbf{v}_i$. If C is a convolutional code, then its Euclidean dual code $C^\perp$ is defined by

$$C^\perp = \{\mathbf{u}(D) \in \mathbb{F}_q[D]^n \mid \langle \mathbf{u}(D) \mid \mathbf{v}(D) \rangle = 0, \ \forall \ \mathbf{v}(D) \in C\}.$$

Similarly, we define here the Hermitian dual of a convolutional code.

Definition 7.1.5 The Hermitian inner product is defined as $\langle \mathbf{u}(D) \mid \mathbf{v}(D) \rangle_h = \sum_i \mathbf{u}_i \cdot \mathbf{v}_i^q$, where $\mathbf{u}_i, \mathbf{v}_i \in \mathbb{F}_{q^2}^n$ and $\mathbf{v}_i^q = (v_{1i}^q, \ldots, v_{ni}^q)$. The Hermitian dual of a code C is defined by

$$C^{\perp_h} = \{\mathbf{u}(D) \in \mathbb{F}_{q^2}[D]^n \mid \langle \mathbf{u}(D) \mid \mathbf{v}(D) \rangle_h = 0 \ \forall \ \mathbf{v}(D) \in C\}.$$

We will now show how to construct a convolutional code by means of a linear block code. This technique is due to Piret [130]. The initial point is to consider $C \subseteq \mathbb{F}_q^n$ as an $[n, k, d]_q$ linear block code with parity check matrix H. In order to construct a convolutional code derived from C, we proceed as follows:

- Split H into $\mu + 1$ disjoint submatrices H_i such that $H = \begin{bmatrix} H_0 \\ H_1 \\ \vdots \\ H_\mu \end{bmatrix}$, where each H_i has n columns.
- Construct the polynomial matrix $G(D) = \tilde{H}_0 + \tilde{H}_1 D + \tilde{H}_2 D^2 + \cdots + \tilde{H}_\mu D^\mu$, where $\tilde{H}_i$, for all $1 \leq i \leq \mu$, are derived from the respective matrices H_i by adding zero-rows at the bottom such that $\tilde{H}_i$ has κ rows in total, where κ is the maximal number of rows among the matrices H_i.
- The matrix $G(D)$ generates a convolutional code V.

The following theorem due to Aly et al. is a powerful tool in order to construct convolutional codes. This method is, in fact, a generalization of Piret's technique [130] to nonbinary alphabets.

Theorem 7.1.1 ([6, Theorem 3]) *Let $C \subseteq \mathbb{F}_q^n$ be a linear code with parameters $[n, k, d]_q$. Assume that $H \in \mathbb{F}_q^{(n-k) \times n}$ is a parity check matrix for C partitioned into submatrices $H_0, H_1, \ldots, H_\mu$ as above such that $\kappa = \mathrm{rk}\, H_0$ and $\mathrm{rk}\, H_i \leq \kappa$ for $1 \leq i \leq \mu$.*

(a) *The matrix $G(D)$ is a reduced basic generator matrix.*

(b) *If d_f and $d_f^\perp$ denote the free distances of V and $V^\perp$, respectively, d_i denote the minimum distance of the code $C_i = \{\mathbf{v} \in \mathbb{F}_q^n \mid \mathbf{v}\tilde{H}_i^t = 0\}$ and $d^\perp$ is the minimum distance of $C^\perp$, then one has $\min\{d_0 + d_\mu, d\} \leq d_f^\perp \leq d$ and $d_f \geq d^\perp$.*

(c) *If C contains its Euclidean dual $C^\perp$, respectively, its Hermitian dual $C^{\perp_h}$, then the convolutional code $V = \{\mathbf{v}(D) = \mathbf{u}(D)G(D) \mid \mathbf{u}(D) \in F_q^{n-k}[D]\}$ is contained in its dual $V^\perp$, respectively, its Hermitian dual $V^{\perp_h}$.*

7.2 Defining QCCs

Researches dealing with constructions of quantum convolutional codes (QCCs) have received few attention when compared with its classical counterpart [5–7, 32, 41, 57–59, 66, 98, 99, 123, 124, 154, 160, 161, 170]. This is natural because the quantum coding theory has gained strength only in the last three decades, whereas theory of classical convolutional codes was already established in the literature a long time ago. The precursors in the investigation of the structure of quantum convolutional codes were Ollivier and Tillich [123, 124]. Almeida and Palazzo Jr. [32] constructed the first concatenated [(4, 1, 3)] quantum convolutional code.

To put the reader into context, Grassl and Rötteler [58, 59] generated quantum convolutional codes as well as they provide algorithms to obtain non-catastrophic encoders. Forney et al., constructed rate $(n - 2)/n$ quantum convolutional codes. Wilde and Brun [160, 161] generated entanglement-assisted quantum convolutional coding. Houshmand et al. [66] investigated constructions of quantum convolutional encoders with minimal-memory. In the papers [98, 99], we constructed families of optimal (MDS) quantum convolutional codes derived from BCH codes and from negacyclic codes, respectively. Klappenecker et al. [5–7] constructed several families of QCCs derived from Reed–Solomon, Reed–Muller and BCH codes. Additionally, they derived the generalized quantum Singleton bound for quantum convolutional codes. Zhu et al. [170] constructed optimal QCCs derived from classical constacyclic codes.

We will briefly expose the background on QCCs. The readers who are interested to learn more about QCCs will find a detailed description of this class of (quantum) codes in the works [123, 124].

Definition 7.2.1 A quantum convolutional code is defined by means of its stabilizer which is a subgroup of the infinite version of the Pauli group, consisting of tensor products of generalized Pauli matrices acting on a semi-infinite stream of qudits. The stabilizer can be defined by a stabilizer matrix of the form

$$S(D) = (X(D) \mid Z(D)) \in \mathbb{F}_q[D]^{(n-k) \times 2n}$$

satisfying the symplectic orthogonality condition given by

$$X(D)Z(1/D)^t - Z(D)X(1/D)^t = 0.$$

The *constraint length* and the *degree* of a QCC are defined similarly as in the case of convolutional codes.

Definition 7.2.2 Let $\mathcal{Q}$ be a QCC defined by a full rank (as always) stabilizer matrix $S(D)$. The constraint length of $\mathcal{Q}$ is defined to be

$$\gamma_i = \max_{1 \leq j \leq n} \{\max\{\deg X_{ij}(D), \deg Z_{ij}(D)\}\},$$

and the overall constraint length is defined by

$$\gamma = \sum_{i=1}^{n-k} \gamma_i.$$

If γ has the smallest value among all basic generator matrices, then γ is the called *degree* of the code.

The memory of QCCs is defined in the sequence.

Definition 7.2.3 The memory μ of $\mathcal{Q}$ is defined as

$$\mu = \max_{1 \leq i \leq n-k, 1 \leq j \leq n} \{\max\{\deg X_{ij}(D), \deg Z_{ij}(D)\}\},$$

and the free distance is defined analogously as to classical convolutional codes, i.e., the minimum of the weights of nonzero codewords in $\mathcal{Q}$.

The parameters of a quantum convolutional code is denoted by

$$[(n, k, \mu; \gamma, d_f)]_q,$$

where n is the frame size, k is the number of logical qudits per frame, μ is the memory, γ is the degree and d_f is the free distance of the code.

A quantum convolutional code can be also described in terms of a semi-infinite stabilizer matrix S with entries in $\mathbb{F}_q \times \mathbb{F}_q$ in the following way. If $S(D) = \sum_{i=0}^{\mu} G_i D^i$, where each matrix G_i for all $i = 0, \ldots, \mu$, is a matrix of size $(n-k) \times n$, then the semi-infinite matrix is defined as

$$S = \begin{bmatrix} G_0 & G_1 & \ldots & G_\mu & 0 & \ldots & \ldots & \ldots \\ 0 & G_0 & G_1 & \ldots & G_\mu & 0 & \ldots & \ldots \\ 0 & 0 & G_0 & G_1 & \ldots & G_\mu & 0 & \ldots \\ \vdots & \vdots & \vdots & \vdots & \vdots & \vdots & \vdots & \vdots \end{bmatrix}.$$

We will utilize the CSS-like construction (i.e., an analogous version of the CSS code construction to convolutional codes) to generate families of QCCs.

Theorem 7.2.1 ([25, 58]) (CSS-like Construction) *Let C_1 and C_2 be two classical convolutional codes with parameters $(n, k_1)_q$ and $(n, n - k_2)_q$, respectively, such that $C_2^\perp \subset C_1$. The stabilizer matrix is given by*

$$\begin{pmatrix} H_2(D) & | & 0 \\ 0 & | & H_1(D) \end{pmatrix} \in \mathbb{F}_q[D]^{(n-k_1+k_2) \times 2n},$$

where $H_1(D)$ and $H_2(D)$ denote parity check matrices of C_1 and C_2, respectively. Then there exists an $[(n, K = k_1 - k_2, (d_z)_f/(d_x)_f)]_q$ convolutional stabilizer code, where $(d_x)_f = \min\{\text{wt}(C_1\backslash C_2^\perp), \text{wt}(C_2\backslash C_1^\perp)\}$ and $(d_z)_f = \max\{\text{wt}(C_1\backslash C_2^\perp), \text{wt}(C_2\backslash C_1^\perp)\}$.

Remark 7.2.1 To avoid stress of notation, we assume throughout this subsection that if $(d_x)_f > (d_z)_f$, then the values are changed.

7.3 Constructions of Optimal QCCs

In the previous subsection we defined a quantum convolutional code. Here, we construct a family of unit-memory quantum MDS convolutional codes derived from BCH codes. The content of this part can be found in our paper [98].

The first family of QCCs that we construct is a family of MDS convolutional codes with parameters

$$(n, n - 2i, 2; 1, 2i + 3)_q,$$

where $1 \le i \le \frac{q}{2} - 1, q = 2^t, t \ge 3$ and $n = q + 1$. After this, we utilize these convolutional codes to construct our MDS QCCs with parameters

$$[(n, n - 4i, 1; 2, 2i + 3)]_q,$$

where $2 \le i \le \frac{q}{2} - 2, q = 2^t, t \ge 3$ and $n = q^2 + 1$.

It is interesting to observe that the order between the degree and the memory is changed when comparing the parameters of classical and quantum convolutional codes. This is usual in some papers in the literature (see for instance [6]).

The following lemma states that self-orthogonal convolutional codes can by utilized to construct quantum convolutional codes.

Lemma 7.3.1 ([6, Proposition 2]) *Let C be an $(n, (n - k)/2, \gamma; m)_{q^2}$ convolutional code such that $C \subseteq C^{\perp_h}$. Then there exists an $[(n, k, m; \gamma, d_f)]_q$ convolutional stabilizer code, where $d_f = wt(C^{\perp_h} \backslash C)$.*

Let C be an $[(n, k, m; \gamma, d_f)]_q$ QCC. Recall that C is *pure* if does not exist errors of weight less than d_f in the stabilizer of C. The following result is an analogous to the Singleton bound for quantum convolutional codes.

Theorem 7.3.1 ([7]) (Quantum Singleton bound) *The free distance of an $[(n, k, m; \gamma, d_f)]_q$ F_{q^2}-linear pure convolutional stabilizer code is bounded by*

$$d_f \le \frac{n - k}{2} \left(\left\lfloor \frac{2\gamma}{n + k} \right\rfloor + 1 \right) + \gamma + 1.$$

Remark 7.3.1 It is interesting to note that this is one of the few bounds presenting in the literature concerning quantum convolutional codes. Due to this, we recommend for interested readers to investigate asymptotic bounds, Hamming bound and other types of structures concerning convolutional stabilizer codes.

7.3.1 Optimal Convolutional Codes

In this section, we construct a family of (classical) convolutional MDS codes that will be utilized in the construction of quantum MDS convolutional codes exhibited in the next subsection (Sect. 7.3.2).

We begin by recalling the following well-known result.

Lemma 7.3.2 ([114, Theorem 9, Chap. 11]) *Suppose that $q = 2^t$, where $t \geq 2$ is an integer, $n = q + 1$ and consider that $a = \frac{q}{2}$. Then one has*

(i) *With exception of coset $C_0 = \{0\}$, each one of the other q-cosets is of the form $C_{a-i} = \{a - i, a + i + 1\}$, where $0 \leq i \leq a - 1$;*
(ii) *The q-ary cosets $C_{a-i} = \{a - i, a + i + 1\}$, where $0 \leq i \leq a - 1$, are mutually disjoint.*

We are now able to construct optimal convolutional codes.

Theorem 7.3.2 *Assume that $q = 2^t$, where $t \geq 3$ is an integer, $n = q + 1$ and consider that $a = \frac{q}{2}$. Then there exists a classical MDS convolutional code with parameters $(n, n - 2i, 2; 1, 2i + 3)_q$, where $1 \leq i \leq a - 1$.*

Proof We first note that $\gcd(n, q) = 1$ and $ord_n(q) = 2$. The proof consists of two steps. We first construct suitable BCH (block) codes and after constructing convolutional codes derived from them.

Let C_2 be the BCH code of length n over $\mathbb{F}_q$ generated by the product of the minimal polynomials

$$C_2 = \langle g_2(x) \rangle = \langle M^{(a-i)}(x) M^{(a-i+1)}(x) \cdot \cdots \cdot M^{(a-1)}(x) M^{(a)}(x) \rangle.$$

A parity check matrix of C_2 is obtained from the matrix

$$H_{2i+3,a-i} = \begin{bmatrix} 1 & \alpha^{(a-i)} & \alpha^{2(a-i)} & \cdots & \alpha^{(n-1)(a-i)} \\ 1 & \alpha^{(a-i+1)} & \alpha^{2(a-i+1)} & \cdots & \alpha^{(n-1)(a-i+1)} \\ \vdots & \vdots & \vdots & \vdots & \vdots \\ 1 & \alpha^{(a-1)} & \cdots & \cdots & \alpha^{(n-1)(a-1)} \\ 1 & \alpha^{a} & \cdots & \cdots & \alpha^{(n-1)a} \end{bmatrix}$$

by expanding each entry as a column vector (containing 2 rows) with respect to some $\mathbb{F}_q$-basis β of $\mathbb{F}_{q^2}$ and then removing any linearly dependent rows. This new matrix H_{C_2} is a parity check matrix of C_2 and it has $2i + 2$ rows. Since the dimension of C_2 is equal to $n - 2(i + 1)$ (as proved in the paragraph below), so there is no linearly dependent rows in H_{C_2}.

From Lemma 7.3.2, each one of the q-ary cyclotomic cosets C_{a-i}, where $0 \leq i \leq a - 1$ (corresponding to the minimal polynomials $M^{(a-i)}(x)$), has two elements and they are mutually disjoint. Since the degree of the generator polynomial $g_2(x)$ of the code C_2 equals the cardinality of its defining set, then one has $\deg(g_2(x)) =$

$2(i + 1)$, so the dimension k_{C_2} of C_2 equals $k_{C_2} = n - \deg(g_2(x)) = n - 2(i + 1)$. Moreover, the defining set of the code C_2 consists of the sequence $\{a - i, a - i + 1, \ldots, a, a + 1, \ldots, a + i + 1\}$ of $2i + 2$ consecutive integers, so, from the BCH bound, the minimum distance d_{C_2} of C_2 satisfies $d_{C_2} \geq 2i + 3$. Thus, C_2 is a MDS code with parameters $[n, n - 2i - 2, 2i + 3]_q$ and, consequently, its (Euclidean) dual code has dimension $2i + 2$.

Let C_1 be the BCH code of length n over $\mathbb{F}_q$ generated by $g_1(x)$ given by

$$g_1(x) = M^{(a-i+1)}(x) M^{(a-i+2)}(x) \cdots \cdot M^{(a-1)}(x) M^{(a)}(x).$$

We know that C_1 has a parity check matrix derived from the matrix

$$H_{2i+1,a-i+1} = \begin{bmatrix} 1 & \alpha^{(a-i+1)} & \alpha^{2(a-i+1)} & \cdots & \alpha^{(n-1)(a-i+1)} \\ 1 & \alpha^{(a-i+2)} & \alpha^{2(a-i+2)} & \cdots & \alpha^{(n-1)(a-i+2)} \\ \vdots & \vdots & \vdots & \vdots & \vdots \\ 1 & \alpha^{(a-1)} & \cdots & \cdots & \alpha^{(n-1)(a-1)} \\ 1 & \alpha^{a} & \cdots & \cdots & \alpha^{(n-1)a} \end{bmatrix}$$

by expanding each entry as a column vector (containing 2 rows) with respect to β (already done, since $H_{2i+1,a-i+1}$ is a submatrix of $H_{2i+3,a-i}$). After performing the expansion for all entries, such new matrix is denoted by H_{C_1} (H_{C_1} is a submatrix of H_{C_2}). Applying Lemma 7.3.2 and proceeding in the same way as above, it follows that C_1 is an $[n, n - 2i, 2i + 1]_q$ MDS code.

Let us now consider C as the BCH code of length n over $\mathbb{F}_q$ generated by the minimal polynomial $M^{(a-i)}(x)$; C is an $[n, n - 2, d \geq 2]_q$ code. A parity check matrix H_C of C is given by expanding each entry of the matrix

$$H_{2,a-i} = \begin{bmatrix} 1 & \alpha^{(a-i)} & \alpha^{2(a-i)} & \cdots & \alpha^{(n-1)(a-i)} \end{bmatrix}$$

with respect to β (already done, since $H_{2,a-i}$ is a submatrix of $H_{2i+3,a-i}$). Because C has dimension $n - 2$, H_C has rank 2 (H_C is also a submatrix of H_{C_2}).

We here rearrange the rows of H_{C_2} in the form

$$H = \begin{bmatrix} 1 & \alpha^{a} & \cdots & \cdots & \alpha^{(n-1)a} \\ 1 & \alpha^{(a-1)} & \cdots & \cdots & \alpha^{(n-1)(a-1)} \\ \vdots & \vdots & \vdots & \vdots & \vdots \\ 1 & \alpha^{(a-i+1)} & \alpha^{2(a-i+1)} & \cdots & \alpha^{(n-1)(a-i+1)} \\ 1 & \alpha^{(a-i)} & \alpha^{2(a-i)} & \cdots & \alpha^{(n-1)(a-i)} \end{bmatrix},$$

(to simplify the notation we write H in terms of powers of α, although it is clear from the context that this matrix has entries in $\mathbb{F}_q$, which are derived by expanding each entry with respect to β already performed).

We then split H into two disjoint submatrices H_0 and H_1 of the forms

$$H_0 = \begin{bmatrix} 1 & \alpha^a & \cdots & \cdots & \alpha^{(n-1)a} \\ 1 & \alpha^{(a-1)} & & \cdots & \alpha^{(n-1)(a-1)} \\ \vdots & \vdots & \vdots & \vdots & \vdots \\ 1 & \alpha^{(a-i+1)} & \alpha^{2(a-i+1)} & \cdots & \alpha^{(n-1)(a-i+1)} \end{bmatrix}$$

and

$$H_1 = \begin{bmatrix} 1 & \alpha^{(a-i)} & \alpha^{2(a-i)} & \cdots & \alpha^{(n-1)(a-i)} \end{bmatrix},$$

respectively, where H_0 is obtained from the matrix H_{C_1} by rearranging rows and H_1 is derived from H_C also by rearranging rows. Hence, it follows that $\mathrm{rk}(H_0) \geq \mathrm{rk}(H_1)$.

Let V be the convolutional code generated by the reduced basic (according to Theorem 7.1.1 Item (a)) generator matrix

$$G(D) = \tilde{H}_0 + \tilde{H}_1 D,$$

where $\tilde{H}_0 = H_0$ and $\tilde{H}_1$ is obtained from H_1 by adding zero-rows at the bottom such that $\tilde{H}_1$ has the number of rows of H_0 in total. By construction, V is a unit-memory convolutional code of dimension $2i$ and degree $\delta_V = 2$.

Let $V^\perp$ be the Euclidean dual of V. We know that $V^\perp$ has dimension $n - 2i$ and degree 2. By Theorem 7.1.1 Item (c), the free distance of $V^\perp$ is bounded by $\min\{d_0 + d_1, d\} \leq d_f^\perp \leq d$, where d_i is the minimum distance of the code $C_i = \{\mathbf{v} \in \mathbb{F}_q^n \mid \mathbf{v}\tilde{H}_i' = 0\}$. From construction one has $d = 2i + 3$, $d_0 = 2i + 1$ and $d_1 \geq 2$, so $V^\perp$ is an $(n, n - 2i, 2; 1, 2i + 3)_q$ code.

Recall that the generalized (classical) Singleton bound [145] of an $(n, k, \gamma; m, d_f)_q$ convolutional code is given by

$$d_f \leq (n - k)[\lfloor \gamma/k \rfloor + 1] + \gamma + 1.$$

Replacing the values of the parameters of $V^\perp$ in the above inequality it follows that $V^\perp$ is a MDS convolutional code. The proof is complete. $\qquad\square$

Let us now present an illustrative example.

Example 7.3.1 According to Theorem 7.3.2, let $q = 16$, $n = q + 1 = 17$ and $a = 8$. Assume C_2 is an $[17, 11, 7]_{16}$ MDS code generated by $M^{(8)}(x)\, M^{(7)}(x) M^{(6)}(x)$. The cosets of C_2 are $\{8, 9\}$, $\{7, 10\}$ and $\{6, 11\}$. Let C_1 be the (cyclic) MDS code generated by the product of the minimal polynomials $M^{(8)}(x) M^{(7)}(x)$; C_1 has parameters $[17, 13, 5]_{16}$. Finally, suppose C is $[17, 15, d \geq 2]_{16}$ cyclic code generated by $M^{(6)}(x)$. In this case, $i = 2$. We then form the convolutional code V generated by $G(D) = \tilde{H}_0 + \tilde{H}_1 D$, where $\tilde{H}_0 = H_0$ and $\tilde{H}_1$ is obtained from H_1 by adding zero-rows at the bottom such that $\tilde{H}_1$ has the number of rows of H_0 in total. The matrix H_0 is the parity check matrix of C_1 (up to permutation of rows) and H_1 is the parity check matrix of C. We know that V is an $(17, 4, 2; 1, d_f)_{16}$ code; $V^\perp$ is an $(17, 13, 2; 1, d_f^\perp)_{16}$ code, where $\min\{d_0 + d_1, d\} \leq d_f^\perp \leq d$, where $d_0 = 5$, $d_1 \geq 2$

and $d = 7$. Therefore $V^\perp$ has parameters $(17, 13, 2; 1, 7)_{16}$. Applying the generalized Singleton bound one has $7 = 4(\lfloor 2/13 \rfloor + 1) + 2 + 1$, so $V^\perp$ is maximum-distance-separable code.

7.3.2 Construction of Optimal QCCs

Here we apply Theorem 7.3.2 in order to obtain family of optimal QCCs. To do this, let us recall a result shown in [91].

Lemma 7.3.3 *Assume that $q = 2^t$, where t is an integer $t \geq 1$. Let $n = q^2 + 1$ and $a = \frac{q^2}{2}$. If C is the cyclic code whose defining set Z is given by $Z = C_{a-i} \cup \cdots \cup C_a$, where $0 \leq i \leq \frac{q}{2} - 1$, then C is Hermitian dual-containing.*

Proof See [91, Lemma 4.2].   □

In the following theorem, we present a family of quantum convolutional MDS codes.

Theorem 7.3.3 *Assume $q = 2^t$, where $t \geq 3$ is an integer, $n = q^2 + 1$ and consider that $a = \frac{q^2}{2}$. Then there exists a quantum MDS convolutional code with parameters $[(n, n - 4i, 1; 2, 2i + 3)]_q$, where $2 \leq i \leq \frac{q}{2} - 2$.*

Proof We consider the same notation utilized in Theorem 7.3.2. We know that $\gcd(n, q^2) = 1$. From Theorem 7.3.2, there exists a classical convolutional MDS code with parameters $(n, n - 2i, 2; 1, 2i + 3)_{q^2}$, for each $2 \leq i \leq \frac{q}{2} - 2$. This code is the Euclidean dual $V^\perp$ of the convolutional code V whose parameters are given by $(n, 2i, 2; 1, d_f)_{q^2}$. The codes $V^\perp$ and $V^{\perp_h}$ have the same degree as code (see the proof of Theorem 7 in [7]). Additionally, it is straightforward to check that $\text{wt}(V^\perp) = \text{wt}(V^{\perp_h})$, so $V^{\perp_h}$ is an $(n, n - 2i, 2; m^*, 2i + 3)_{q^2}$ code. From Lemma 7.3.3 and from Theorem 7.1.1 Item (b), one has $V \subset V^{\perp_h}$. Applying Lemma 7.3.1, there exists an $[(n, n - 4i, 1; 2, d_f \geq 2i + 3)]_q$ convolutional stabilizer code, for each $2 \leq i \leq \frac{q}{2} - 2$. Replacing the parameters of the previously constructed codes in the quantum generalized Singleton bound (Theorem 7.3.1) one has the equality $2i + 3 = 2i \left(\lfloor \frac{4}{2n-4i} \rfloor + 1 \right) + 2 + 1$. The proof is complete.   □

Example 7.3.2 To illustrate an application Theorem 7.3.3, assume that $q = 8$, $n = 65$ and $i = 2$. Applying Theorem 7.3.3, there exists an $[(65, 57, 1; 2, 7)]_8$ MDS convolutional stabilizer code.

Taking $q = 16$, $n = 257$ and $i = 2, 3, 4, 5$, one has an optimal QCC with parameters $[(257, 249, 1; 2, 7)]_{16}$, $[(257, 245, 1; 2, 9)]_{16}$, $[(257, 241, 1; 2, 11)]_{16}$, $[(257, 237, 1; 2, 13)]_{16}$, respectively, and so on.

7.4 More QCCs from BCH Codes

We describe in this part how to obtain families of unit-memory as well as multi-memory convolutional stabilizer codes with good parameters. The content presented here can be found in the paper [93]. The technique of construction employed in the sequence is similar as the utilized in Sect. 7.3. Our constructions differ from those shown given in [6] at least in two aspects: (1) we construct unit-memory and also multi-memory QCCs whereas in [6] only unit-memory QCCs were constructed; (2) we make use directly of minimal polynomials to define classical BCH codes utilized in our quantum code constructions.

The first construction generates quantum convolutional codes of length $n = q^4 - 1$, ($q \geq 3$ is a prime power), with parameters

- $[(n, n - 4(i - 2) - 2, 1; 2, d_f \geq i + 1)]_q$, $3 \leq i \leq q^2 - 1$;
- $[(n, n - 4i - 2, 1; 2j, d_f \geq i + j + 2)]_q$, $1 \leq i = j$ and $2 \leq i + j \leq q^2 - 2$.

In the second, we show how to obtain quantum convolutional codes of length $n = q^{2m} - 1$, where $q \geq 4$ is a prime power and $m = ord_n(q^2) \geq 3$, derived from the Hermitian construction. Our codes have parameters:

- $[(n, n - 2m(2q^2 - 3) - 2, 1; m, d_f \geq 2q^2 + 2)]_q$;
- $[(n, n - 2mi - 2, 1; mj, d_f \geq i + j + 2)]_q$, where $1 \leq i = j \leq q^2 - 2$;
- $[(n, n - 2m(i - 1) - 2, 1; m, d_f \geq i + 2)]_q$, where $1 \leq i < q^2 - 1$;
- $[(n, n - 2m(q^2 - 2) - 2, 1; m, d_f \geq q^2 + 2)]_q$;
- $[(n, n - 2m(i + q^2 - 2) - 2, 1; m, d_f \geq i + q^2 + 2)]_q$, where $1 \leq i < q^2 - 1$;
- $[(n, n - 2m(i - 2) - 2, 2; 2m, d_f \geq i + 2)]_q$, where $3 \leq i < q^2 - 1$;
- $[(n, n - 2m(i - \mu) - 2, \mu; m\mu, d_f \geq i - \mu + 4)]_q$, where $\mu \geq 3$ and $\mu + 1 \leq i < q^2 - 1$.

Finally, the third construction proposed provides convolutional stabilizer codes of length $n = q^m - 1$, ($q \geq 4$ is a prime power and $m = ord_n(q) \geq 3$), derived from the Euclidean construction, with parameters

- $[(n, n - 2m(c - 1) - 2, 1; m, d_f \geq c + 2)]_q$, where $2 \leq c = i + j \leq q - 2$ and $i, j \geq 1$;
- $[(n, n - 2mi - 2, 1; mj, d_f \geq i + j + 2)]_q$, where $1 \leq i = j \leq q - 2$;
- $[(n, n - 2m(q - 2) - 2, 1; m, d_f \geq q + 2)]_q$;
- $[(n, n - 2m(q - 1), 1; m + 1, d_f \geq q + 3)]_q$;
- $[(n, n - 2m(q - 1) - 2, 1; mj, d_f \geq q + j + 2)]_q$, where $1 \leq j < q - 1$;
- $[(n, n - 2m(2q - 3), 1; m, d_f \geq 2q + 1)]_q$.

As we can see above, as families of unit-memory as well as multi-memory QCCs will be constructed, although we focus the attention on the construction of unit-memory codes.

Let us recall the following two useful lemmas.

Lemma 7.4.1 ([6, Proposition 1]) *Let C be an $(n, (n-k)/2, \gamma; \mu)_q$ convolutional code such that $C \subset C^{\perp}$. Then there exists an $[(n, k, \mu; \gamma, d_f)]_q$ convolutional stabilizer code, where $d_f = wt(C^{\perp}\backslash C)$.*

Lemma 7.4.2 ([6, Proposition 2]) *Let C be an $(n, (n-k)/2, \gamma; \mu)_{q^2}$ convolutional code such that $C \subset C^{\perp_h}$. Then there exists an $[(n, k, \mu; \gamma, d_f)]_q$ convolutional stabilizer code, where $d_f = wt(C^{\perp_h}\backslash C)$.*

7.4.1 Construction I

Here we focus on the construction of convolutional stabilizer codes of length $q^4 - 1$ over $\mathbb{F}_{q^2}$. To proceed further we need some results available in [89].

Lemma 7.4.3 *Let $q \geq 3$ be a prime power and $n = q^4 - 1$. Consider the $q^2 - 1$ q^2-ary cosets modulo n given by*

$$\mathbb{C}_{[q^2+1]},$$
$$\mathbb{C}_{[q^2+2]} = \{q^2 + 2, \ 1 + 2q^2\},$$
$$\vdots$$
$$\mathbb{C}_{[2q^2-1]} = \{2q^2 - 1, \ 1 + (q^2 - 1)q^2\}.$$

Then the following hold:

(a) the q^2-ary coset $\mathbb{C}_{[q^2+1]}$ contains only one element;
(b) each of the other cosets contains two elements;
(c) all these q^2-cosets are mutually disjoints.

Proof The proof can be found in [89, Lemma 3.2]. □

Lemma 7.4.4 *Let $n = q^4 - 1$ where $q \geq 3$ is a prime power. Let C be the cyclic code of length n over $\mathbb{F}_{q^2}$ generated by the product of the minimal polynomials*

$$M^{(q^2+1)}(x)M^{(q^2+2)}(x)\ldots M^{(q^2+j)}(x),$$

$1 \leq j \leq q^2 - 1$. *Then C is Hermitian self-orthogonal.*

Proof See [89, Theorem III.1]. □

At this point we are ready to show Theorem 7.4.1.

Theorem 7.4.1 *Let $n = q^4 - 1$, where $q \geq 3$ is a prime power. Then there exists an $[(n, n - 4(i-2) - 2, 1; 2, d_f \geq i + 1)]_q$ quantum convolutional code, where $3 \leq i \leq q^2 - 1$.*

Proof We know the equalities $\gcd(q, n) = 1$ and $ord_n(q) = 2$ hold. Assume first that C is the BCH code of length $n = q^4 - 1$ over $\mathbb{F}_{q^2}$, generated by the product of the minimal polynomials

$$C = \langle M^{(q^2+1)}(x)M^{(q^2+2)}(x) \cdots \cdots M^{(q^2+i-1)}(x)M^{(q^2+i)}(x) \rangle,$$

where $3 \leq i \leq q^2 - 1$. A parity check matrix of C is obtained from the matrix

$$H_{i+1,q^2+1} = \begin{bmatrix} 1 & \alpha^{(q^2+1)} & \alpha^{2(q^2+1)} & \cdots & \alpha^{(n-1)(q^2+1)} \\ 1 & \alpha^{(q^2+2)} & \alpha^{2(q^2+2)} & \cdots & \alpha^{(n-1)(q^2+2)} \\ \vdots & \vdots & \vdots & \vdots & \vdots \\ 1 & \alpha^{(q^2+i-1)} & \cdots & \cdots & \alpha^{(n-1)(q^2+i-1)} \\ 1 & \alpha^{(q^2+i)} & \cdots & \cdots & \alpha^{(n-1)(q^2+i)} \end{bmatrix}$$

by expanding each entry as a column vector (in this case, containing 2 rows) over some $\mathbb{F}_{q^2}$-basis β of $\mathbb{F}_{q^4}$ and then removing any linearly dependent rows. We denote this new matrix by H. From Lemma 7.4.3, C has parameters $[n, n - 2(i - 1) - 1, d > i + 1]_{q^2}$. Moreover, since C has dimension $n - 2(i - 1) - 1$, H has $2(i - 1) + 1$ linearly independent rows.

We next consider that C_0 is the BCH code of length $n = q^4 - 1$, over $\mathbb{F}_{q^2}$, generated by the product of the minimal polynomials

$$C_0 = \langle M^{(q^2+1)}(x)M^{(q^2+2)}(x) \cdots \cdots M^{(q^2+i-2)}(x)M^{(q^2+i-1)}(x) \rangle.$$

Analogously, C_0 has a parity check matrix derived from the matrix

$$H_{i,q^2+1} = \begin{bmatrix} 1 & \alpha^{(q^2+1)} & \alpha^{2(q^2+1)} & \cdots & \alpha^{(n-1)(q^2+1)} \\ 1 & \alpha^{(q^2+2)} & \alpha^{2(q^2+2)} & \cdots & \alpha^{(n-1)(q^2+2)} \\ \vdots & \vdots & \vdots & \vdots & \vdots \\ 1 & \alpha^{(q^2+i-2)} & \cdots & \cdots & \alpha^{(n-1)(q^2+i-2)} \\ 1 & \alpha^{(q^2+i-1)} & \cdots & \cdots & \alpha^{(n-1)(q^2+i-1)} \end{bmatrix}$$

by expanding each entry as a column vector (containing 2 rows) over some $\mathbb{F}_{q^2}$-basis β of $\mathbb{F}_{q^4}$, then removing any linearly dependent rows. After these operations the new matrix is denoted by H_0. Applying again Lemma 7.4.3, the code C_0 has parameters $[n, n - 2(i - 2) - 1, d_0 \geq i]_{q^2}$. Since C_0 has dimension $n - 2(i - 2) - 1$, H_0 has $2(i - 2) + 1$ linearly independent rows.

Let C_1 be the BCH code of length $n = q^4 - 1$ over $\mathbb{F}_{q^2}$, generated by $M^{(q^2+i)}(x)$; C_1 has parameters $[n, n - 2, d_1 \geq 2]_{q^2}$ and a parity check matrix of C_1 is given by expanding each entry of the matrix

$$H_{2,q^2+i} = \begin{bmatrix} 1 & \alpha^{(q^2+i)} & \alpha^{2(q^2+i)} & \cdots & \alpha^{(n-1)(q^2+i)} \end{bmatrix}$$

with respect to β. The new matrix is denoted by H_1 and since C_1 has dimension $n - 2$, H_1 has 2 linearly independent rows.

We know that rk $H_0 \geq$ rk H_1. Consider the convolutional code V generated by

$$G(D) = \tilde{H}_0 + \tilde{H}_1 D,$$

where $\tilde{H}_0 = H_0$ and $\tilde{H}_1$ is obtained from H_1 by adding zero-rows at the bottom such that $\tilde{H}_1$ has the number of rows of H_0 in total. By construction, V has dimension $2(i - 2) + 1$ and degree $\delta_V = 2$; hence, V is an $(n, 2(i - 2) + 1, 2; 1, d_{f*})_{q^2}$ code. The Euclidean dual $V^\perp$ of V has dimension $n - 2(i - 2) - 1$ and degree 2. Let us now compute the free distance $d_f^\perp$ of $V^\perp$. By Theorem 7.1.1 Item (b), the free distance of $V^\perp$ is bounded by $\min\{d_0 + d_1, d\} \leq d_f^\perp \leq d$, where d_i is the minimum distance of the code $C_i = \{\mathbf{v} \in F_q^n \mid \mathbf{v} H_i^t = 0\}$. From construction one has $d \geq i + 1, d_0 \geq i$ and $d_1 \geq 2$, so $d_f^\perp \geq i + 1$ and $V^\perp$ has parameters $(n, n - 2(i - 2) - 1, 2; \mu, d_f^\perp \geq i + 1)_{q^2}$ for each $3 \leq i \leq q^2 - 1$. The codes $V^\perp$ and $V^{\perp_h}$ have the same degree as code (see the proof of Theorem 7 in [7]). Since wt$(V^\perp) =$ wt$(V^{\perp_h})$, the convolutional code $V^{\perp_h}$ has parameters $(n, n - 2(i - 2) - 1, 2; m^*, d_f^{\perp_h} \geq i + 1)_{q^2}$. From Lemma 7.4.4 and from Theorem 7.1.1 Item (c), we have $V \subset V^{\perp_h}$. Applying Lemma 7.3.1, there exists an $[(n, n - 4(i - 2) - 2, 1; 2, d_f \geq i + 1)]_q$ convolutional stabilizer code, for each $3 \leq i \leq q^2 - 1$. $\square$

The next theorem generates more new QCCs:

Theorem 7.4.2 *Let $n = q^4 - 1$ where $q \geq 3$ is a prime power. Then there exists an $[(n, n - 4i - 2, 1; 2j, d_f \geq i + j + 2)]_q$ QCC, where $1 \leq i = j$ and $2 \leq i + j \leq q^2 - 2$.*

Proof Let C be the BCH code generated by

$$M^{(q^2+1)}(x) M^{(q^2+2)}(x) \cdots \cdots \cdots M^{(q^2+i+j)}(x) M^{(q^2+i+j+1)}(x),$$

where $1 \leq i = j$ and $2 \leq i + j \leq q^2 - 2$. C has a parity check matrix H. Suppose that C_0 is the BCH code generated by

$$M^{(q^2+1)} \cdots M^{(q^2+i)}(x) M^{(q^2+i+1)}(x),$$

where $1 \leq i = j$ and $2 \leq i + j \leq q^2 - 2$; C_0 has parity check matrix H_0. Assume also that C_1 is the BCH code generated by

$$M^{(q^2+i+2)} \cdots M^{(q^2+i+j+1)}(x),$$

where $1 \leq i = j$ and $2 \leq i + j \leq q^2 - 2$; C_1 has parity check matrix H_1. Applying Lemma 7.4.3 one can easily verify that C has parameters $[n, n - 2(i + j) - 1, d \geq$

$i + j + 2]_{q^2}$, C_0 has parameters $[n, n - 2i - 1, d_0 \geq i + 2]_{q^2}$ and C_1 has parameters $[n, n - 2j, d_1 \geq j + 1]_{q^2}$.

The convolutional code V generated by the reduced basic generator matrix $G(D) = \tilde{H}_0 + \tilde{H}_1 D$, is a unit-memory convolutional code of dimension $2i + 1$ and degree $\delta_V = 2j$, so V has parameters $(n, 2i + 1, 2j; 1, d_{f*})_{q^2}$. The convolutional code $V^{\perp_h}$ has parameters $(n, n - 2i - 1, 2j; \mu, d_f^{\perp_h} \geq i + j + 2)_{q^2}$, where the lower bound for the free distance $i + j + 2$ was found by applying Theorem 7.1.1 Item (b). From Lemma 7.4.4 and from Theorem 7.1.1 Item (b), one has $V \subset V^{\perp_h}$. Applying Lemma 7.3.1, there exists an $[(n, n - 4i - 2, 1; 2j, d_f \geq i + j + 2)]_q$ QCC, for each $1 \leq i = j$ and $2 \leq i + j \leq q^2 - 2$. □

Example 7.4.1 In Theorem 7.3.2 consider that $q = 5$ and $i = 8$. Let C be the BCH code of length 624 over $\mathbb{F}_{25}$, generated by

$$C = \langle M^{(26)}(x)M^{(27)}(x) \ldots M^{(32)}(x)M^{(33)}(x)\rangle,$$

C_0 be the BCH code of length 624 over $\mathbb{F}_{25}$, generated by

$$C_0 = \langle M^{(26)}(x)M^{(27)}(x) \ldots M^{(31)}(x)M^{(32)}(x)\rangle,$$

and suppose also that C_1 is the BCH code of length 624 over $\mathbb{F}_{25}$, generated by $M^{(33)}(x)$. Applying Theorem 7.4.1 one has an $[(624, 598, 1; 2, d_f \geq 9)]_5$ quantum convolutional code.

Analogously, in Theorem 7.4.2, let us consider that $q = 5$ and $i = j = 3$. Let C be the BCH code generated by

$$M^{(26)}(x)M^{(27)}(x) \ldots M^{(32)}(x),$$

C_0 be the BCH code generated by

$$M^{(26)}M^{(27)}(x)M^{(28)}(x)M^{(29)}(x).$$

Suppose that C_1 is the BCH code generated by

$$M^{(30)} \ldots M^{(32)}(x).$$

Applying Theorem 7.4.2, we obtain an $[(624, 610, 1; 6, d_f \geq 8)]_5$ QCC.

7.4.2 Construction II

In this subsection, we apply similar technique which was developed in the previous subsection in order to obtain more convolutional stabilizer codes. Lemmas 7.4.5 and 7.4.6 are essentials to our constructions.

Lemma 7.4.5 *Suppose that* $n = q^{2m} - 1$, *where* $q \geq 4$ *is a prime power and* $m = \text{ord}_n(q^2) \geq 3$. *Let* $s = \sum_{i=0}^{m-1} (q^2)^i$. *Then the following hold:*

(a) *the* q^2-*coset* $\mathbb{C}_{[s]}$ *has only one element;*
(b) *the* q^2-*cosets* $\mathbb{C}_{[s+i]}$ *are mutually disjoints, where* $1 \leq i \leq q^2 - 1$;
(c) *the* q^2-*cosets* $\mathbb{C}_{[s-j]}$ *are mutually disjoints, where* $1 \leq j \leq q^2 - 1$;
(d) *the* q^2-*cosets of the forms* $\mathbb{C}_{[s+i]}$ *and* $\mathbb{C}_{[s-j]}$ *are mutually disjoints, where* $1 \leq i, j \leq q^2 - 1$;
(e) *the cosets of the form* $\mathbb{C}_{[s+i]}$, *where* $1 \leq i \leq q^2 - 1$, *contain m elements;*
(f) *the cosets of the form* $\mathbb{C}_{[s-j]}$ *contain m elements, where* $1 \leq j \leq q^2 - 1$.

Proof All these results can be found in [89, Lemmas III.3, III.4, and III.5].   □

Lemma 7.4.6 *Let* $q \geq 4$ *be a prime power and* $n = q^{2m} - 1$. *Assume also that* $\gcd(q^2, n) = 1$ *and* $m = \text{ord}_n(q^2) \geq 3$. *Let* $s = \sum_{i=0}^{m-1} (q^2)^i$. *If C is the cyclic code generated by*

$$M^{(s)}(x)M^{(s+1)}(x)\ldots M^{(s+i)}(x) \cdot M^{(s-1)}(x)\ldots M^{(s-j)}(x),$$

for all $1 \leq i, j \leq q^2 - 1$, *then C is Hermitian self-orthogonal.*

Proof See [89, Lemma III.6].   □

Keeping these results in mind we are able to prove Theorems 7.4.3 and 7.4.4 and their respective corollaries.

Theorem 7.4.3 *Let* $n = q^{2m} - 1$, *where* $q \geq 4$ *is a prime power and* $m = \text{ord}_n(q^2) \geq 3$. *Then there exists an* $[(n, n - 2m(2q^2 - 3) - 2, 1; m, d_f \geq 2q^2 + 2)]_q$ *QCC.*

Proof Clearly one has $\gcd(q, n) = 1$. Consider first that C is the BCH code of length $n = q^{2m} - 1$ over $\mathbb{F}_{q^2}$, generated by the product of the minimal polynomials

$$M^{(s)}(x)M^{(s+1)}(x)\ldots M^{(s+q^2-2)}(x)M^{(s+q^2-1)}(x) \cdot$$
$$\cdot M^{(s-1)}(x)\ldots M^{(s-q^2+1)}(x),$$

where $s = \sum_{i=0}^{m-1} (q^2)^i$. A parity check matrix of C is obtained from the matrix

$$H_{2q^2+2,\,s-q^2+1} = \begin{bmatrix} 1 & \alpha^{(s-q^2+1)} & \alpha^{2(s-q^2+1)} & \cdots & \alpha^{(n-1)(s-q^2+1)} \\ 1 & \alpha^{(s-q^2+2)} & \alpha^{2(s-q^2+2)} & \cdots & \alpha^{(n-1)(s-q^2+2)} \\ \vdots & \vdots & \vdots & \vdots & \vdots \\ 1 & \alpha^{(s-1)} & \cdots & \cdots & \alpha^{(n-1)(s-1)} \\ 1 & \alpha^{(s)} & \cdots & \cdots & \alpha^{(n-1)(s)} \\ 1 & \alpha^{(s+1)} & \cdots & \cdots & \alpha^{(n-1)(s+1)} \\ \vdots & \vdots & \vdots & \vdots & \vdots \\ 1 & \alpha^{(s+q^2-2)} & \cdots & \cdots & \alpha^{(n-1)(s+q^2-2)} \\ 1 & \alpha^{(s+q^2-1)} & \cdots & \cdots & \alpha^{(n-1)(s+q^2-1)} \end{bmatrix}$$

by expanding each entry as a column vector over some $\mathbb{F}_{q^2}$-basis β of $\mathbb{F}_{q^{2m}}$ and then removing any linearly dependent rows. We denote this new matrix by H. From Lemma 7.4.5, C has parameters $[n, n - 2m(q^2 - 1) - 1, d \geq 2q^2 + 2]_{q^2}$. Moreover, since C has dimension $n - 2m(q^2 - 1) - 1$, H has $2m(q^2 - 1) + 1$ linearly independent rows.

Let C_0 be the BCH code generated by

$$M^{(s)}(x)M^{(s+1)}(x)\ldots M^{(s+q^2-2)}(x)\ldots M^{(s-1)}(x)\ldots M^{(s-q^2+1)}(x).$$

Analogously, C_0 has a parity check matrix derived from the matrix

$$H_{2q^2,\,s-q^2+1} = \begin{bmatrix} 1 & \alpha^{(s-q^2+1)} & \alpha^{2(s-q^2+1)} & \cdots & \alpha^{(n-1)(s-q^2+1)} \\ 1 & \alpha^{(s-q^2+2)} & \alpha^{2(s-q^2+2)} & \cdots & \alpha^{(n-1)(s-q^2+2)} \\ \vdots & \vdots & \vdots & \vdots & \vdots \\ 1 & \alpha^{(s-1)} & \cdots & \cdots & \alpha^{(n-1)(s-1)} \\ 1 & \alpha^{(s)} & \cdots & \cdots & \alpha^{(n-1)(s)} \\ 1 & \alpha^{(s+1)} & \cdots & \cdots & \alpha^{(n-1)(s+1)} \\ \vdots & \vdots & \vdots & \vdots & \vdots \\ 1 & \alpha^{(s+q^2-2)} & \cdots & \cdots & \alpha^{(n-1)(s+q^2-2)} \end{bmatrix}$$

by expanding each entry as a column vector over some $\mathbb{F}_{q^2}$-basis β of $\mathbb{F}_{q^{2m}}$ and then removing any linearly dependent rows. After performing these operations the obtained matrix is denoted by H_0. Applying again Lemma 7.4.5, the code C_0 has parameters $[n, n - m(2q^2 - 3) - 1, d_0 \geq 2q^2]_{q^2}$. Since C_0 has dimension $n - m(2q^2 - 3) - 1$, H_0 has $m(2q^2 - 3) + 1$ linearly independent rows.

Let C_1 be the BCH code generated by $M^{(s+q^2-1)}(x)$; C_1 has parameters $[n, n - m, d_1 \geq 2]_{q^2}$. A parity check matrix of C_1 is given by expanding each entry of the matrix

$$H_{2,\,s+q^2-1} = \begin{bmatrix} 1 & \alpha^{(s+q^2-1)} & \cdots\cdots\cdots & \alpha^{(n-1)(s+q^2-1)} \end{bmatrix}$$

with respect to β. The new matrix is denoted by H_1 and since C_1 has dimension $n - m$, H_1 has m linearly independent rows. Let V generated by $G(D) = \tilde{H}_0 + \tilde{H}_1 D$, where $\tilde{H}_0 = H_0$ and $\tilde{H}_1$ is obtained from H_1 by adding zero-rows at the bottom such that $\tilde{H}_1$ has the number of rows of H_0 in total. By construction, V has parameters $(n, m(2q^2 - 3) + 1, m; 1, d_{f^*})_{q^2}$. The Hermitian dual $V^{\perp_h}$ of the convolutional code V has dimension $n - m(2q^2 - 3) - 1$ and degree m.

From construction one has $d \geq 2q^2 + 2$, $d_0 \geq 2q^2$ and $d_1 \geq 2$; so, by Theorem 7.1.1 Item (b), the free distance of $V^{\perp_h}$ satisfies $d_f^\perp \geq 2q^2 + 2$. Thus $V^{\perp_h}$ has parameters $(n, n - m(2q^2 - 3) - 1, m; \mu, d_f^{\perp_h} \geq 2q^2 + 2)_{q^2}$. We know that $V \subset V^{\perp_h}$. Applying Lemma 7.3.1, there exists an $[(n, n - 2m(2q^2 - 3) - 2, 1; m, d_f \geq 2q^2 + 2)]_q$ QCC. The proof is complete. $\square$

Theorem 7.4.4 also generates good QCCs.

Theorem 7.4.4 *Let* $n = q^{2m} - 1$, *where* $q \geq 4$ *is a prime power and* $m = \mathrm{ord}_n$ $(q^2) \geq 3$. *Then there exist quantum convolutional codes with parameters* $[(n, n - 2mi - 2, 1; mj, d_f \geq i + j + 2)]_q$, *for each* $1 \leq i = j \leq q^2 - 2$.

Proof Left to the reader. $\square$

Corollary 7.4.1 *Let* $n = q^{2m} - 1$, *where* $q \geq 4$ *is a prime power and* $m = \mathrm{ord}_n$ $(q^2) \geq 3$. *Then there exist convolutional stabilizer codes with parameters*

(a) $[(n, n - 2m(i - 1) - 2, 1; m, d_f \geq i + 2)]_q$, *for each* $1 \leq i < q^2 - 1$;
(b) $[(n, n - 2m(q^2 - 2) - 2, 1; m, d_f \geq q^2 + 2)]_q$;
(c) $[(n, n - 2m(j + q^2 - 2) - 2, 1; m, d_f \geq j + q^2 + 2)]_q$, *for each* $1 \leq j < q^2 - 1$.

Exercise 7.4.1 Show Theorem 7.4.4 and Corollary 7.4.1.

Until now we only have constructed unit-memory convolutional stabilizer codes. However, the technique utilized here can be also applied to generate multi-memory convolutional codes. These constructions are possible due to Lemma 7.4.5, since such lemma provides the exact parameters of the corresponding classical block codes utilized in the proposed construction. Let us next present the constructions of families of multi-memory QCCs.

Theorem 7.4.5 (memory two QCCs) *Let* $n = q^{2m} - 1$, *where* $q \geq 4$ *is a prime power and* $m = \mathrm{ord}_n(q^2) \geq 3$. *Then there exists an* $[(n, n - 2m(i - 2) - 2, 2; 2m, d_f \geq i + 2)]_q$ *QCC, for each* $3 \leq i < q^2 - 1$.

Proof Let C be the BCH code generated by

$$M^{(s)}(x)M^{(s+1)}(x)\ldots M^{(s+i-2)}(x)M^{(s+i-1)}(x)M^{(s+i)}(x),$$

where $3 \le i < q^2 - 1$. We know from Lemma 7.4.5 that C has parameters $[n, n - mi - 1, d \ge i + 2]_{q^2}$, where $3 \le i < q^2 - 1$. A parity check matrix of C is the matrix H.

Let C_0 be the BCH code generated by

$$M^{(s)}(x)M^{(s+1)}(x)\ldots M^{(s+i-2)}(x),$$

where $3 \le i < q^2 - 1$. From Lemma 7.4.5, C_0 has parameters $[n, n - m(i - 2) - 1, d_0 \ge i]_{q^2}$, $3 \le i < q^2 - 1$. A parity check matrix of C_0 is H_0.

Assume also that C_1 is the BCH code generated by $M^{(s+i-1)}(x)$, and let C_2 be the BCH code generated by $M^{(s+i)}(x)$, where $3 \le i < q^2 - 1$. We know that C_1 and C_2 have parameters $[n, n - m, d_1 \ge 2]_{q^2}$ and $[n, n - m, d_2 \ge 2]_{q^2}$, respectively. A parity check matrices of C_1 and C_2 are, respectively, H_1 and H_2.

The convolutional code V generated by the matrix $G(D) = \tilde{H}_0 + \tilde{H}_1 D + \tilde{H}_2 D^2$ has parameters $(n, m(i - 2) + 1, 2m; 2, d_{f*})_{q^2}$. Note that $\gamma = 2m$ because, from Lemma 7.4.5, since the q^2-ary coset $\mathbb{C}_{[s+i]}$ contains m elements, it follows that H_2 have m linearly independent rows, and each of the first m linearly independent rows of $\tilde{H}_2$ has degree 2.

We know that rk $H_0 \ge$ rk H_1 and rk $H_0 \ge$ rk H_2. The code $V^{\perp_h}$ is an $(n, n - m(i - 2) - 1, 2m; \mu, d_f^{\perp_h} \ge i + 2)_{q^2}$ code, where we compute the free distance $d_f^{\perp_h}$ by applying Theorem 7.1.1 (Item (b)). From Lemma 7.4.6 and by Theorem 7.1.1, Item (b), one has $V \subset V^{\perp_h}$. Applying Lemma 7.3.1, there exists an $[(n, n - 2m(i - 2) - 2, 2; 2m, d_f \ge i + 2)]_q$ QCC, for each $3 \le i < q^2 - 1$. $\square$

Theorem 7.4.5 can be generalized as follows.

Theorem 7.4.6 (multi-memory QCCs) *Let* $n = q^{2m} - 1$, *where* $q \ge 4$ *is a prime power and* $m = \mathrm{ord}_n(q^2) \ge 3$. *Then there exists an*

$$[(n, n - 2m(i - \mu) - 2, \mu; m\mu, d_f \ge i - \mu + 4)]_q$$

QCC, where $\mu \ge 3$ *and* $\mu + 1 \le i < q^2 - 1$.

Proof Let C be the BCH code generated by

$$M^{(s)}(x)M^{(s+1)}(x)\ldots M^{(s+i)}(x),$$

where $\mu \ge 3$ and $\mu + 1 \le i < q^2 - 1$. Assume that C_0 is the BCH code generated by

$$M^{(s)}(x)M^{(s+1)}(x)\ldots M^{(s+i-\mu)}(x),$$

where $\mu + 1 \le i < q^2 - 1$. Assume also that C_j, for $j = 1, \ldots, \mu$, is the BCH code generated by

$$M^{(s+i-\mu+j)}(x),$$

where $\mu + 1 \leq i < q^2 - 1$. Proceeding similarly as in the proof of Theorem 7.4.5, the results follows.   □

Exercise 7.4.2 Complete the proof of Theorem 7.4.6.

Remark 7.4.1 Note that in Theorem 7.4.6, we can also consider that C is generated by

$$M^{(s)}(x)M^{(s+1)}(x)\ldots M^{(s+i)}(x) \cdot M^{(s-1)}(x)\ldots M^{(s-j)}(x),$$

for all $1 \leq i, j \leq q^2 - 1$, because from Lemma 7.4.6, C is Hermitian self-orthogonal. After this we choose suitable range for μ (greater than displayed in Theorem 7.4.6), generating therefore more QCCs. Consequently, also in this case, the proposed construction method holds.

7.4.3 Construction III

In this subsection, we construct QCCs with respect to the Euclidean inner product. Let us recall some results proved in [89].

Lemma 7.4.7 *Suppose that* $n = q^m - 1$, *where* $q \geq 4$ *is a prime power and* $m = \mathrm{ord}_n(q) \geq 3$. *Let* $s = \sum_{i=0}^{m-1} q^i$. *Then the following hold:*

(a) *The* q-coset $\mathbb{C}_{[s]}$ *has only one element;*
(b) *Each one of the* q-ary cosets $\mathbb{C}_{[s+i]}$ *are mutually disjoints, where* $1 \leq i \leq q - 1$;
(c) *Each one of the* q-ary cosets $\mathbb{C}_{[s-j]}$ *are mutually disjoints, where* $1 \leq j \leq q - 1$;
(d) *The* q-cosets of the forms $\mathbb{C}_{[s+i]}$ *and* $\mathbb{C}_{[s-j]}$ *are mutually disjoints, where* $1 \leq i, j \leq q - 1$;
(e) *The cosets of the form* $\mathbb{C}_{[s+i]}$, *where* $1 \leq i \leq q - 1$, *have* m *elements;*
(f) *The cosets of the form* $\mathbb{C}_{[s-j]}$ *have* m *elements, where* $1 \leq j \leq q - 1$.

Proof See [89, Lemmas III.7, III.8, and III.9].   □

Lemma 7.4.8 *Let* $q \geq 4$ *be a prime power and* $n = q^m - 1$. *Let* $m = \mathrm{ord}_n(q) \geq 3$ *and* $s = \sum_{i=0}^{m-1} q^i$. *If* C *is the cyclic code generated by*

$$M^{(s)}(x)M^{(s+1)}(x)\ldots M^{(s+j)}(x) \cdot M^{(s-1)}(x)\ldots M^{(s-j)}(x),$$

where $1 \leq j \leq q - 1$, *then* C *is Euclidean self-orthogonal.*

Proof See [89, Lemma III.10].   □

Applying Lemmas 7.4.7 and 7.4.8 we obtain more QCCs.

Theorem 7.4.7 *Let $q \geq 4$ be a prime power and $n = q^m - 1$. Assume that $m = \mathrm{ord}_n(q) \geq 3$. Then there exists an $[(n, n - 2m(c - 1) - 2, 1; m, d_f \geq c + 2)]_q$ quantum convolutional code, where $2 \leq c = i + j \leq q - 2$ and $i, j \geq 1$.*

Proof Left to the reader. □

Exercise 7.4.3 Prove Theorem 7.4.7.

Theorem 7.4.8 *Let $n = q^m - 1$, where $q \geq 4$ is a prime power and assume that $m = \mathrm{ord}_n(q) \geq 3$. Then there exists an $[(n, n - 2mi - 2, 1; mj, d_f \geq i + j + 2)]_q$ QCC, where $1 \leq i = j \leq q - 2$.*

Proof Left to the reader. □

Theorem 7.4.9 *Suppose that $n = q^m - 1$, where $q \geq 4$ is a prime power and $m = \mathrm{ord}_n(q) \geq 3$. Then there exist quantum convolutional codes with parameters*

(a) $[(n, n - 2m(q - 2) - 2, 1; m, d_f \geq q + 2)]_q$;
(b) $[(n, n - 2m(q - 1), 1; m + 1, d_f \geq q + 3)]_q$;
(c) $[(n, n - 2m(q - 1) - 2, 1; mj, d_f \geq q + j + 2)]_q$, *for each* $1 \leq j < q - 1$;
(d) $[(n, n - 2m(2q - 3), 1; m, d_f \geq 2q + 1)]_q$.

Exercise 7.4.4 Prove Theorems 7.4.8 and 7.4.9.

7.4.4 Code Comparison

In this section, we compare the parameters of the QCCs constructed here with the ones available in [6]. The parameters $[(n, k, \mu; \gamma, d_f)]_q$ denote the parameters of QCCs shown in that paper.

The parameters of our code shown in Table 7.1 are obtained from Construction I (see Theorem 7.4.2), Construction II (see Theorem 7.4.4) and from Construction III (see Theorem 7.4.8), respectively.

The criterion adopted to compare the codes is the usual: if the codes have the same code length and the same lower bound for the free distance, the code with greater dimension is better than the other. For example, our $[(624, 598, 1; 12, d_f \geq 14)]_5$ code is better than the $[(624, 592, 1; \gamma, d_f \geq 14)]_5$ code shown in [6] since these two codes have same code length (624) and same lower bound for the free distance (14), but the new code has greater dimension (598) than the dimension (592) of the $[(624, 592, 1; \gamma, d_f \geq 14)]_5$ code. According to the criterion established, we can see in Table 7.1 that the our codes are better than the ones shown in [6].

Table 7.1 Code comparison

The new codes	Codes shown in [6]
$[(n, n - 4i - 2, 1; 2j, d_f \geq i + j + 2)]_q$	$[(n, k, \mu; \gamma, d_f)]_q$
$[(624, 614, 1; 4, d_f \geq 6)]_5$	$[(624, 612, 1; \gamma, d_f \geq 6)]_5$
$[(624, 610, 1; 6, d_f \geq 8)]_5$	$[(624, 608, 1; \gamma, d_f \geq 8)]_5$
$[(624, 606, 1; 8, d_f \geq 10)]_5$	$[(624, 604, 1; \gamma, d_f \geq 9)]_5$
$[(624, 602, 1; 10, d_f \geq 12)]_5$	$[(624, 596, 1; \gamma, d_f \geq 12)]_5$
$[(624, 598, 1; 12, d_f \geq 14)]_5$	$[(624, 592, 1; \gamma, d_f \geq 14)]_5$
$[(n, n - 2mi - 2, 1; mj, d_f \geq i + j + 2)]_q$	$[(n, k, \mu; \gamma, d_f)]_q$
$[(4095, 4081, 1; 6, d_f \geq 6)]_4$	$[(4095, 4077, 1; \gamma, d_f \geq 6)]_4$
$[(4095, 4075, 1; 9, d_f \geq 8)]_4$	$[(4095, 4071, 1; \gamma, d_f \geq 8)]_4$
$[(4095, 4069, 1; 12, d_f \geq 10)]_4$	$[(4095, 4065, 1; \gamma, d_f \geq 9)]_4$
$[(4095, 4063, 1; 15, d_f \geq 12)]_4$	$[(4095, 4053, 1; \gamma, d_f \geq 12)]_4$
$[(4095, 4057, 1; 18, d_f \geq 14)]_4$	$[(4095, 4047, 1; \gamma, d_f \geq 14)]_4$
$[(4095, 4051, 1; 21, d_f \geq 16)]_4$	$[(4095, 4041, 1; \gamma, d_f \geq 15)]_q$
$[(4095, 4045, 1; 24, d_f \geq 18)]_4$	$[(4095, 4029, 1; \gamma, d_f \geq 18)]_4$
$[(4095, 4039, 1; 27, d_f \geq 20)]_4$	$[(4095, 4023, 1; \gamma, d_f \geq 20)]_4$
$[(n, n - 2mi - 2, 1; mj, d_f \geq i + j + 2)]_q$	$[(n, k, \mu; \gamma, d_f)]_q$
$[(64, 49, 1; 6, d_f \geq 6)]_4$	$[(64, 45, 1; \gamma, d_f \geq 6)]_4$
$[(124, 110, 1; 6, d_f \geq 6)]_5$	$[(124, 106, 1; \gamma, d_f \geq 6)]_5$
$[(124, 104, 1; 9, d_f \geq 8)]_5$	$[(124, 100, 1; \gamma, d_f \geq 8)]_5$
$[(342, 328, 1; 6, d_f \geq 6)]_7$	$[(328, 324, 1; \gamma, d_f \geq 6)]_7$
$[(342, 322, 1; 9, d_f \geq 8)]_7$	$[(328, 318, 1; \gamma, d_f \geq 8)]_7$
$[(342, 316, 1; 12, d_f \geq 10)]_7$	$[(328, 312, 1; \gamma, d_f \geq 9)]_7$
$[(342, 310, 1; 15, d_f \geq 12)]_7$	$[(328, 306, 1; \gamma, d_f \geq 12)]_7$
$[(728, 714, 1; 6, d_f \geq 6)]_9$	$[(728, 710, 1; \gamma, d_f \geq 6)]_9$
$[(728, 708, 1; 9, d_f \geq 8)]_9$	$[(728, 704, 1; \gamma, d_f \geq 8)]_9$
$[(728, 702, 1; 12, d_f \geq 10)]_9$	$[(728, 698, 1; \gamma, d_f \geq 9)]_9$
$[(728, 696, 1; 15, d_f \geq 12)]_9$	$[(728, 686, 1; \gamma, d_f \geq 12)]_9$
$[(728, 690, 1; 18, d_f \geq 14)]_9$	$[(728, 680, 1; \gamma, d_f \geq 14)]_9$
$[(728, 684, 1; 21, d_f \geq 16)]_9$	$[(728, 680, 1; \gamma, d_f \geq 15)]_9$
$[(2400, 2382, 1; 8, d_f \geq 6)]_7$	$[(2400, 2376, 1; \gamma, d_f \geq 6)]_7$
$[(2400, 2374, 1; 12, d_f \geq 8)]_7$	$[(2400, 2368, 1; \gamma, d_f \geq 8)]_7$
$[(2400, 2366, 1; 16, d_f \geq 10)]_7$	$[(2400, 2360, 1; \gamma, d_f \geq 9)]_7$
$[(2400, 2358, 1; 20, d_f \geq 12)]_7$	$[(2400, 2352, 1; \gamma, d_f \geq 12)]_7$

7.5 QCCs from Negacyclic Codes

In this subsection, we utilize the class of negacyclic codes [14, 16, 18, 87] in order to construct classical and quantum MDS convolutional codes. More specifically, we apply the famous method proposed by Piret [130] (generalized to nonbinary alphabets in [6]), which consists in the construction of classical convolutional codes derived from block codes. After this, we utilize our classical convolutional codes constructed here to obtain our optimal (MDS) quantum convolutional codes. The interested reader can consult the paper [98] for more details.

By applying this algebraic technique, it is possible to construct families of classical and quantum convolutional codes and not by only few codes with specific parameters, in contrast with many works where only exhaustively computational search or even specific codes are constructed.

Our classical convolutional MDS codes constructed have parameters

- $(n, n - 2i + 1, 2; 1, 2i + 2)_{q^2}$, where $q \equiv 1 \pmod 4$ is a power of an odd prime, $n = q^2 + 1$ and $2 \leq i \leq n/2 - 1$;
- $(n, n - 2i + 2, 2; 1, 2i + 1)_{q^2}$, where q is a power of an odd prime, $n = (q^2 + 1)/2$ and $2 \leq i \leq (n - 1)/2$;
- $(n, n - 2i + 1, 2; 1, 2i + 2)_{q^2}$, where $q \geq 5$ is a power of an odd prime, $n = (q^2 + 1)/2$ and $2 \leq i \leq (n - 1)/2 - 1$.

The quantum convolutional MDS codes constructed here have parameters

- $[(n, n - 4i + 2, 1; 2, 2i + 2)]_q$, where $q \equiv 1 \pmod 4$ is a power of an odd prime, $n = q^2 + 1$ and $2 \leq i \leq (q - 1)/2$;
- $[(n, n - 4i + 4, 1; 2, 2i + 1)]_q$, where $q \geq 7$ is a power of an odd prime, $n = (q^2 + 1)/2$ and $2 \leq i \leq (q - 1)/2$.

7.5.1 Negacyclic Codes

The class of negacyclic codes [14, 16, 18, 76, 87] has been studied in the literature. This class of codes is a particular class of a more general class of constacyclic codes [18]. In this subsection, we review basic concepts on these codes.

As usual, we assume that the length and the cardinality of q of the field $\mathbb{F}_q$ are relatively prime. Let us consider the quotient ring $R_n = \mathbb{F}_q[x]/(x^n + 1)$. A *negacyclic* code is a principal ideal of R_n under the usual correspondence

$$\mathbf{c} = (c_0, c_1, \dots, c_{n-1}) \longrightarrow c_0 + c_1 x + \dots + c_{n-1}x^{n-1} \quad \mod (x^n + 1).$$

The generator polynomial $g(x)$ of a negacyclic code C divides the polynomial $(x^n + 1)$. The roots of $(x^n + 1)$ are the roots of $(x^{2n} - 1)$ which are not roots of $(x^n - 1)$ in some extension field of $\mathbb{F}_{q^2}$ (since we will work with codes endowed with the Hermitian inner product).

Let us consider that $m = ord_{2n}(q^2)$ and let β be a primitive $2n$th root of unity in $\mathbb{F}_{q^{2m}}$; hence, $\alpha = \beta^2 \in \mathbb{F}_{q^{2m}}$ is a primitive nth root of unity. The roots of $x^n + 1$ are given by $\beta^{2i+1}, 0 \le i \le n - 1$. Put $\mathbb{O}_{2n} = \{1, 3, \ldots, 2n - 1\}$.

Definition 7.5.1 The defining set of a negacyclic code C of length n generated by $g(x)$ is the set

$$\mathcal{Z} = \{i \in \mathbb{O}_{2n} | \beta^i \text{ is root of } g(x)\}.$$

The defining set is a union of q^2-ary cosets defined by

$$C_i = \{i, iq^2, \ldots, iq^{2(m_i - 1)}\},$$

where m_i is the smallest positive integer such that $iq^{2(m_i)} \equiv i \mod 2n$.

Definition 7.5.2 The minimal polynomial $M^{(j)}(x)$, over $\mathbb{F}_{q^2}$, of $\beta^j \in \mathbb{F}_{q^{2m}}$ is given by

$$M^{(j)}(x) = \prod_{j \in C_i} (x - \beta^j).$$

The dimension of C is equal to $n - |\mathcal{Z}|$.

The BCH bound for constacyclic codes (see for example [14, 87]) asserts, that is, C is a q^2-ary negacyclic code of length n with generator polynomial $g(x)$, and if $g(x)$ has the elements $\{\beta^{2i+1} | 0 \le i \le d - 2\}$ as roots, where β is a primitive $2n$th root of unity, then C has minimum distance d_C lower bounded by d, that is, $d_C \ge d$.

A definition of a negacyclic BCH code is given below.

Definition 7.5.3 (*Negacyclic BCH codes*) Let q be a power of an odd prime with $\gcd(n, q) = 1$. Let β be a primitive $2n$th root of unity in $\mathbb{F}_{q^m}$. A negacyclic code C of length n over $\mathbb{F}_q$ is a BCH code with designed distance δ if, for some odd integer $b \ge 1$, we have

$$g(x) = \text{lcm}\{M^{(b)}(x), M^{(b+2)}(x), \ldots, M^{[b+2(\delta-2)]}(x)\},$$

that is, $g(x)$ is the monic polynomial of smallest degree over $\mathbb{F}_q$ having $\alpha^b, \alpha^{b+2}, \ldots,$ $\alpha^{[b+2(\delta-2)]}$ as zeros.

Therefore, $c \in C$ if and only if $c(\alpha^b) = c(\alpha^{(b+2)}) = \cdots = c(\alpha^{[b+2(\delta-2)]}) = 0$. Thus the code has a string of $\delta - 1$ consecutive odd powers of β as zeros.

7.5.2 Convolutional MDS Codes

In this subsection, families of convolutional codes are constructed. First, we recall some results shown in the literature.

Lemma 7.5.1 ([77, Lemma 4.1]) *Let* $n = q^2 + 1$, *where* $q \equiv 1 \mod 4$ *is a power of an odd prime and suppose that* $s = n/2$. *Then the* q^2-*ary cosets modulo* $2n$ *are given by* $C_s = \{s\}$, $C_{3s} = \{3s\}$ *and* $C_{s-2i} = \{s - 2i, s + 2i\}$, *where* $1 \leq i \leq s - 1$.

Lemma 7.5.2 ([77, Lemma 4.4]) *Let* $n = (q^2 + 1)/2$, *where* q *is a power of an odd prime. Then the* q^2-*ary cosets modulo* $2n$ *containing all odd integers from* 1 *to* $2n - 1$ *are given by:* $C_n = \{n\}$, *and* $C_{2i-1} = \{2i - 1, 1 - 2i\}$, *where* $1 \leq i \leq (n-1)/2$.

Remark 7.5.1 Let $\mathcal{B} = \{b_1, \ldots, b_l\}$ be a basis of $\mathbb{F}_{q^l}$ over $\mathbb{F}_q$. If $u = (u_1, \ldots, u_n) \in \mathbb{F}_{q^l}^n$, then we can write the vectors u_i, $1 \leq i \leq n$, as linear combinations of the elements of $\mathcal{B}$, i.e., $u_i = u_{i1}b_1 + \cdots + u_{il}b_l$. Consider that $u^{(j)} = (u_{1j}, \ldots, u_{nj})$ are vectors in $\mathbb{F}_q^n$, where $1 \leq j \leq l$. Then, if $v \in \mathbb{F}_q^n$, it follows that $v \cdot u = 0$ if and only if $v \cdot u^{(j)} = 0$ for all $1 \leq j \leq l$.

In the following theorem we construct a parity check matrix for negacyclic codes:

Theorem 7.5.1 *Assume that* q *is a power of an odd prime,* $\gcd(n, q) = 1$, *and* $m = ord_{2n}(q)$. *Let* β *be a primitive* $2n$th *root of unity in* $\mathbb{F}_{q^m}$. *Let* b *be an odd positive integer with* $1 \leq b \leq 2n - 1$. *Then a parity check matrix for the negacyclic BCH code* C *of length* n *and designed distance* δ, *generated by the polynomial*

$$g(x) = \text{lcm}\{M^{(b)}(x), M^{(b+2)}(x), \ldots, M^{[b+2(\delta-2)]}(x)\}$$

is the matrix

$$H_{\delta,b} =$$
$$= \begin{bmatrix} 1 & \beta^b & \beta^{2b} & \cdots & \beta^{(n-1)b} \\ 1 & \beta^{(b+2)} & \beta^{2(b+2)} & \cdots & \beta^{(n-1)(b+2)} \\ 1 & \beta^{(b+4)} & \beta^{2(b+4)} & \cdots & \beta^{(n-1)(b+4)} \\ \vdots & \vdots & \vdots & \vdots & \vdots \\ 1 & \beta^{[b+2(\delta-2)]} & \beta^{2[b+2(\delta-2)]} & \cdots & \beta^{(n-1)[b+2(\delta-2)]} \end{bmatrix},$$

where each entry is replaced by the corresponding column of m *elements from* $\mathbb{F}_q$ *and then removing any linearly dependent rows.*

Proof Assume that $\mathbf{c} = (c_0, c_1, \ldots, c_{n-1}) \in C$. We then have

$$\mathbf{c}(\beta^b) = \mathbf{c}(\beta^{b+2}) = \mathbf{c}(\beta^{b+4}) = \cdots = \mathbf{c}(\beta^{[b+2(\delta-2)]}) = 0;$$

hence,

$$\begin{bmatrix} 1 & \beta^b & \beta^{2b} & \cdots & \beta^{(n-1)b} \\ 1 & \beta^{(b+2)} & \beta^{2(b+2)} & \cdots & \beta^{(n-1)(b+2)} \\ 1 & \beta^{(b+4)} & \beta^{2(b+4)} & \cdots & \beta^{(n-1)(b+4)} \\ \vdots & \vdots & \vdots & \vdots & \vdots \\ 1 & \beta^{[b+2(\delta-2)]} & \beta^{2[b+2(\delta-2)]} & \cdots & \beta^{(n-1)[b+2(\delta-2)]} \end{bmatrix} \cdot \begin{bmatrix} c_0 \\ c_1 \\ c_2 \\ \vdots \\ c_{n-1} \end{bmatrix} = \begin{bmatrix} 0 \\ 0 \\ \vdots \\ 0 \end{bmatrix}_{(\delta-1,1)}.$$

From Remark 7.5.1 and from the definition of negacyclic BCH codes, the result follows. The proof is complete.   □

We are now ready to show one of the main results of this subsection.

Theorem 7.5.2 Let $n = q^2 + 1$, where $q \equiv 1 \mod 4$ is a power of an odd prime and suppose that $s = n/2$. Then there exists an $(n, n - 2i + 1, 2; 1, 2i + 2)_{q^2}$ MDS convolutional code, where $2 \leq i \leq n/2 - 1$.

Proof First, note that $\gcd(n, q) = 1$ and $ord_{2n}(q^2) = 2$. Let β be a primitive $2n$th root of unity in $\mathbb{F}_{q^{2m}}$. Consider that C_2 is the negacyclic BCH code of length n over $\mathbb{F}_{q^2}$ generated by the product of the minimal polynomials

$$C_2 = \langle g_2(x) \rangle = \langle M^{(s)}(x) M^{(s+2)}(x) \cdots M^{(s+2i)}(x) \rangle,$$

where $2 \leq i \leq s - 1$.

By Theorem 7.5.1, a parity check matrix of C_2 is obtained from the matrix

$$H_2 = \begin{bmatrix} 1 & \beta^s & \beta^{2s} & \cdots & \beta^{(n-1)s} \\ 1 & \beta^{(s+2)} & \beta^{2(s+2)} & \cdots & \beta^{(n-1)(s+2)} \\ 1 & \beta^{(s+4)} & \beta^{2(s+4)} & \cdots & \beta^{(n-1)(s+4)} \\ \vdots & \vdots & \vdots & \vdots & \vdots \\ 1 & \beta^{(s+2i)} & \beta^{2(s+2i)} & \cdots & \beta^{(n-1)(s+2i)} \end{bmatrix}$$

by expanding each entry as a column vector (containing 2 rows) with respect to some $\mathbb{F}_{q^2}$-basis β of $\mathbb{F}_{q^4}$ and then removing one linearly dependent row. From Lemma 7.5.1, this new matrix H_{C_2} has rank $2i + 1$; so, C_2 has dimension $n - 2i - 1$. From the BCH bound for negacyclic codes, it follows that the minimum distance d_2 of C_2 satisfies $d_2 \geq 2i + 2$. Thus, from the (classical) Singleton bound, one concludes that C_2 is a MDS code with parameters $[n, n - 2i - 1, 2i + 2]_{q^2}$ and, consequently, its Hermitian dual code has dimension $2i + 1$.

Next we assume that C_1 is the negacyclic BCH code of length n over $\mathbb{F}_{q^2}$ generated by the product of the minimal polynomials

$$C_1 = \langle g_1(x) \rangle = \langle M^{(s)}(x) M^{(s+2)}(x) \cdots M^{[s+2(i-1)]}(x) \rangle,$$

Similarly, by Theorem 7.5.1, C_1 has a parity check matrix derived from the matrix

$$H_1 = \begin{bmatrix} 1 & \beta^s & \beta^{2s} & \cdots & \beta^{(n-1)s} \\ 1 & \beta^{(s+2)} & \beta^{2(s+2)} & \cdots & \beta^{(n-1)(s+2)} \\ 1 & \beta^{(s+4)} & \beta^{2(s+4)} & \cdots & \beta^{(n-1)(s+4)} \\ \vdots & \vdots & \vdots & \vdots & \vdots \\ 1 & \beta^{[s+2(i-1)]} & \beta^{2[s+2(i-1)]} & \cdots & \beta^{(n-1)[s+2(i-1)]} \end{bmatrix}$$

by expanding each entry as a column vector with respect to some $\mathbb{F}_{q^2}$-basis β of $\mathbb{F}_{q^4}$ (already done, since H_1 is a submatrix of H_2) and then removing one linearly dependent row. From Lemma 7.5.1, this new matrix H_{C_1} has rank $2i - 1$, so C_1 has dimension $n - 2i + 1$. From the BCH bound for negacyclic codes, the minimum distance d_1 of C_1 satisfies $d_1 \geq 2i$, so C_1 is an $[n, n - 2i + 1, 2i]_{q^2}$ MDS code. Hence, its Hermitian dual code has dimension $2i - 1$.

Let C_0 be the negacyclic BCH code of length n over $\mathbb{F}_{q^2}$ generated by the minimal polynomial $M^{(s+2i)}(x)$. Then C_0 is an $[n, n - 2, d_0 \geq 2]_{q^2}$ code. A parity check matrix H_{C_0} of C_0 is given by expanding the entries of the matrix

$$H_0 = \left[\, 1 \;\; \alpha^{(s+2i)} \;\; \alpha^{2(s+2i)} \; \cdots \; \alpha^{(n-1)(s+2i)} \,\right]$$

with respect to β (already done, since H_0 is a submatrix of H_2).

Further, let us construct the convolutional code V generated by the reduced basic (according to Theorem 7.1.1 Item (a)) generator matrix

$$G(D) = \tilde{H}_{C_1} + \tilde{H}_{C_0} D,$$

where $\tilde{H}_{C_1} = H_{C_1}$ and $\tilde{H}_{C_0}$ is obtained from H_{C_0} by adding zero-rows at the bottom such that $\tilde{H}_{C_0}$ has the number of rows of H_{C_1} in total. By construction, V is a unit-memory convolutional code of dimension $2i - 1$ and degree $\delta_V = 2$. We know that the Hermitian dual $V^{\perp_h}$ of V has dimension $n - 2i + 1$ and degree 2. By Theorem 7.1.1 Item (c), the free distance of $V^{\perp_h}$ is bounded by $\min\{d_0 + d_1, d_2\} \leq d_f^{\perp_h} \leq d_2$, where d_i is the minimum distance of the code $C_i = \{\mathbf{v} \in \mathbb{F}_q^n \mid \mathbf{v}\tilde{H}_{C_i}^t = 0\}$. From construction one has $d_2 = 2i + 2$, $d_1 = 2i$ and $d_0 \geq 2$, so $V^{\perp_h}$ has parameters $(n, n - 2i + 1, 2; 1, 2i + 2)_{q^2}$. It is easy to see that $V^{\perp_h}$ is MDS. $\qquad\square$

We can also construct optimal convolutional codes of length $n = (q^2 + 1)/2$, where q is an odd prime power.

Theorem 7.5.3 *Let $n = (q^2 + 1)/2$, where q is a power of an odd prime. Then there exists an $(n, n - 2i + 2, 2; 1, 2i + 1)_{q^2}$ MDS convolutional code, where $2 \leq i \leq (n - 1)/2$.*

Proof Left to exercise. $\qquad\square$

Exercise 7.5.1 Show Theorem 7.5.3.

Theorem 7.5.4 *Let $n = (q^2 + 1)/2$, where $q \geq 5$ is a power of an odd prime. Then there exists a MDS convolutional code with parameters $(n, n - 2i + 1, 2; 1, 2i + 2)_{q^2}$, where $2 \leq i \leq \frac{(n-1)}{2} - 1$.*

Proof Consider that C_2, C_1 and C_0 are negacyclic BCH codes of length n over $\mathbb{F}_{q^2}$ generated, respectively, by $\langle g_2(x) \rangle = \langle M^{(n)}(x)M^{(n+2)}(x) \cdots M^{(n+2i)}(x) \rangle$, $\langle g_1(x) \rangle = \langle M^{(n)}(x)M^{(n+2)}(x) \cdots M^{(n+2i-2)}(x) \rangle$ and $\langle g_0(x) \rangle = \langle M^{(n+2i)}(x) \rangle$. Applying the same procedure given in the proofs of Theorems 7.5.2 and 7.5.3, the result follows. $\qquad\square$

7.5.3   More Optimal QCCs

We propose in this subsection the construction of MDS convolutional stabilizer codes derived from the convolutional codes constructed in Sect. 7.5.2. We begin by recalling some results that we will need in our constructions.

Lemma 7.5.3 ([77]) *Let* $n = q^2 + 1$, *where* $q \equiv 1$ *(mod 4) is a power of an odd prime and suppose that* $s = n/2$. *If* C *is a* q^2-*ary negacyclic code of length* n *with defining set* $\mathcal{Z} = \cup_{i=0}^{\delta} C_{s-2i}$, *where* $0 \leq \delta \leq (q-1)/2$, *then* $C^{\perp_h} \subseteq C$.

Lemma 7.5.4 ([77]) *Let* $n = (q^2 + 1)/2$, *where* q *is a power of an odd prime. If* C *is a* q^2-*ary negacyclic code of length* n *with defining set* $\mathcal{Z} = \cup_{i=1}^{\delta} C_{2i-1}$, *where* $1 \leq \delta \leq (q-1)/2$, *then* $C^{\perp_h} \subseteq C$.

Keeping these results in mind, we are able to show how to obtain families of optimal quantum convolutional codes.

Theorem 7.5.5 *Let* $n = q^2 + 1$, *where* $q \equiv 1$ *(mod 4) is a power of an odd prime and suppose that* $s = n/2$. *Then there exists a quantum MDS convolutional code with parameters* $[(n, n - 4i + 2, 1; 2, 2i + 2)]_q$, *where* $2 \leq i \leq (q-1)/2$.

Proof We consider the same notation utilized in Theorem 7.5.2. From Theorem 7.5.2, there exists a classical convolutional MDS code $V^{\perp_h}$ with parameters $(n, n - 2i + 1, 2; 1, 2i + 2)_{q^2}$, for each $2 \leq i \leq n/2 - 1$. This code is the Hermitian dual of the code V with parameters $(n, 2i - 1, 2; 1, d_f)_{q^2}$. From Lemma 7.5.3 and from Theorem 7.1.1 Item (b), one has $V \subset V^{\perp_h}$. Applying Lemma 7.3.1, there exists an $[(n, n - 4i + 2, 1; 2, d_f \geq 2i + 2)]_q$ convolutional stabilizer code $\mathcal{Q}$, for each $2 \leq i \leq (q-1)/2$. Replacing the parameters of $\mathcal{Q}$ in Theorem 7.3.1, the result follows.   □

Theorem 7.5.6 *Let* $n = (q^2 + 1)/2$, *where* $q \geq 7$ *is a power of an odd prime. Then there exists a quantum MDS convolutional code with parameters* $[(n, n - 4i + 4, 1; 2, 2i + 1)]_q$, *where* $2 \leq i \leq (q-1)/2$.

Proof Left to the reader.   □

Exercise 7.5.2 Prove Theorem 7.5.6.

Table 7.2 given in the sequence, contains the parameters of some optimal quantum convolutional codes constructed here.

7.6   QCCs from AG Codes

This subsection is devoted to the construction of families of quantum convolutional codes derived from algebraic geometry codes. All the results presented here can be found in our paper [127].

Table 7.2 Quantum MDS

Our convolutional stabilizer codes
$[(n, n - 4i + 2, 1; 2, 2i + 2)]_q, q \equiv 1(\bmod 4), n = q^2 + 1, 2 \le i \le (q - 1)/2$
$[(26, 20, 2; 1, 6)]_5$
$[(82, 80, 2; 1, 4)]_9$
$[(82, 76, 2; 1, 6)]_9$
$[(82, 72, 2; 1, 8)]_9$
$[(82, 68, 2; 1, 10)]_9$
$[(170, 168, 2; 1, 4)]_{13}$
$[(170, 164, 2; 1, 6)]_{13}$
$[(170, 160, 2; 1, 8)]_{13}$
$[(170, 156, 2; 1, 10)]_{13}$
$[(170, 152, 2; 1, 12)]_{13}$
$[(170, 148, 2; 1, 14)]_{13}$
$[(n, n - 4i + 4, 2; 1, 2i + 1)]_q, n = (q^2 + 1)/2, 2 \le i \le (q - 1)/2$
$[(25, 21, 2; 1, 5)]_7$
$[(25, 17, 2; 1, 7)]_7$
$[(61, 57, 2; 1, 5)]_{11}$
$[(61, 53, 2; 1, 7)]_{11}$
$[(61, 49, 2; 1, 9)]_{11}$
$[(61, 45, 2; 1, 11)]_{11}$
$[(145, 141, 2; 1, 5)]_{17}$
$[(145, 137, 2; 1, 7)]_{17}$
$[(145, 133, 2; 1, 9)]_{17}$
$[(145, 129, 2; 1, 11)]_{17}$
$[(145, 125, 2; 1, 13)]_{17}$
$[(145, 121, 2; 1, 15)]_{17}$
$[(145, 117, 2; 1, 17)]_{17}$

We first construct (classical) unit-memory convolutional codes (CCs) derived from (block) algebraic geometry (AG) codes, after deriving families of QCCs. After this, we also generate families of CCs by applying the techniques of puncturing, extending, expanding and by the direct product code construction applied to AG codes. Additionally, utilizing such CCs, we also obtain families of QCCs. In particular, in the case of CCs, a family of at least almost near maximum-distance-separable (MDS) (i.e., the difference between the generalized Singleton bound and the real free distance of the convolutional code is equal to two) is obtained. In the quantum case, we construct a family of MDS quantum convolutional codes.

7.6.1 Convolutional AG Code

In this subsection, we present the constructions of convolutional codes from AG codes.

Theorem 7.6.1 *Assume that $F/\mathbb{F}_q$ is an algebraic function field of genus g. Consider the AG code $C_\Omega(D, G)$ shown in Definition 4.5.13. Then there exists an $(n, k - r, r; 1, d_f \geq d)_q$ unit-memory CC, where $r \leq k/2$, $k = l(G) - l(G - D)$ and $d \geq n - \deg(G)$.*

Proof Let $C_\Omega(D, G)$ be the AG code with parity check matrix

$$
H_\Omega = \begin{bmatrix} x_1(P_1) \ x_1(P_2) \ \dots \ x_1(P_n) \\ \vdots \qquad \vdots \qquad \qquad \vdots \\ x_k(P_1) \ x_k(P_2) \ \dots \ x_k(P_n) \end{bmatrix},
$$

where $\{x_1, \dots, x_k\}$ is a basis of $\mathcal{L}(G)$. Let $C_{\mathcal{L}}(D, G)$ be the Euclidean dual of $C_\Omega(D, G)$. The matrix H_Ω is a generator matrix for $C_{\mathcal{L}}(D, G)$. We know that $C_{\mathcal{L}}(D, G)$ is an $[n, k = l(G) - l(G - D), d \geq n - \deg(G)]_q$ AG code, where $n = \deg(D)$. We define a convolutional code with generator matrix $M(D) = H_0 + \tilde{H}_1 D$, where H_0 is the submatrix of H consisting of the $k - r$ first rows and $\tilde{H}_1$ is the matrix consisting of the last r rows of H_Ω by adding zero-rows at the bottom such that the matrix $\tilde{H}_1$ has $k - r$ rows in total. From hypothesis, it follows that rk $H_0 \geq$ rk $\tilde{H}_1$. From Theorem 7.1.1, $M(D)$ is reduced and basic matrix. Again, from Theorem 7.1.1, it follows that the CC generated by $M(D)$ is a unit-memory code with dimension $k - r$, degree r and free distance $d_f \geq d$, and the code follows. □

Applying Proposition 4.5.1 in Theorem 7.6.1, we derive more families of CCs:

Corollary 7.6.1 *Assume all the hypotheses of Theorem 7.6.1 hold. If $2g - 2 < \deg(G) < n$, then there exists an $(n, k - r, r; 1, d_f \geq d)_q$, where $n = \deg(D)$, $k = \deg(G) + 1 - g$, $r \leq k/2$ and $d \geq n - \deg(G)$.*

Corollary 7.6.2 *Assume all the hypotheses of Theorem 7.6.1 hold. If $2g - 2 < \deg(G) < n$, then there exists an $(n, k - 1, 1; 1, d_f \geq d)_q$, where $k = \deg(G) + 1 - g$ and $d \geq n - \deg(G)$.*

Remark 7.6.1 Applying the generalized Singleton bound to the CCs constructed in Corollary 7.6.2, it follows the free distance of such codes satisfy $d_f \leq n - k + 3$, where n and k are the parameters of $C_{\mathcal{L}}(D, G)$. Further, $d_f \geq n - k + 1 - g$, i.e., $n - k + 1 - g \leq d_f \leq n - k + 3$. In particular, if the genus of the function fields is zero, these convolutional codes are almost near MDS or near MDS or MDS, because the Singleton defect is at most two.

Corollary 7.6.3 *Assume that $F = \mathbb{F}_q(z)$ is a rational function field. For $\beta \in \mathbb{F}_q$, let P_β be the zero of $z - \beta$ and P_∞ be the pole of z in $\mathbb{F}_q(z)$. Then there exists an $(q, t, 1; 1, d_f \geq q - t)_q$, where $1 \leq t \leq q - 1$.*

Table 7.3 Some CCs

CCs from Theorems 7.6.2 and 7.6.3
$(32, 15, 1; 1, d_f \geq 15)_{16}$
$(32, 1, 1; 1, d_f \geq 29)_{16}$
$(128, 64, 1; 1, d_f \geq 60)_{64}$
$(176, 64, 1; 1, d_f \geq 105)_{64}$
$(512, 128, 1; 1, d_f \geq 376)_{256}$

Table 7.4 Some CCs

CCs from Corollary 7.6.3
$(q, t, 1; 1, d_f \geq q - t)_q, 1 \leq t \leq q - 1$
$(8, 2, 1; 1, d_f \geq 6)_8$
$(37, 17, 1; 1, d_f \geq 20)_{37}$
$(71, 68, 1; 1, d_f \geq 3)_{71}$
$(128, 64, 1; 1, d_f \geq 64)_{128}$
$(256, 128, 1; 1, d_f \geq 128)_{256}$

Proof Let $C_{\mathcal{L}}(D, G)$ be the AG code with $D = \sum_{\beta \in \mathbb{F}_q} P_\beta$ and $G = t P_\infty$, $1 \leq t \leq q - 1$. We know that $C_{\mathcal{L}}(D, G)$ has parameters $n = q$, $k = t + 1$ and $d \geq n - r$. Applying Theorem 7.1.1 to $C_{\mathcal{L}}^{\perp}(D, G)$, the code follows. $\qquad\square$

In Table 7.4 we exhibit some examples of CCs from Corollary 7.6.3.

Theorem 7.6.2 *Let $q = 2^t$, where $t \geq 1$ is an integer. Then there exists an $(2q^2, r - q/2, 1; 1, d_f \geq 2q^2 - r)_{q^2}$ CC, where $q - 2 < r < 2q^2$.*

Proof It follows from the fact that in the function field $F = \mathbb{F}_q(x, y)$, defined by $y^2 + y = x^{q+1}$, it is possible to construct an $[2q^2, r - q/2 + 1, d \geq 2q^2 - r]_{q^2}$ AG code, where $q - 2 < r < 2q^2$ (see [150]). $\qquad\square$

Theorem 7.6.3 *Let $q = 2^t$, where $t \geq 1$ is an odd integer. Then there exists an $(3q^2 - 2q, r - q + 1, 1; 1, d_f \geq 3q^2 - 2q - r)_{q^2}$ CC, where $2q - 4 < r < 3q^2 - 2q$.*

Proof Let us consider F as the function field over $\mathbb{F}_{q^2}$ defined by $y^q + y = x^3$. We know that the genus of F is $g = q - 1$ and the number of places of degree one is equal to $3q^2 - 2q + 1$ (see [71]). Let $D = P_1 + \cdots + P_n$ be a divisor, where $n = 3q^2 - 2q$ and $G = r P_{3q^2 - 2q + 1} = r P_\infty$, $2q - 4 < r < 3q^2 - 2q$. From Theorem 7.1.1, there exists an $(3q^2 - 2q, r - q + 1, 1; 1, d_f \geq 3q^2 - 2q - r)_{q^2}$ CC, as desired. $\qquad\square$

The following result shows us how to construct CCs by puncturing AG codes.

Theorem 7.6.4 *Assume the same notation of Theorem 7.6.1. Suppose also that the code $C_{\mathcal{L}}(D, G)$ has no minimum weight codeword with a nonzero jth coordinate.*

Then there exists an $(n - 1, k - r, r; 1, d_f)_q$ CC, where $d_f \geq d$, $k = \deg(G) + 1 - g$ and $d \geq n - \deg(G)$.

Proof Let $C_{\mathcal{L}}(D, G)$ be the AG with parameters $[n, k, d]_q$ considered in Theorem 7.6.1, where $D = P_1 + \cdots + P_n$. Assume that $D^* = D - P_j$, where $j \in \{1, 2, \ldots, n\}$. Define the puncture code $C_{\mathcal{L}}(D^*, G)$ derived from $C_{\mathcal{L}}(D, G)$ which is also an AG code (see [126]). Since the supports of D^* and G are disjoint, the definition of the code $C_{\mathcal{L}}(D^*, G)$ makes sense. We know that $C_{\mathcal{L}}(D^*, G)$ is an $[n - 1, k, d]_q$. Proceeding similarly as in the proof of Theorem 7.6.1, one has an $(n - 1, k - r, r; 1, d_f)_q$ CC. The proof is complete.   □

Theorem 7.6.5 *Assume the same notation of Theorem 7.6.1. Then there exists an* $(n + 1, k - r, r; 1, d_f \geq d^e)_q$ *CC, where* $d^e = d$ *or* $d^e = d + 1$, $k = \deg(G) + 1 - g$, $r \leq k/2$ *and* $d \geq n - \deg(G)$.

Proof Let $C_{\mathcal{L}}(D, G)$ be the AG with parameters $[n, k, d]_q$ considered in Theorem 7.6.1. We construct a new code $C_{\mathcal{L}}^e(D, G)$ by extending $C_{\mathcal{L}}(D, G)$; this new code is an $[n, k, d^e]_q$, where d^e is specified in the hypotheses. Proceeding similarly as in the proof of Theorem 7.6.1, we have the code.   □

Applying code expansion we have more families of CCs.

Theorem 7.6.6 *Assume the same notation of Theorem 7.6.1, where* $C_{\mathcal{L}}(D, G)$ *is an AG code over* $\mathbb{F}_{q^r}$. *Then there exists an* $(rn, rk - t, t; 1, d_f \geq d)_q$ *CC, where* $k = \deg(G) + 1 - g$, $t \leq k/2$ *and* $d \geq n - \deg(G)$.

Proof The proof is left to the reader.   □

Theorem 7.6.7 *Assume the same notation of Theorem 7.6.1. Then there exists an* $(n^2, k^2 - r, r; 1, d_f \geq d^2)_q$ *CC, where* $k = \deg(G) + 1 - g$, $r \leq k/2$ *and* $d \geq n - \deg(G)$.

Proof Left to the reader.   □

7.6.2 Quantum Convolutional AG Codes

Fixing the Notation. As usual p denotes a prime number, q is a prime power, $\mathbb{F}_q$ is the finite field with q elements and $F/\mathbb{F}_q$ denotes an algebraic function field over $\mathbb{F}_q$ of genus g.

In this subsection, we construct several families of quantum convolutional codes derived from AG codes. Additionally, the family of QCCs constructed in Theorems 7.6.9 consists of maximum-distance-separable (MDS) codes.

In our construction, we need to utilize the following result from [6].

Lemma 7.6.1 ([6, Props. 1 and 2]) *Let* C *be an* $(n, (n-k)/2, \gamma; m)_{q^2}$ *convolutional code satisfying* $C \subseteq C^{\perp_h}$ *(respectively,* $C \subseteq C^{\perp}$, *where* C *is an* $(n, (n-k)/2, \gamma; m)_q)$. *There there exists an* $[(n, k, m; \gamma, d_f)]_q$ *quantum convolutional code, where* $d_f = \mathrm{wt}(C^{\perp_h} \setminus C)$ *(respectively,* $d_f = \mathrm{wt}(C^{\perp} \setminus C))$. *If does not exist errors of weight less than* d_f *in the stabilizer of* C, *then* C *is said to be pure.*

Theorem 7.6.8 *Let* $C_{\mathcal{L}}(D, G)$ *be an one-point self-orthogonal code (as given in Definition 4.5.11) with parameters* $[n, k, d]_q$ *and generator matrix* M. *If* d_1 *is the minimum distance of the code with parity check matrix* $\tilde{M}_1$ *and* $C_{\mathcal{L}}(D, G^*)$ *is an AG code generated by the divisor* G^*, *where* $\dim(\mathcal{L}(G^*)) = k - l$, *then there exists an* $[(n, n - 2(k-l), 1; l, d_f)]$ *QCC, where* $d_f \geq \min\{d(C_{\mathcal{L}}(D, G^*)) + d_1, d(C_{\Omega}(D, G))\}$. *In particular, if* $2g - 2 \leq \deg(G) \leq n/2 + g$, *then one has* $d_f \geq \deg(G) - (2g - 2)$.

Proof Let us consider $S(Q) = \{0 = \rho_1 < \rho_2 < \cdots\}$ be the Weierstrass semigroup of Q. We construct the *Weierstrass basis* of the Riemann–Roch space $\mathcal{L}(rQ)$, where $r \geq 0$, with dimension $l(rQ) = k$ as follows. The first element of the basis of $\mathcal{L}(rQ)$ is a vector $x_1 \in F/\mathbb{F}_q$ such that $\nu_Q(x_1) = \rho_1 = 0$. The choice for the second vector follows the same idea: we take a vector $x_2 \in F/\mathbb{F}_q$ with $\nu_Q(x_2) = \rho_2$, and so on. From construction, it follows that each of such vectors has different valuation in the place Q; hence, the set $\{x_1, x_2, \ldots, x_k\}$ contains k LI vectors. Additionally, these vectors belong to the space $\mathcal{L}(rQ)$. In other words, $\{x_1, x_2, \ldots, x_k\}$ is a basis of $\mathcal{L}(rQ)$ (called a Weierstrass basis of $\mathcal{L}(rQ)$). Hence, the set $\{x_1, x_2, \ldots, x_{k-1}\}$ is also a basis of a Riemann–Roch space $\mathcal{L}(r^*Q)$, where $\mathcal{L}(r^*Q) \subset \mathcal{L}(rQ)$ and $l(r^*Q) = k - 1$.

Let $C_{\mathcal{L}}(D, G)$ be an $[n, k, d]_q$ one-point self-orthogonal code (as in Definition 4.5.11) with generator matrix M with lines $\{m_1, \ldots, m_k\}$. Let $\{x_1, \ldots, x_k\}$ be the basis of $\mathcal{L}(G)$ constructed above. Applying Theorem 7.1.1, the corresponding convolutional code is also self-orthogonal and it has parameters $(n, k - l, 1; 1, d_f \geq d(C_{\mathcal{L}}(D, G)))_q$. Applying Lemma 7.6.1 for this code and since the set of vectors $\{x_1, \ldots, x_{k-l}\}$ generates another Riemann–Roch space associated with the divisor $G^* = \rho_{k-l}Q$, where ρ_{k-l} is the $(k-l)$th element of $S(Q)$, we obtain an $[(n, n - 2(k-l), 1; l, d_f)]$ QCC, where $d_f \geq \min\{d(C_{\mathcal{L}}(D, G^*)) + d_1, d(C_{\Omega}(D, G))\}$. $\square$

Remark 7.6.2 It is interesting to observe that our construction can be utilized over any curve that satisfies the conditions of Definition 4.5.11.

Remark 7.6.3 Note that, in Theorem 7.6.8, by applying the generalized quantum Singleton bound, it follows that the free distance of the QCCs previously constructed are bounded by $d_f \leq \deg(G) - g + 2$ (where $2g - 2 \leq \deg(G) \leq n/2 + g$). Further, $d_f \geq \deg(G) - 2g + 2$; then the free distance d_f is bounded by $\deg(G) - 2g + 2 \leq d_f \leq \deg(G) - g + 2$. In particular, for function fields $F/\mathbb{F}_q$ with $g = 0$, the new QCCs are MDS. In the other cases, the generalized quantum Singleton defect of the QCCs is at most equal to the genus of the curve utilized to construct the corresponding AG code.

Let us see how to obtain optimal QCCs.

Table 7.5 Some MDS QCCs

QCCs from Theorem 7.6.9
$[(q, q - 2r, 1; 1, d_f = r + 2)]_q, 1 \leq r \leq (q - 2)/2$
$[(4, 2, 1; 1, d_f = 3)]_4$
$[(7, 3, 1; 1, d_f = 4)]_7$
$[(32, 2, 1; 1, d_f = 17)]_{32}$
$[(64, 2, 1; 1, d_f = 33)]_{64}$
$[(128; 126; 1; 1; d_f = 3)]_{128}$
$[(256, 254, 1; 1, d_f = 3)]_{256}$

Theorem 7.6.9 *Assume that all the hypotheses of Theorem 7.6.8 hold and let* $F = \mathbb{F}_q(z)$ *be a rational function field. Then there exists an* $[(q, q - 2r, 1; 1, d_f)]_q$ *maximum-distance-separable QCC, where* $1 \leq r \leq (q - 2)/2$ *and* $d_f = r + 2$.

Proof Proceeding similarly as in Corollary 7.6.3, a rational function field is utilized to construct a CC. For $r \leq (q - 2)/2$ (see [152]), we can get a self-orthogonal AG code, which implies that the CC derived from such code is also self-orthogonal. Applying Theorem 7.6.8, we obtain the desired code. Note that the parameters of this code satisfy the Singleton bound with equality, i.e., it is a MDS code. □

Table 7.5 shows some examples of maximum-distance-separable QCCs obtained from Theorem 7.6.9.

In Theorems 7.6.10 and 7.6.11 we show how to construct more families of QCCs.

Theorem 7.6.10 *Let us consider the* $(2q^2, r - q/2, 1; 1, d_f \geq 2q^2 - r)_{q^2}$ *CC constructed in Theorem 7.6.2, where* $q = 2^t$ *and* $t \geq 3$. *If* $q - 2 \leq r \leq q^2 + q/2 - 1$, *then there exists an* $[(2q^2, 2q^2 + q - 2r, 1; 1, d_f)]_{q^2}$ *QCC, where* $d_f \geq r - q + 2$. *On the other hand, if* $q - 2 \leq r \leq 2q - 2$, *we can also construct an* $[(2q^2, 2q^2 + q - 2r, 1; 1, d_f)]_q$ *QCC, where* $d_f \geq r - q + 2$.

Proof Let $F/\mathbb{F}_q$ be the function field of Theorem 7.6.2. For $q - 2 \leq r \leq q^2 + q/2 - 1$, the AG code constructed over $F/\mathbb{F}_q$ is Euclidean self-orthogonal, and for $r \leq 2q - 2$ is Hermitian self-orthogonal (see [71]). Thus, applying Theorem 7.6.8, one can get an $[(2q^2, 2q^2 + q - 2r, 1; 1, d_f)]_{q^2}$ and an $[(2q^2, 2q^2 + q - 2r, 1; 1, d_f)]_q$ quantum convolutional codes. □

Theorem 7.6.11 *Let* $(3q^2 - 2q, r - q + 1, 1; 1, d_f \geq 3q^2 - 2q - r)_q$ *be the CC constructed from Theorem 7.6.3, where* $q = 2^t$ *and* $t \geq 3$ *is an odd integer. If* $2q - 4 \leq r \leq \frac{3q^2}{2} - 2$, *then there exists an* $[(3q^2 - 2q, 3q^2 - 2(r + 1), 1; 1, d_f)]_{q^2}$ *QCC, where* $d_f \geq r - 2q + 4$. *On the other hand, if* $2q - 4 \leq r \leq 3q - 4$, *one has an* $[(3q^2 - 2q, 3q^2 - 2(r + 1), 1; 1, d_f)]_q$ *QCC.*

Table 7.6 Some QCCs

QCCs from Theorem 7.6.10	Singleton defect
$[(2q^2, 2q^2 + q - 2r, 1; 1, d_f \geq r - q + 2)]_{q^2}$, $q - 2 \leq r \leq 2q - 2$ and $q = 2^t$ with $t \geq 3$	
$[(128, 118, 1; 1, d_f \geq 3)]_8$	4
$[(2048, 2014, 1; 1, d_f \geq 3)]_{32}$	16
$[(8192, 8066, 1; 1, d_f \geq 33)]_{64}$	32

Table 7.7 Some QCCs

QCCs from Theorem 7.6.11	Singleton defect
$[(3q^2 - 2q; 3q^2 - 2(r + 1); 1; 1; d_f \geq r - 2q + 4)]_q$ $2q - 4 \leq r \leq 3q - 4$ and $q = 2^t, t \geq 3$	
$[(176, 160, 1; 1, d_f \geq 3)]_8$	7
$[(3008, 2944, 1; 1, d_f \geq 3)]_{32}$	31

Proof Let $F/\mathbb{F}_q$ be the function field of Theorem 7.6.3. For $r \leq \frac{3q^2}{2} - 2$, the corresponding AG code over $F/\mathbb{F}_q$ is Euclidean self-orthogonal, and for $r \leq 3q - 4$ is Hermitian self-orthogonal (see [71]). Applying Theorem 7.6.8, the result follows. $\square$

7.7 AQCCs

In this section, we present the construction of the first families of asymmetric quantum convolutional codes (AQCCs) [102]. Unfortunately, this is the unique work exhibited in the literature addressing such construction. Therefore, much more investigations must be done: Singleton and Hamming bounds, asymptotic-type bounds such as Gilbert–Varshamov bound among others; constructions of more families of AQCCs with good parameters, weight enumerator and so on. If the reader is interested, we think it is a good area of research to be developed.

Let us return our attention to the code construction to be presented. In order to perform it, we first construct families of good (classical) convolutional codes, i.e., convolutional codes with large dimensions and free distances, after applying a version of the CSS quantum code construction to these (classical) convolutional codes, which we call *CSS-type construction*. As it is desirable, our AQCCs have non-catastrophic (see Definition 7.1.3) generator matrices.

Definition 7.7.1 An asymmetric quantum convolutional code is a quantum code defined over quantum channels where qudit-flip errors and phase-shift errors may

Table 7.8 Some new codes

MDS QCCs	QCCs
$[(q, q - 2m, 1; 1, d_f = m + 2)]_q$	$[(2q^2, 2q^2 + q - 2m, 1; 1, d_f \geq m - q + 2)]_{q^2}$
$1 \leq m \leq (q - 2)/2$	$q - 2 \leq m \leq 2q - 2$ and $q = 2^t$ with $t \geq 3$
$[(4, 2, 1; 1, d_f = 3)]_4$	—
$[(7, 5, 1; 1, d_f = 3)]_7$	—
$[(7, 3, 1; 1, d_f = 4)]_7$	—
$[(8, 6, 1; 1, d_f = 3)]_8$	$[(128, 118, 1; 1, d_f \geq 3)]_8$
$[(8, 2, 1; 1, d_f = 5)]_8$	$[(128, 114, 1; 1, d_f \geq 5)]_8$
$[(32, 30, 1; 1, d_f = 3)]_{32}$	$[(2048, 2014, 1; 1, d_f \geq 3)]_{32}$
$[(32, 2, 1; 1, d_f = 17)]_{32}$	$[(2048, 1986, 1; 1, d_f \geq 17)]_{32}$
$[(64, 62, 1; 1, d_f = 3)]_{64}$	$[(8192, 8126, 1; 1, d_f \geq 3)]_{64}$
$[(64, 2, 1; 1, d_f = 33)]_{64}$	$[(8192, 8066, 1; 1, d_f \geq 33)]_{64}$
$[(128; 126; 1; 1; d_f = 3)]_{128}$	$[(32768; 32638; 1; 1; d_f \geq 3)]_{128}$
$[(128; 2; 1; 1; d_f = 65)]_{128}$	
$[(256, 254, 1; 1, d_f = 3)]_{256}$	
$[(256, 2, 1; 1, d_f = 129)]_{256}$	
$[(512; 510; 1; 1; d_f = 3)]_{512}$	
$[(512; 2; 1; 1; d_f = 257)]_{512}$	
$[(1024, 1022, 1; 1, d_f = 3)]_{1024}$	
$[(1024, 2, 1; 1, d_f = 513)]_{1024}$	

have different probabilities. The parameters of an AQCC are denoted by $[(n, k, \mu^*; \gamma, [d_z]_f/[d_x]_f)]_q$, where n is the frame size, k is the number of logical qudits per frame, m is the memory, $[d_z]_f$ and $[d_x]_f$ are the free distance corresponding to phase-shift and qudit-flip errors, respectively, and γ is the degree of the code.

In this subsection, we present the first families of AQQCs exhibited in the literature [102]. Our asymmetric quantum convolutional codes have parameters given in the following:

- $[(n, 2i - 4, \mu^*; 6, [d_z]_f/[d_x]_f)]_q$,
  where $q = 2^t$, $t \geq 4$, $n = q + 1$, $(d_z)_f \geq n - 2i - 1$ and $(d_x)_f \geq 3$, for all $3 \leq i \leq \frac{q}{2} - 1$;
- $[(n, 2i - 2t - 2, \mu^*; 6, [d_z]_f \geq n - 2i - 1/[d_x]_f \geq 2t + 3)]_q$,
  where $q = 2^l$, $l \geq 4$, $n = q + 1$, t integer with $1 \leq t \leq i - 2$, $3 \leq i \leq \frac{q}{2}$;
- $[(n, 2i - 2t, \mu^*; 4, [d_z]_f/[d_x]_f)]_q$,
  where $(d_z)_f \geq n - 2i - 1$ and $(d_x)_f \geq 2t + 3$, $q = 2^l$, $l \geq 4$, $n = q + 1$, t integer with $1 \leq t \leq i - 1$, $2 \leq i \leq \frac{q}{2}$;
- $[(n, 2i - 2t - 2, \mu^*; 6, [d_z]_f/[d_x]_f)]_q$,
  where $q = p^l$, p is an odd prime, $l \geq 2$, $n = q + 1$, $(d_z)_f \geq n - 2i$ and $(d_x)_f \geq 2t + 2$, for all $1 \leq t \leq i - 2$, where $3 \leq i \leq \frac{n}{2} - 1$;

- $[(n, 2i - 2t, \mu^*; 4, [d_z]_f/[d_x]_f)]_q$,
  where $q = p^l$, p is an odd prime, $l \geq 2$, $n = q + 1$, $(d_z)_f \geq n - 2i$ and $(d_x)_f \geq 2t + 2$, for all $1 \leq t \leq i - 1$, with $2 \leq i \leq \frac{n}{2} - 1$;
- $[(q - 1, i - t - 1, \mu^*; 3, [d_z]_f/[d_x]_f)]_q$,
  where $q \geq 8$ is a prime power $(d_z)_f \geq q - i - 1$ and $(d_x)_f \geq t + 2$, for all $1 \leq t \leq i - 2$, where $3 \leq i \leq q - 3$;
- $[(q - 1, i - t, \mu^*; 2, [d_z]_f/[d_x]_f)]_q$,
  where $(d_z)_f \geq q - i - 1$, $(d_x)_f \geq t + 2$, for all $1 \leq t \leq i - 1$, where $2 \leq i \leq q - 3$;
- $[(n, n - t - k - 2, \mu^*; 3, [d_z]_f/[d_x]_f)]_q$,
  where $(d_z)_f \geq t + 2$ and $(d_x)_f \geq k + 1$, where $q \geq 5$ is a prime power, $k \geq 1$ and n are integers such that $5 \leq n \leq q$ and $k \leq n - 4$ and t is an integer with $1 \leq t \leq n - k - 2$;
- $[(n, n - t - k - 1, \mu^*; 2, [d_z]_f/[d_x]_f)]_q$,
  where $q \geq 5$ is a prime power, $k \geq 1$, $n \geq 5$ are integers such that $n \leq q$, $k \leq n - 4$, $1 \leq t \leq n - k - 1$, $(d_z)_f \geq t + 2$ and $(d_x)_f \geq k + 1$.

For propaedeutic purposes we divide into three different types of constructions. Construction I, shown in Sect. 7.7.1, is a general construction of AQCCs, i.e., we do not consider any specific class of codes. In Construction II (Sect. 7.7.2) we utilize the class of BCH codes to derive families of AQCCs and in Construction III (Sect. 7.7.3) we show how to derive families of AQCCs from Reed–Solomon (RS) and generalized Reed–Solomon (GRS) codes.

7.7.1 Construction I—General Construction

Theorem 7.7.1 establishes the first construction method.

Theorem 7.7.1 (General Construction) *Let q be a prime power and n be a positive integer. Then there exists asymmetric quantum convolutional codes with parameters*

$$[(n, \text{rk } H_0, \mu^*; \gamma_1 + \gamma_2, (d_z)_f/(d_x)_f)]_q,$$

where $\gamma_1 = \mu(\text{rk } H_\mu + \text{rk } H'_\mu) + \sum_{i=1}^{\mu-1}(\mu - i)[\text{rk } H'_{(\mu-i)} - \text{rk } H'_{(\mu-i+1)}]$, $\gamma_2 = \mu(\text{rk }$

$H'_\mu) + \sum_{i=1}^{\mu-1}(\mu - i)[\text{rk } H'_{(\mu-i)} - \text{rk } H'_{(\mu-i+1)}]$, $(d_x)_f \geq (d_1)_f \geq d^\perp$ and $(d_z)_f \geq (d_2)^\perp_f$,

where $(d_1)_f, d^\perp, (d_2)^\perp_f$ *and the matrices* $H_0, H'_0, H_1, H'_1, \ldots, H_\mu, H'_\mu$, *are constructed below.*

Proof Consider a set of $m < n$ linearly independent (LI) vectors $\mathbf{v}_i \in \mathbb{F}_q^n$, $i = 1, 2, \ldots, m$. Let

$$\mathcal{H} = \begin{bmatrix} H_0 \\ H_0' \\ H_1 \\ H_1' \\ \vdots \\ H_\mu \\ H_\mu' \end{bmatrix}$$

be the matrix whose rows are the vectors v_i, $i = 1, 2, \ldots, m$. The matrices H_0, H_0', H_1, $H_1', \ldots, H_\mu, H_\mu'$, are mutually disjoint. The matrices H_i, $i = 0, 1, \ldots, \mu$, are chosen in such a way that

$$\mathrm{rk}\ H_i = \mathrm{rk}\ H_j,$$

for all $i, j = 0, 1, \ldots, \mu$ (the choice of the vectors in each H_i is arbitrary).

In order to compute the degree of the convolutional code constructed in the sequence, we assume that H_0' has full rank and also $\mathrm{rk}\ H_0' \geq \mathrm{rk}\ H_1' \geq \cdots \geq \mathrm{rk}\ H_\mu'$. The matrices $\tilde{H}_i'$ with $1 \leq i \leq \mu$ are obtained from the respective matrices H_i' by adding zero-rows at the bottom such that $\tilde{H}_i'$ has $\mathrm{rk}\ H_0'$ rows in total.

Let $\mathcal{H}$ be a parity check matrix of a linear block code $C = [n, k, d]_q$, where $k = n - m$. Consider the linear block code $C^* = [n, k^*, d^*]_q$ with parity check matrix

$$H^* = \begin{bmatrix} H_0' \\ H_1' \\ \vdots \\ H_\mu' \end{bmatrix}.$$

Next, we construct a matrix $G_1(D)$ as follows:

$$G_1(D) = \begin{bmatrix} H_0 \\ -- \\ H_0' \end{bmatrix} + \begin{bmatrix} H_1 \\ -- \\ \tilde{H}_1' \end{bmatrix} D + \begin{bmatrix} H_2 \\ -- \\ \tilde{H}_2' \end{bmatrix} D^2 + \cdots + \begin{bmatrix} H_\mu \\ -- \\ \tilde{H}_\mu' \end{bmatrix} D^\mu.$$

Further, let us consider the submatrices $G_0(D)$ and $G_2(D)$ of $G_1(D)$, given, respectively, by

$$G_0(D) = H_0 + H_1 D + H_2 D^2 + \cdots + H_\mu D^\mu$$

and

$$G_2(D) = H_0' + \tilde{H}_1' D + \tilde{H}_2' D^2 + \cdots + \tilde{H}_\mu' D^\mu.$$

We know that $G_1(D)$ has full rank $\kappa = \mathrm{rk}\, H_0 + \mathrm{rk}\, H_0'$ and $G_2(D)$ has full rank $k_0' = \mathrm{rk}\, H_0'$. From construction, it follows that $G_1(D)$ and $G_2(D)$ are reduced basic generator matrices of the convolutional codes V_1 and V_2, respectively. Both convolutional codes have memory μ. Applying a similar idea as in the proof of [6, Theorem 3], the free distance $(d_1)_f$ of the convolutional code V_1 and the free distance $(d_1)_f^\perp$ of its Euclidean dual $V_1^\perp$ satisfy

$$\min\{D_0 + D_\mu, d\} \leq (d_1)_f^\perp \leq d$$

and

$$(d_1)_f \geq d^\perp,$$

where D_0 is the minimum distance of the code with parity check matrix

$$\left[\begin{array}{c} H_0 \\ -- \\ H_0' \end{array} \right],$$

and D_μ is the minimum distance of the code with parity check matrix

$$\left[\begin{array}{c} H_\mu \\ -- \\ \tilde{H}_\mu' \end{array} \right].$$

Similarly, the free distance $(d_2)_f$ of V_2 and the free distance $(d_2)_f^\perp$ of $V_2^\perp$ satisfy

$$\min\{d_0' + d_\mu', d^*\} \leq (d_2)_f^\perp \leq d^*$$

and

$$(d_2)_f \geq (d^\perp)^*,$$

where d_0' is the minimum distance of the code C_0' with parity check matrix H_0' and d_μ' is the minimum distance of the code with parity check matrix $\tilde{H}_\mu'$. The degree γ_2 of the code V_2 is equal to

$$\gamma_2 = \mu(\mathrm{rk}\, H_\mu') + \sum_{i=1}^{\mu-1}(\mu - i)[\mathrm{rk}\, H_{(\mu-i)}' - \mathrm{rk}\, H_{(\mu-i+1)}'].$$

The dual code $V_2^\perp$ also has degree γ_2.

Similarly, the degree γ_1 of V_1 is

$$\gamma_1 = \mu(\text{rk } H_\mu + \text{rk } H'_\mu) + \sum_{i=1}^{\mu-1} (\mu - i)[\text{rk } H'_{(\mu-i)} - \text{rk } H'_{(\mu-i+1)}],$$

so $V_1^\perp$ has also degree γ_1.

We know that $V_2 \subset V_1$. The corresponding CSS-type code derived from V_1 and V_2 has frame size n, $k = \text{rk } H_0$ logical qudits per frame, degree $\gamma = \gamma_1 + \gamma_2$, $(d_x)_f \geq (d_1)_f \geq d^\perp$ and $(d_z)_f \geq (d_2)_f^\perp$, where $\min\{d_0' + d_\mu', d^*\} \leq (d_2)_f^\perp \leq d^*$. Thus one can get an $[(n, \text{rk } H_0, \mu^*; \gamma_1 + \gamma_2, (d_z)_f/(d_x)_f)]_q$ AQCC. If $H_1(D)$ is a generator matrix of the code $V_1^\perp$ then a stabilizer matrix of our AQCC is given by

$$\begin{pmatrix} G_2(D) & 0 \\ 0 & H_1(D) \end{pmatrix}.$$

Other variant of this construction can be obtained by considering a CSS-type code derived from the pair of classical convolutional codes $V_1^\perp \subset V_2^\perp$. The proof is complete. $\qquad\square$

7.7.2 Construction II

In this subsection, we utilize Bose–Chaudhuri–Hocquenghem (BCH) codes to construct families of AQCCs. To proceed with Construction II, we fix some notation. As always, q is a prime power and n a positive integer such that $\gcd(q, n) = 1$. Let α be a primitive nth root of unity in some extension field.

Recall that a cyclic code C of length n over $\mathbb{F}_q$ is a BCH code with designed distance δ if, for some integer $b \geq 0$, we have

$$g(x) = \text{l. c. m.}\{M^{(b)}(x), M^{(b+1)}(x), \ldots, M^{(b+\delta-2)}(x)\},$$

i.e., $g(x)$ is the monic polynomial of smallest degree over $\mathbb{F}_q$ having $\alpha^b, \alpha^{b+1}, \ldots, \alpha^{b+\delta-2}$ as zeros.

Therefore, $c \in C$ if and only if $c(\alpha^b) = c(\alpha^{b+1}) = \cdots = c(\alpha^{b+\delta-2}) = 0$. Thus the code has a string of $\delta - 1$ consecutive powers of α as zeros. It is well known that the minimum distance of a BCH code is greater than or equal to its designed distance δ. A parity check matrix for C is given by

$$H_{\delta,b} = \begin{bmatrix} 1 & \alpha^b & \alpha^{2b} & \cdots & \alpha^{(n-1)b} \\ 1 & \alpha^{(b+1)} & \alpha^{2(b+1)} & \cdots & \alpha^{(n-1)(b+1)} \\ \vdots & \vdots & \vdots & \vdots & \vdots \\ 1 & \alpha^{(b+\delta-2)} & \cdots & \cdots & \alpha^{(n-1)(b+\delta-2)} \end{bmatrix},$$

where each entry is replaced by the corresponding column of l elements from $\mathbb{F}_q$, where $l = \operatorname{ord}_n(q)$, and then removing any linearly dependent rows. The rows of the resulting matrix over $\mathbb{F}_q$ are the parity checks satisfied by C.

We need to utilize useful results shown in [98].

Theorem 7.7.2 ([98, Theorem 4.2]) *Assume that $q = 2^t$, where $t \geq 3$ is an integer, $n = q + 1$ and consider that $a = \frac{q}{2}$. Then there exists an $(n, n - 2i, 2; 1, 2i + 3)_q$, classical MDS convolutional code, where $1 \leq i \leq a - 1$.*

Theorem 7.7.3 establishes conditions to construct AQCCs derived from BCH codes.

Theorem 7.7.3 *Let $q = 2^t$, where $t \geq 4$ and consider that $n = q + 1$ and $a = \frac{q}{2}$. Then there exists an $[(n, 2i - 4, \mu^*; 6, [d_z]_f/[d_x]_f)]_q$ AQCC, where $(d_z)_f \geq n - 2i - 1$ and $(d_x)_f \geq 3$, for all $3 \leq i \leq a - 1$.*

Proof Consider the parity check $\mathbb{F}_q$-matrix of the BCH code C given by

$$
\mathcal{H} = \begin{bmatrix}
1 & \alpha^a & \cdots & \cdots & \alpha^{(n-1)a} \\
1 & \alpha^{(a-1)} & \cdots & \cdots & \alpha^{(n-1)(a-1)} \\
\vdots & \vdots & \vdots & \vdots & \vdots \\
1 & \alpha^{(a-i+1)} & \alpha^{2(a-i+1)} & \cdots & \alpha^{(n-1)(a-i+1)} \\
1 & \alpha^{(a-i)} & \alpha^{2(a-i)} & \cdots & \alpha^{(n-1)(a-i)}
\end{bmatrix},
$$

whose entries are expanded with respect to some $\mathbb{F}_q$-basis $\mathcal{B}$ of $\mathbb{F}_{q^2}$, after removing the linearly dependent rows. This BCH code was constructed in the proof of [98, Theorem 4.2] (more precisely, it is the code C_2 constructed there); C is a MDS code with parameters $[n, n - 2i - 2, 2i + 3]_q$. Its (Euclidean) dual code $C^\perp$ is also a MDS code with parameters $[n, 2i + 2, n - 2i - 1]_q$.

We next construct a classical convolutional code V_1 generated by the reduced basic matrices

$$
G_1(D) = \begin{bmatrix}
1 & \alpha^{(a-i+2)} & \alpha^{2(a-i+2)} & \cdots & \alpha^{(n-1)(a-i+2)} \\
- & - & - & - & - \\
1 & \alpha^a & \cdots & \cdots & \alpha^{(n-1)a} \\
1 & \alpha^{(a-1)} & \cdots & \cdots & \alpha^{(n-1)(a-1)} \\
\vdots & \vdots & \vdots & \vdots & \vdots \\
1 & \alpha^{(a-i+3)} & \alpha^{2(a-i+3)} & \cdots & \alpha^{(n-1)(a-i+3)}
\end{bmatrix} +
$$

$$
\begin{bmatrix}
1 & \alpha^{(a-i+1)} & \alpha^{2(a-i+1)} & \cdots & \alpha^{(n-1)(a-i+1)} \\
- & - & - & - & - \\
1 & \alpha^{(a-i)} & \alpha^{2(a-i)} & \cdots & \alpha^{(n-1)(a-i)} \\
0 & 0 & 0 & 0 & 0 \\
\vdots & \vdots & \vdots & \vdots & \vdots \\
0 & 0 & 0 & 0 & 0
\end{bmatrix} D
$$

and

$$G_2(D) = \left[\, 1\ \alpha^{(a-i+2)}\ \alpha^{2(a-i+2)}\ \cdots\ \alpha^{(n-1)(a-i+2)} \,\right] +$$
$$\left[\, 1\ \alpha^{(a-i+1)}\ \alpha^{2(a-i+1)}\ \cdots\ \alpha^{(n-1)(a-i+1)} \,\right] D.$$

The code V_1, generated by $G_1(D)$, is a unit-memory code of dimension $k_1 = 2(i - 1)$ and degree $\gamma_1 = 4$; V_1 is an $(n, 2[i - 1], 4; 1, [d_1]_f \geq n - 2i - 1)_q$ code. Its Euclidean dual code $V_1^{\perp}$ has parameters $(n, n - 2[i - 1], 4; \mu_1^{\perp}, [d_1]_f^{\perp} \geq 2i + 2)_q$. The code V_2, generated by $G_2(D)$, is an $(n, 2, 2; 1, [d_2]_f)_q$ code, so $V_2^{\perp}$ has parameters $(n, n - 2, 2; \mu_2^{\perp}, [d_2]_f^{\perp} \geq 3)_q$. From construction, it follows that $V_2 \subset V_1$, so $V_1^{\perp} \subset V_2^{\perp}$. Consider the stabilizer matrix given by

$$\begin{pmatrix} H_1(D) \mid & 0 \\ 0 & \mid G_2(D) \end{pmatrix},$$

where $H_1(D)$ is a parity check matrix of the code $V_1^{\perp}$. The corresponding CSS-type code has $K = 2i - 4$, $\gamma = 6$, $(d_z)_f \geq n - 2i - 1$ and $(d_x)_f \geq 3$. Thus there exists an $[(n, 2i - 4, \mu^*; 6, [d_z]_f/[d_x]_f)]_q$ AQCC.   □

Remark 7.7.1 It is interesting to note that the idea of construction of the matrix $G_2(D)$ shown in the proof of Theorem 7.7.3 is distinct from that given in Theorem 7.7.1.

Theorem 7.7.4 *Let $q = 2^l$, where $l \geq 4$ and consider that $n = q + 1$ and $a = \frac{q}{2}$. Then there exist AQCCs with parameters*

(a) $[(n, 2i - 2t - 2, \mu^*; 6, [d_z]_f/[d_x]_f)]_q$, *where* $(d_z)_f \geq n - 2i - 1$, $(d_x)_f \geq 2t + 3$, *i and t are positive integers such that* $1 \leq t \leq i - 2$ *and* $3 \leq i \leq a - 1$;
(b) $[(n, 2i - 2t, \mu^*; 4, [d_z]_f/[d_x]_f)]_q$, *where* $(d_z)_f \geq n - 2i - 1$, $(d_x)_f \geq 2t + 3$, *i and t are positive integers such that* $1 \leq t \leq i - 1$ *and* $2 \leq i \leq a - 1$.

Proof We only show Item (a) since Item (b) is similar. The notation and the matrix $\mathcal{H}$ is the same as in the proof of Theorem 7.7.3. We split $\mathcal{H}$ into disjoint submatrices in order to construct a reduced basic generator matrix $G_1(D)$ of V_1 given by

$$G_1(D) = \begin{bmatrix} 1 & \alpha^{[a-(t+1)]} & \alpha^{2[a-(t+1)]} & \cdots & \alpha^{(n-1)[a-(t+1)]} \\ 1 & \alpha^a & \cdots & \cdots & \alpha^{(n-1)a} \\ 1 & \alpha^{(a-1)} & \cdots & \cdots & \alpha^{(n-1)(a-1)} \\ \vdots & \vdots & \vdots & \vdots & \vdots \\ 1 & \alpha^{[a-(t-1)]} & \alpha^{2[a-(t-1)]} & \cdots & \alpha^{(n-1)[a-(t-1)]} \\ - & - & - & - & - \\ 1 & \alpha^{[a-(t+2)]} & \alpha^{2[a-(t+2)]} & \cdots & \alpha^{(n-1)[a-(t+2)]} \\ \vdots & \vdots & \vdots & \vdots & \vdots \\ 1 & \alpha^{[a-(i-2)]} & \alpha^{2[a-(i-2)]} & \cdots & \alpha^{(n-1)[a-(i-2)]} \\ 1 & \alpha^{[a-(i-1)]} & \alpha^{2[a-(i-1)]} & \cdots & \alpha^{(n-1)[a-(i-1)]} \end{bmatrix} +$$

$$
\begin{bmatrix}
1 & \alpha^{(a-i)} & \alpha^{2(a-i)} & \cdots & \alpha^{(n-1)(a-i)} \\
1 & \alpha^{(a-t)} & \alpha^{2(a-t)} & \cdots & \alpha^{(n-1)(a-t)} \\
0 & 0 & 0 & 0 & 0 \\
\vdots & \vdots & \vdots & \vdots & \vdots \\
0 & 0 & 0 & 0 & 0 \\
\hline
0 & 0 & 0 & 0 & 0 \\
\vdots & \vdots & \vdots & \vdots & \vdots \\
0 & 0 & 0 & 0 & 0
\end{bmatrix} D.
$$

Assume that V_2 is the convolutional code generated by

$$
G_2(D) =
\begin{bmatrix}
1 & \alpha^a & \cdots & \cdots & \alpha^{(n-1)a} \\
1 & \alpha^{(a-1)} & \cdots & \cdots & \alpha^{(n-1)(a-1)} \\
\vdots & \vdots & & \vdots & \vdots \\
1 & \alpha^{[a-(t-1)]} & \alpha^{2[a-(t-1)]} & \cdots & \alpha^{(n-1)[a-(t-1)]}
\end{bmatrix} +
$$

$$
\begin{bmatrix}
1 & \alpha^{(a-t)} & \alpha^{2(a-t)} & \cdots & \alpha^{(n-1)(a-t)} \\
0 & 0 & 0 & 0 & 0 \\
\vdots & \vdots & \vdots & \vdots & \vdots \\
0 & 0 & 0 & 0 & 0
\end{bmatrix} D.
$$

It is easy to see that the code V_1 has parameters

$$(n, 2i - 2, 4; 1, [d_1]_f \geq n - 2i - 1)_q$$

and $V_1^{\perp}$ has parameters

$$(n, n - 2i + 2, 4; \mu_1^{\perp}, [d_1]_f^{\perp})_q.$$

The code V_2 is an

$$(n, 2t, 2; 1, [d_2]_f)_q$$

convolutional code; so the dual code $V_2^{\perp}$ has parameters

$$(n, n - 2t, 2; \mu_2^{\perp}, [d_2]_f^{\perp} \geq 2t + 3)_q.$$

Since $V_2 \subset V_1$, it follows that $V_1^{\perp} \subset V_2^{\perp}$. Hence, there exists an

$$[(n, 2i - 2t - 2, \mu^*; 6, [d_z]_f / [d_x]_f)]_q$$

AQCC, where $(d_z)_f \geq n - 2i - 1$ and $(d_x)_f \geq 2t + 3$. This completes the proof. $\square$

Example 7.7.1 We now list the parameters of some AQCCs obtaining by applying Theorem 7.7.4: $[(17, 6, \mu^*; 6, [d_z]_f \geq 6/[d_x]_f \geq 5)]_{16}$,
$[(17, 8, \mu^*; 4, [d_z]_f \geq 6/[d_x]_f \geq 5)]_{16}$, $[(17, 8, \mu^*; 6, [d_z]_f \geq 5/[d_x]_f \geq 4)]_{16}$,
$[(17, 10, \mu^*; 4, [d_z]_f \geq 5/[d_x]_f \geq 4)]_{16}$, $[(33, 24, \mu^*; 6, [d_z]_f \geq 5/[d_x]_f \geq 4)]_{32}$,
$[(33, 26, \mu^*; 4, [d_z]_f \geq 5/[d_x]_f \geq 4)]_{32}$, $[(33, 22, \mu^*; 6, [d_z]_f \geq 6/[d_x]_f \geq 5)]_{32}$,
$[(33, 24, \mu^*; 4, [d_z]_f \geq 6/[d_x]_f \geq 5)]_{32}$, $[(33, 20, \mu^*; 6, [d_z]_f \geq 8/[d_x]_f \geq 5)]_{32}$,
$[(33, 22, \mu^*; 4, [d_z]_f \geq 8/[d_x]_f \geq 5)]_{32}$.

More families of AQCCs can be obtained by applying Theorem 7.7.5.

Theorem 7.7.5 *Assume that $q = p^l$, where p is an odd prime and $l \geq 2$. Consider that $n = q + 1$ and $a = \frac{n}{2}$. Then there exist AQCCs with parameters*

(a) $[(n, 2i - 2t - 2, \mu^*; 6, [d_z]_f/[d_x]_f)]_q$, *where* $(d_z)_f \geq n - 2i$ *and* $(d_x)_f \geq 2t + 2$, *for all* $1 \leq t \leq i - 2$, *where* $3 \leq i \leq a - 1$;
(b) $[(n, 2i - 2t, \mu^*; 4, [d_z]_f/[d_x]_f)]_q$, *where* $(d_z)_f \geq n - 2i$ *and* $(d_x)_f \geq 2t + 2$, *for all* $1 \leq t \leq i - 1$, *where* $2 \leq i \leq a - 1$.

Proof Analogous to that of Theorem 7.7.4.   □

Exercise 7.7.1 Show Theorem 7.7.5.

Remark 7.7.2 We call the attention that the idea of construction of the matrix $G_2(D)$ is different for each of Theorems 7.7.1, 7.7.3, 7.7.4 and 7.7.5.

7.7.3 Construction III

We here are interested in the construction of AQCCs derived from Reed–Solomon (RS) and generalized Reed–Solomon (GRS) codes. We first deal with RS codes.

Recall that a RS code over $\mathbb{F}_q$ is a BCH code, of length $n = q - 1$, with parameters $[n, n - d + 1, d]_q$, where $2 \leq d \leq n$. A parity check matrix of a RS code is given by

$$H_{\delta,b} = \begin{bmatrix} 1 & \alpha^b & \alpha^{2b} & \cdots & \alpha^{(n-1)b} \\ 1 & \alpha^{(b+1)} & \alpha^{2(b+1)} & \cdots & \alpha^{(n-1)(b+1)} \\ \vdots & \vdots & \vdots & \vdots & \vdots \\ 1 & \alpha^{(b+d-2)} & \cdots & \cdots & \alpha^{(n-1)(b+d-2)} \end{bmatrix}.$$

In Theorem 7.7.6 presented in the sequence, we construct AQCCs derived from RS codes:

Theorem 7.7.6 *Assume that $q \geq 8$ is a prime power. Then there exist AQCCs with parameters*

(a) $[(q - 1, i - t - 1, \mu^*; 3, [d_z]_f/[d_x]_f)]_q$, *where* $(d_z)_f \geq q - i - 1$, $(d_x)_f \geq t + 2$, *for all* $1 \leq t \leq i - 2$, *where* $3 \leq i \leq q - 3$;

(b) $[(q-1, i-t, \mu^*; 2, [d_z]_f/[d_x]_f)]_q$, *where* $(d_z)_f \geq q-i-1$, $(d_x)_f \geq t+2$, *for all* $1 \leq t \leq i-1$, *where* $2 \leq i \leq q-3$.

Proof We only show Item (a) (Item (b) is similar). The construction is the same as in the proof of Theorem 7.7.4, although the codes have distinct parameters. More specifically, starting from a parity check matrix $\mathcal{H}$

$$\mathcal{H} = \begin{bmatrix} 1 & \alpha^a & \cdots & \cdots & \alpha^{(n-1)a} \\ 1 & \alpha^{(a-1)} & \cdots & \cdots & \alpha^{(n-1)(a-1)} \\ \vdots & \vdots & \vdots & \vdots & \vdots \\ 1 & \alpha^{(a-i+1)} & \alpha^{2(a-i+1)} & \cdots & \alpha^{(n-1)(a-i+1)} \\ 1 & \alpha^{(a-i)} & \alpha^{2(a-i)} & \cdots & \alpha^{(n-1)(a-i)} \end{bmatrix},$$

of an $[q-1, q-i-2, i+2]_q$ RS code, we construct generator matrices $G_1(D)$ and $G_2(D)$ for codes V_1 and V_2, respectively, as per Theorem 7.7.4. In this context, it is easy to see that V_1 has parameters

$$(q-1, i-1, 2; 1, [d_1]_f \geq q-i-1)_q,$$

and $V_1^\perp$ has parameters

$$(q-1, q-i, 2; \mu_1^\perp, [d_1]_f^\perp)_q.$$

Similarly, it is easy to see that V_2 is an

$$(q-1, t, 1; 1, |d_2]_f)_q$$

code and $V_2^\perp$ has parameters

$$(q-1, q-t-1, 1; \mu_2^\perp, [d_1]_f^\perp \geq t+2)_q.$$

Hence, the corresponding CSS-type code is an

$$[(q-1, i-t-1, \mu^*; 3, [d_z]_f/[d_x]_f)]_q,$$

where $(d_z)_f \geq q-i-1$ and $(d_x)_f \geq t+2$. The proof is complete. $\square$

Example 7.7.2 Applying Theorem 7.7.6, more AQCCs can be constructed:
$[(10, 4, \mu^*; 3, [d_z]_f \geq 4/[d_x]_f \geq 3)]_{11}$, $\quad [(10, 5, \mu^*; 3, [d_z]_f \geq 3/[d_x]_f \geq 3)]_{11}$,
$[(10, 2, \mu^*; 3, [d_z]_f \geq 6/[d_x]_f \geq 3)]_{11}$, $[(10, 1, \mu^*; 3, [d_z]_f \geq 6/[d_x]_f \geq 4)]_{11}$.

In the following, we will construct AQQCs derived from generalized RS codes. Let n be an integer with $1 \leq n \leq q$. Choose an n-tuple $\zeta = (\zeta_0, \ldots, \zeta_{n-1})$ of distinct elements of $\mathbb{F}_q$. Assume that $\mathbf{v} = (v_0, \ldots, v_{n-1})$ is an n-tuple of nonzero (not necessary distinct) elements of the field $\mathbb{F}_q$. For any integer k, where $1 \leq k \leq n$, let us consider the set of polynomials of degree less than k, in $\mathbb{F}_q[x]$, denoted by $\mathcal{P}_k$. Assuming all these facts, we can define a GRS code.

Definition 7.7.2 The GRS code is defined as

$$\mathrm{GRS}_k(\zeta, \mathbf{v}) = \{(v_0 f(\zeta_0), v_1 f(\zeta_1), \ldots, v_{n-1} f(\zeta_{n-1})) | f \in \mathcal{P}_k\}.$$

The generalized RS code $\mathrm{GRS}_k(\zeta, \mathbf{v})$ is a MDS code with parameters $[n, k, n - k + 1]_q$. It is not difficult to see that the Euclidean dual $\mathrm{GRS}_k^{\perp}(\zeta, \mathbf{v})$ of the GRS code $\mathrm{GRS}_k(\zeta, \mathbf{v})$ is also a GRS code and $\mathrm{GRS}_k^{\perp}(\zeta, \mathbf{w}) = \mathrm{GRS}_{n-k}(\zeta, \mathbf{v})$ for some n-tuple $\mathbf{w} = (w_0, \ldots, w_{n-1})$ of nonzero elements of $\mathbb{F}_q$.

Exercise 7.7.2 Show that $\mathrm{GRS}_k(\zeta, \mathbf{v})$ is a maximum-distance-separable code.

A generator matrix of $\mathrm{GRS}_k(\zeta, \mathbf{v})$ is given as

$$G = \begin{bmatrix} v_0 & v_1 & \cdots & v_{n-1} \\ v_0\zeta_0 & v_1\zeta_1 & \cdots & v_{n-1}\zeta_{n-1} \\ v_0\zeta_0^2 & v_1\zeta_1^2 & \cdots & v_{n-1}\zeta_{n-1}^2 \\ \vdots & \vdots & \vdots & \vdots \\ v_0\zeta_0^{k-1} & v_1\zeta_1^{k-1} & \cdots & v_{n-1}\zeta_{n-1}^{k-1} \end{bmatrix}.$$

A parity check matrix for this code is the matrix

$$H = \begin{bmatrix} w_0 & w_1 & \cdots & w_{n-1} \\ w_0\zeta_0 & w_1\zeta_1 & \cdots & w_{n-1}\zeta_{n-1} \\ w_0\zeta_0^2 & w_1\zeta_1^2 & \cdots & w_{n-1}\zeta_{n-1}^2 \\ \vdots & \vdots & \vdots & \vdots \\ w_0\zeta_0^{n-k-1} & w_1\zeta_1^{n-k-1} & \cdots & w_{n-1}\zeta_{n-1}^{n-k-1} \end{bmatrix}.$$

Exercise 7.7.3 Show that G and H are, in fact, a generator and a parity check matrix for $\mathrm{GRS}_k(\zeta, \mathbf{v})$, respectively.

In the next result, we construct new AQCCs derived from GRS codes.

Theorem 7.7.7 *Let $q \geq 5$ be a prime power. Assume that $k \geq 1$ and $n \geq 5$ are integers such that $n \leq q$ and $k \leq n - 4$. Choose an n-tuple $\zeta = (\zeta_0, \ldots, \zeta_{n-1})$ of distinct elements of $\mathbb{F}_q$ and an n-tuple $\mathbf{v} = (v_0, \ldots, v_{n-1})$ of nonzero elements of $\mathbb{F}_q$. Then there exists an $[(n, n - t - k - 2, \mu^*; 3, [d_z]_f/[d_x]_f)]_q$ AQCC, where $(d_z)_f \geq t + 2$ and $(d_x)_f \geq k + 1$, $1 \leq t \leq n - k - 2$.*

Proof Let

$$
H = \begin{bmatrix}
w_0 & w_1 & \cdots & w_{n-1} \\
w_0\zeta_0 & w_1\zeta_1 & \cdots & w_{n-1}\zeta_{n-1} \\
w_0\zeta_0^2 & w_1\zeta_1^2 & \cdots & w_{n-1}\zeta_{n-1}^2 \\
\vdots & \vdots & \vdots & \vdots \\
w_0\zeta_0^{n-k-1} & w_1\zeta_1^{n-k-1} & \cdots & w_{n-1}\zeta_{n-1}^{n-k-1}
\end{bmatrix}
$$

be a parity check matrix of an $\mathrm{GRS}_k(\zeta, \mathbf{v})$ code. We split H to form polynomial matrices $G_1(D)$ and $G_2(D)$ of codes V_1 and V_2, respectively, as follows:

$$
G_1(D) = \left[\begin{array}{cccc}
w_0\zeta_0^{n-k-3} & w_1\zeta_1^{n-k-3} & \cdots & w_{n-1}\zeta_{n-1}^{n-k-3} \\
w_0 & w_1 & \cdots & w_{n-1} \\
w_0\zeta_0 & w_1\zeta_1 & \cdots & w_{n-1}\zeta_{n-1} \\
\vdots & \vdots & \vdots & \vdots \\
w_0\zeta_0^{t-1} & w_1\zeta_1^{t-1} & \cdots & w_{n-1}\zeta_{n-1}^{t-1} \\
\hline
w_0\zeta_0^{t+1} & w_1\zeta_1^{t+1} & \cdots & w_{n-1}\zeta_{n-1}^{t+1} \\
\vdots & \vdots & \vdots & \vdots \\
w_0\zeta_0^{n-k-2} & w_1\zeta_1^{n-k-2} & \cdots & w_{n-1}\zeta_{n-1}^{n-k-1}
\end{array}\right] +
$$

$$
\left[\begin{array}{cccc}
w_0\zeta_0^{n-k-1} & w_1\zeta_1^{n-k-1} & \cdots & w_{n-1}\zeta_{n-1}^{n-k-1} \\
w_0\zeta_0^t & w_1\zeta_1^t & \cdots & w_{n-1}\zeta_{n-1}^t \\
0 & 0 & 0 & 0 \\
\vdots & \vdots & \vdots & \vdots \\
0 & 0 & 0 & 0 \\
\hline
0 & 0 & 0 & 0 \\
\vdots & \vdots & \vdots & \vdots \\
0 & 0 & 0 & 0
\end{array}\right] D
$$

and

$$
G_2(D) = \begin{bmatrix}
w_0 & w_1 & \cdots & w_{n-1} \\
w_0\zeta_0 & w_1\zeta_1 & \cdots & w_{n-1}\zeta_{n-1} \\
w_0\zeta_0^2 & w_1\zeta_1^2 & \cdots & w_{n-1}\zeta_{n-1}^2 \\
\vdots & \vdots & \vdots & \vdots \\
w_0\zeta_0^{t-1} & w_1\zeta_1^{t-1} & \cdots & w_{n-1}\zeta_{n-1}^{t-1}
\end{bmatrix} +
$$

$$
\begin{bmatrix}
w_0\zeta_0^t & w_1\zeta_1^t & \cdots & w_{n-1}\zeta_{n-1}^t \\
0 & 0 & 0 & 0 \\
\vdots & \vdots & \vdots & \vdots \\
0 & 0 & 0 & 0
\end{bmatrix} D,
$$

where $\mathbf{w} = (w_0, \ldots, w_{n-1})$ is a vector such that $GRS_k^{\perp}(\zeta, \mathbf{w}) = GRS_{n-k}(\zeta, \mathbf{v})$. The code V_1 has parameters $(n, n-k-2, 2; 1, [d_1]_f \geq k+1)_q$ and $V_1^{\perp}$ has parameters $(n, k+2, 2; \mu_1^{\perp}, [d_1]_f^{\perp})_q$. Similarly, V_2 is an $(n, t, 1; 1, [d_2]_f)_q$ code and $V_2^{\perp}$ is an $(n, n-t, 1; \mu_2^{\perp}, [d_1]_f^{\perp} \geq t+2)_q$ code. Then there exists an $[(n, n-t-k-2, \mu^*; 3, [d_z]_f/[d_x]_f)]_q$ code, where $(d_z)_f \geq t+2$ and $(d_x)_f \geq k+1$. $\qquad \Box$

Example 7.7.3 From Theorem 7.7.7, we can construct AQCCs with parameters $[(5, 1, \mu^*; 3, [d_z]_f \geq 3/[d_x]_f \geq 2)]_5$, $[(7, 1, \mu^*; 3, [d_z]_f \geq 4/[d_x]_f \geq 3)]_7$, $[(8, 1, \mu^*; 3, [d_z]_f \geq 5/[d_x]_f \geq 3)]_8$, $[(17, 7, \mu^*; 3, [d_z]_f \geq 7/[d_x]_f \geq 4)]_{17}$, $[(17, 7, \mu^*; 3, [d_z]_f \geq 6/[d_x]_f \geq 5)]_{17}$, $[(17, 6, \mu^*; 3, [d_z]_f \geq 7/[d_x]_f \geq 5)]_{17}$, $[(17, 4, \mu^*; 3, [d_z]_f \geq 9/[d_x]_f \geq 5)]_{17}$ and so on.

Theorem 7.7.8 *Let $q \geq 5$ be a prime power. Assume that $k \geq 1$ and $n \geq 5$ are integers such that $n \leq q$ and $k \leq n-4$. Choose an n-tuple $\zeta = (\zeta_0, \ldots, \zeta_{n-1})$ of distinct elements of $\mathbb{F}_q$ and an n-tuple $\mathbf{v} = (v_0, \ldots, v_{n-1})$ of nonzero elements of $\mathbb{F}_q$. Then an $[(n, n-t-k-1, \mu^*; 2, [d_z]_f/[d_x]_f)]_q$ AQCC, where $(d_z)_f \geq t+2$, $(d_x)_f \geq k+1$ and $1 \leq t \leq n-k-1$ can be constructed.*

Proof Similar to that of Theorem 7.7.7. $\qquad \Box$

Example 7.7.4 From Theorem 7.7.8, we obtain AQCCs with parameters $[(5, 1, \mu^*; 2, [d_z]_f \geq 4/[d_x]_f \geq 2)]_5$, $[(7, 2, \mu^*; 2, [d_z]_f \geq 4/[d_x]_f \geq 3)]_7$, $[(7, 2, \mu^*; 2, [d_z]_f \geq 5/[d_x]_f \geq 2)]_7$, $[(7, 1, \mu^*; 2, [d_z]_f \geq 5/[d_x]_f \geq 3)]_7$.

7.7.4 Discussion

Let C be an $[(n, k, \mu; \gamma, d_f)]_q$ quantum convolutional code. Recall that C is said to be *pure* if does not exist errors of weight less than d_f in the stabilizer of C. Let us recall the quantum generalized Singleton bound (GQSB) for quantum convolutional codes.

Theorem 7.7.9 ([7]) *The free distance of an $[(n, k, \mu; \gamma, d_f)]_q$ $\mathbb{F}_{q^2}$-linear pure convolutional stabilizer code is bounded by $d_f \leq \frac{n-k}{2}\left[\left\lfloor \frac{2\gamma}{n+k} \right\rfloor + 1\right] + \gamma + 1$.*

The parameters of our AQCCs are given by $[(n, k, \mu; \gamma, [d_z]_f/[d_x]_f)]_q$, where $(d_x)_f = \min\{\text{wt}(C_1 \backslash C_2^{\perp}), \text{wt}(C_2 \backslash C_1^{\perp})\}$ and $(d_z)_f = \max\{\text{wt}(C_1 \backslash C_2^{\perp}), \text{wt}(C_2 \backslash C_1^{\perp})\}$. In this context, if one puts the constraint of pure codes, the free distance $(d_x)_f$ with respect to qudit-flip errors satisfies the GQSB. However, much research remains to be done in the area of AQCCs. In fact, there is no bound for the respective free distances nor relationships among the parameters of AQCCs. Other impossibility of our comparison is the fact that our codes have parameters quite distinct of the QCCs available in literature. This occurs since our work is the first one with respect to AQCCs. This area of research needs much investigation, since it was introduced recently (see [123]). Additionally, even in the case of constructions of good QCCs, only few works are displayed in literature [7, 41, 98, 99].

Appendix
Review of Basic Algebra

Here, we present some basic definitions of algebra, in order to make this book self-consistent. The first definition is that of group.

Definition A.1 Let G be a set and $* : G \times G \longrightarrow G$ a binary operation on G. We say that $(G, *)$ is a group if the following conditions hold:

(G1)   There exists an element $e \in G$, called identity, such that, for all $x \in G$, $e * x = x * e = x$.

(G2)   The operation $*$ is associative: for all $x, y, z \in G$, $x * (y * z) = (x * y) * z$.

(G3)   For all $x \in G$, there exists $x' \in G$ such that $x * x' = x' * x = e$. The element x' is an inverse of x.

If, in addition, the operation satisfies the commutativity, that is,

(G4)   For all $x, y \in G$, $x * y = y * x$,

then the group is said to be abelian (this name was coined due to Niels Henrik Abel).

Groups satisfy several properties.

Proposition A.1 *Let $(G, *)$ be a group. Then the following hold:*

(I)   *The identity is unique, i.e., there exists a unique element $e \in G$ such that, for all $x \in G$, $x * e = e * x = x$.*

(II)   *For every $x \in G$, its inverse is unique. More precisely, for every $x \in G$, there exists a unique $x^{-1} \in G$ such that $x * x^{-1} = x^{-1} * x = e$.*

(III) *The cancelation laws hold: assume that $x, a, b \in G$; if either $x * a = x * b$ or $a * x = b * x$, then it follows that $a = b$.*

(IV)   *For all $x \in G$, $(x^{-1})^{-1} = x$.*

(V)   *For all $x, y \in G$, $(x * y)^{-1} = y^{-1} * x^{-1}$.*

(VI)   *Let $a, b \in G$. Then the equation $a * x = b$ has a unique solution in G, namely, $x_0 = a^{-1} * b$.*

Proof The proofs are routine and can be found in textbooks of algebra; see, for instance, [137]. $\qquad\square$

G. G. La Guardia, *Quantum Error Correction*, Quantum Science and Technology,
https://doi.org/10.1007/978-3-030-48551-1

Definition A.2 Let $(G, *)$ be a group and $H \subseteq G$. We say that H is a subgroup of G if:

(SG1) $e \in H$, where e is the identity of G.
(SG2)   If $a, b \in H$, then $a * b \in H$.
(SG3)   If $a \in H$, then $a^{-1} \in H$.

A characterization of subgroup is presented in the sequence.

Proposition A.2 *Let $(G, *)$ be a group. A subset $H \subseteq G$ is a subgroup of G if and only if H is nonempty and for all $x, y \in H$, it follows that $x * y^{-1} \in H$.*

Proof See Proposition 2.30 in [137].   □

An interesting and much utilized class of subgroups is the class of normal subgroups.

Definition A.3 Let $(G, *)$ be a group and H be a subgroup of G. We say that H is a normal subgroup of G if, for all $h \in H$ and $g \in G$, imply $g * h * g^{-1} \in H$.

An important feature of normal subgroups is that such groups allow us to define quotient groups.

Definition A.4 Assume that $(G, *)$ is a group and H is a subgroup of G. For all $g \in G$, we define the left coset of g by H as the set $g * H = \{g * h | h \in H\}$.

Proposition A.3 *Let $(G, *)$ be a group, $H \subseteq G$ subgroup of G and $a, b \in G$. Then the following hold:*

(i) $a * H = b * H$ *if and only if* $b^{-1} * a \in H$. *In particular, $a * H = H$ if and only if $a \in H$.*
(ii) *If* $(a * H) \cap (b * H) \neq \emptyset$, *then* $a * H = b * H$.

Proof The proofs follow because the relation on G, defined by $a \sim b \iff b^{-1} * a \in H$, is an equivalence relation whose equivalence classes are the left cosets.   □

If X, Y are nonempty subsets of G, we define the set

$$X * Y = \{x * y | x \in X \text{ and } y \in Y\}.$$

Theorem A.1 *Let $(G, *)$ be a group and $H \subseteq G$ be a normal subgroup of G. Let G/H be the family of all the left cosets of H. Then, for all $a, b \in G$, one has $(a * H)(b * H) = (a * b)H$, and G/H is a group under this operation.*

Proof See Theorem 2.67 in [137].   □

The group G/H is said to be the quotient group G mod H.
In the sequence we define the concept of *ring*.

Definition A.5 Let R be a set and $+ : R \times R \longrightarrow R$ and $\cdot : R \times R \longrightarrow R$ be two binary operations on R. We say that $(R, +, \cdot)$ is a ring if:

(R1) $(R, +)$ is an abelian group. The identity element of $(R, +)$ is called zero and denoted by 0_R or 0 (if there is no possibility of confusion).

(R2)   For all $x, y, z \in R$, $x \cdot (y \cdot z) = (x \cdot y) \cdot z$ (associativity of $\cdot$).

(R3)   For all $x, y, z \in R$, $x \cdot (y + z) = (x \cdot y) + (x \cdot z)$ and $(x + y) \cdot z = (x \cdot z) + (y \cdot z)$ (distributivity).

If $\cdot$ is commutative, that is,

(R4) for all $x, y \in G$, $x \cdot y = y \cdot x$,

then the ring is called commutative.

If there exists an element $1 \in R$ such that

(R5) for every $x \in R$, $1 \cdot x = x \cdot 1 = x$,

we say that $(R, +, \cdot)$ is a ring with unit (and 1 is the unit of R).

If $(R, +, \cdot)$ satisfy the conditions (R1) to (R5), then we say that $(R, +, \cdot)$ is a commutative ring with unit.

Remark A.1 The element $1 \in R$ can be also called *identity* or *one* of the ring.

The following properties are satisfied by rings.

Proposition A.4 *Let $(R, +, \cdot)$ be a commutative ring with unit. Then the following are true:*

(I)   *For all $x \in R$, $0 \cdot x = x \cdot 0 = 0$.*

(II)   *The unit $1 \in R$ is unique.*

(III) *If $1 = 0$, then R consists of the single element 0.*

(IV) *If $x, y \in R$, then $(-x) \cdot y = x \cdot (-y) = -(x \cdot y)$ and $(-x) \cdot (-y) = x \cdot y$.*

(V)   *If $n \in \mathbb{N}$, and $n1 = 0$, then $nx = 0$ for all $x \in R$, where $n1 = \underbrace{1 + 1 + \ldots + 1}_{n \text{ times}}$*

and $\mathbb{N}$ is the set of nonnegative integers.

(VI) *If $x, y \in R$ and $n \in \mathbb{N}$, then one has*

$$(x + y)^n = \sum_{i=0}^{n} \binom{n}{i} x^i \cdot y^{n-i}$$

(Binomial theorem).

Proof See the proof of Proposition 3.2 in [137]. $\qquad\qquad\qquad\qquad\qquad$ $\square$

Definition A.6 Let $(R, +, \cdot)$ be a commutative ring with unit 1, where $1 \neq 0$. We say that $(R, +, \cdot)$ is a integral domain if for all $x, y \in R$ such that $x \cdot y = 0$ it follows that $x = 0$ or $y = 0$.

Remark A.2 Note that we assume in Definition A.6 that $1 \neq 0$, but some authors do not put this constraint (we adopt the notation of [137]).

Another characterization of integral domains is given in the next proposition.

Proposition A.5 *A nonzero commutative ring* $(R, +, \cdot)$ *with unit 1, where* $1 \neq 0$, *is a integral domain if and only if for all* $x, y, z \in R$, *if* $x \cdot y = x \cdot z$, *with* $x \neq 0$, *implies that* $y = z$ *(cancelation law for multiplication).*

Proof See Proposition 3.5 in [137]. □

In the following we define the concept of field.

Definition A.7 Let $(F, +, \cdot)$ be a commutative ring with unit 1, where $1 \neq 0$. We say that $(F, +, \cdot)$ is a field if every nonzero element $x \in F$ has a multiplicative inverse, that is, for all $x \in F$, $x \neq 0$, there exists $x^{-1} \in F$ such that $x \cdot x^{-1} = x^{-1} \cdot x = 1$, where 1 is the unit of the field.

Definition A.8 Let $(R, +, \cdot)$ be a commutative ring. An ideal in R is a nonempty subset $I \subseteq R$ such that:

(I1) For all $x, y \in I$, it follows that $x - y \in I$.
(I2) For all $x \in I$ and for all $a \in R$, one has $a * x \in I$.

In words, an ideal I is a subgroup of the additive (abelian) group $(R, +)$, closed under the multiplication of elements of R by the elements of I.

Ideals have suitable properties, as shown in the following proposition.

Proposition A.6 *Let* I *be an ideal in a commutative ring* $(R, +, \cdot)$. *Then the following hold:*

(i) $0 \in I$.
(ii) *If* $a \in I$, *then* $-a \in I$.
(iii) *If* $a, b \in I$, *then* $a + b \in I$.
(iv) *If the ring has identity* 1 *and if an invertible element of* R *belongs to* I, *then* $I = R$.

Proof The proofs are immediate and left to the reader. □

Let $(R, +, \cdot)$ be a commutative ring with identity and $I \subseteq R$ an ideal of R. We know that I is a normal subgroup of the additive group R, since $(R, +)$ is abelian. We then consider the additive abelian group R/I and define a multiplication on it as follows: for all $a, b \in R$, $(a + I) \cdot (b + I) = (a \cdot b) + I$. Considering what we have said, we can show that R/I with these two operations is a commutative ring with identity (the identity of R/I is the class $1 + I$, where 1 is the identity of R).

Theorem A.2 *If* $(R, +, \cdot)$ *is a commutative ring and* I *an ideal of* R, *then the additive abelian group* R/I *is a commutative ring with identity, where the addition if defined by* $(a + I) + (b + I) = (a + b) + I$ *and the multiplication is given by* $(a + I) \cdot (b + I) = (a \cdot b) + I$, *for all* $a, b \in R$.

Proof See the proof of Theorem 3.110 in [137]. □

References

1. Albuquerque, C.D., Palazzo Jr., R, Silva, E.B.: Topological quantum codes on compact surfaces with genus $g \geq 2$. J. Math. Phys. **50**, 023513-1–023513-20 (2009)
2. Albuquerque, C.D., Palazzo Jr., R, Silva, E.B.: Families of classes of topological quantum codes from tessellations $\{4i + 2, 2i + 1\}$, $\{4i, 4i\}$, $\{8i - 4, 4\}$ and $\{12 - 3, 6\}$. Quantum Inf. Comput. **14**, 1424–1440 (2014)
3. Aly, S.A.: Asymmetric quantum BCH codes. In: Proceedings of the IEEE International Conference on Computer Engineering and Systems (ICCES'08), New Jersey, USA, 25–27 Nov 2008
4. Aly, S.A., Klappenecker, A., Sarvepalli, P.K.: On quantum and classical BCH codes. IEEE Trans. Inf. Theory **53**, 1183–1188 (2007)
5. Aly, S.A., Klappenecker, A., Sarvepalli, P.K.: Quantum convolutional codes derived from Reed-Solomon and Reed-Muller codes. https://arxiv.org/abs/quant-ph/0701037v2 (2007). Accessed 09 Oct 2007
6. Aly, S.A., Grassl, M., Klappenecker, A., Rötteler, M., Sarvepalli, P.K.: Quantum convolutional BCH codes. In: 10th Canadian Workshop on Information Theory (CWIT), Edmonton, AB, Canada, 6–8 June 2007
7. Aly, S.A., Klappenecker, A., Sarvepalli, P.K.: Quantum convolutional codes derived from generalized Reed-Solomon codes. In: IEEE International Symposium on Information Theory (ISIT), Nice, France, 24–29 June 2007
8. Armstrong, M.A.: Basic Topology. Springer (1983)
9. Ashikhmin, A., Barg, A., Knill, E., Litsyn, S.: Quantum error-detection I: statement of the problem. IEEE Trans. Inf. Theory **46**, 778–788 (2000)
10. Ashikhmin, A., Barg, A., Knill, E., Litsyn, S.: Quantum error-detection II: bounds. IEEE Trans. Inf. Theory **46**, 789–800 (2000)
11. Ashikhmin, A., Knill, E.: Non-binary quantum stabilizer codes. IEEE Trans. Inf. Theory **47**, 3065–3072 (2001)
12. Ashikhmin, A., Litsyn, S., Tsfasman, M.A.: Asymptotically good quantum codes. Phys. Rev. A **63**, 1–5 (2001)
13. Atiyah, M.F., MacDonald, I.G.: Introduction to Commutative Algebra. Addison-Wesley Publishing Company Inc., Massachusetts (1969)
14. Aydin, N., Siap, I., Ray-Chaudhuri, D.K.: The structure of 1-generator quasi-twisted codes and new linear codes. Des. Codes Crypt. **24**, 313–326 (2001)
15. Bakshi, G.K., Raka, M.: Minimal cyclic codes of length $p^n q$. Finite Fields Appl. **9**, 432–448 (2003)

© Springer Nature Switzerland AG 2020

G. G. La Guardia, *Quantum Error Correction*, Quantum Science and Technology,
https://doi.org/10.1007/978-3-030-48551-1

16. Berlekamp, E.R.: Negacyclic codes for the Lee metric. In: Symposium on Combinatorial Mathematics and Its Applications, North Carolina, USA, 10–14 April 1967
17. Bierbrauer, J., Edel, Y.: Quantum twisted codes. J. Comb. Des. **8**, 174–188 (2000)
18. Blackford, T.: Negacyclic duadic codes. Finite Fields Appl. **14**, 930–943 (2008)
19. Blahut, R.E.: Algebraic Codes for Data Transmission. Cambridge University Press, New York, USA (2003)
20. Bombin, H., Martin-Delgado, M.A.: Topological quantum error correction with optimal encoding rate. Phys. Rev. A **73**, 062303-1–062303-5 (2006)
21. Bombin, H., Martin-Delgado, M.A.: Topological quantum distillation. Phys. Rev. Lett. **97**, 180501-1-1–180501-4 (2006)
22. Bose, R.C., Ray-Chaudhuri, D.K.: On a class of error correcting binary group codes. Inf. Control **3**, 68–79 (1960)
23. Bose, R.C., Ray-Chaudhuri, D.K.: Further results on error correcting binary group codes. Inf. Control **3**, 279–290 (1960)
24. Buchbinder, S.D., Huang, C.L., Weinstein, Y.S.: Encoding an arbitrary state in a [7, 1, 3] quantum error correction code. Quantum Inf. Process. **12**, 699–719 (2013)
25. Calderbank, A.R., Rains, E.M., Shor, P.W., Sloane, N.J.A.: Quantum error correction via codes over $GF(4)$. IEEE Trans. Inf. Theory **44**, 1369–1387 (1998)
26. Chen, H.: Some good quantum error-correcting codes from algebraic-geometric codes. IEEE Trans. Inf. Theory **47**, 2059–2061 (2001)
27. Chen, H., Ling, S., Xing, C.: Asymptotically good quantum codes exceeding the Ashikhmin-Litsyn-Tsfasman bound. IEEE Trans. Inf. Theory **47**, 2055–2058 (2001)
28. Chen, H., Ling, S., Xing, C.: Quantum codes from concatenated algebraic-geometric codes. IEEE Trans. Inf. Theory **51**, 2915–2920 (2005)
29. Climent, J.J., Herranz, V., Perea.: Linear system modelization of concatenated block and convolutional codes. Linear Algebra Appl. **429**, 1191–1212 (2008)
30. Cohen, G.D., Encheva, S.B., Litsyn, S.: On binary constructions of quantum codes. IEEE Trans. Inf. Theory **45**, 2495–2498 (1999)
31. Curto, J.I., Porras, J.M., Martín, F.J.P., Sotelo, G.S.: Convolutional Goppa codes defined on fibrations. A.A.E.C.C. **23**, 165–178 (2012)
32. de Almeida, A.C.A., Palazzo Jr., R.: A concatenated [(4, 1, 3)] quantum convolutional code. In: IEEE Information Theory Workshop (ITW), San Antonio, Texas, 28–33 Oct 2004
33. Ding, C., Kohel, D., Ling, S.: Elementary 2-group character codes. IEEE Trans. Inf. Theory **46**, 280–284 (2000)
34. Edel, Y.: Table of quantum twisted codes. Electronic address: www.mathi.uni-heidelberg.de/~yves/Matritzen/QTBCH/QTBCHIndex.html
35. Ezerman, M.F., Ling, S., Solé, P.: Additive asymmetric quantum codes. IEEE Trans. Inf. Theory **57**, 5536–5550 (2011)
36. Ezerman, M.F., Ling, S., Yemen, O., Solé, P.: From skew-cyclic codes to asymmetric quantum codes. Adv. Math. Commun. **5**, 41–57 (2011)
37. Ezerman, M.F., Jitman, S., Ling, S.: Pure asymmetric quantum MDS codes from CSS construction: a complete characterization. Int. J. Quantum Inf. **11**(1350027), 1–10 (2013)
38. Feng, K., Ma, Z.: A finite Gilbert-Varshamov bound for pure stabilizer quantum codes. IEEE Trans. Inf. Theory **50**, 3323–3325 (2004)
39. Feng, K., Ling, S., Xing, C.: Asymptotic bounds on quantum codes from algebraic geometry codes. IEEE Trans. Inf. Theory **52**, 986–991 (2006)
40. Forney Jr., G.D.: Convolutional codes I: algebraic structure. IEEE Trans. Inf. Theory **16**, 720–738 (1970)
41. Forney Jr., G.D., Grassl, M., Guha, S.: Convolutional and tail-biting quantum error-correcting codes. IEEE Trans. Inf. Theory **53**, 865–880 (2007)
42. Fujiwara, Y.: Block synchronization for quantum information. Phys. Rev. A **87**, 23–44 (2013)
43. Fujiwara, Y., Tonchev, V.D., Wong, T.W.H.: Algebraic techniques in designing quantum synchronizable codes. Phys. Rev. A **88**, 012318 (2013)

44. Fujiwara, Y., Vandendriessche, P.: Quantum synchronizable codes from finite geometries. IEEE Trans. Inf. Theory **60**, 7345–7354 (2014)
45. Galindo, C., Geil, O., Hernando, F., Ruano, D.: On the distance of stabilizer quantum codes from J-affine variety codes. Quantum Inf. Process. **16**, 111 (2017)
46. Galindo, C., Hernando, F., Ruano, D.: Stabilizer quantum codes from J-affine variety codes and a new Steane-like enlargement. Quantum Inf. Process. **14**, 3211–3231 (2015)
47. Garcia, A., Stichtenoth, H.: On the asymptotic behaviour of some towers of function fields over finite fields. J. Number Theory **61**, 248–273 (1996)
48. Gluesing-Luerssen, H., Rosenthal, J., Smarandache, R.: Strongly MDS convolutional codes. IEEE Trans. Inf. Theory **52**, 584–598 (2006)
49. Gluesing-Luerssen, H., Schmale, W.: Distance bounds for convolutional codes and some optimal codes. e-print arXiv:math/0305135
50. Gluesing-Luerssen, H., Schmale, W.: On doubly-cyclic convolutional codes. AAECC **17**, 151–170 (2006)
51. Gluesing-Luerssen, H., Tsang, F.L.: A matrix ring description for cyclic convolutional codes. Adv. Math. Commun. **2**, 55–81 (2008)
52. Goppa, V.D.: Codes associated with divisors. Problemes Peredachi Informatsii **13**, 33–39 (1977) (English translation in Problems Inform Transmission) **13**, 22–27 (1977)
53. Gottesman, D.: Class of quantum error-correcting codes saturating the quantum Hamming bound. Phys. Rev. A **54**, 1862–1868 (1996)
54. Grassl, M., Beth, Th., Pellizzari, T.: Codes for the quantum erasure channel. Phys. Rev. A **56**, 33–38 (1997)
55. Grassl, M., Beth, T., Rötteler, M.: On optimal quantum codes. Int. J. Quantum Inf. **2**, 55–64 (2004)
56. Grassl, M., Geiselmann, W., Beth, T.: Quantum Reed-Solomon codes. AAECC-13 **1709**, 231–244 (1999)
57. Grassl, M., Rötteler, M.: Quantum block and convolutional codes from self-orthogonal product codes. In: International Symposium on Information Theory (ISIT), Adelaide, Australia, 4–8 Sept 2005
58. Grassl, M., Rötteler, M.: Constructions of quantum convolutional codes. In: IEEE International Symposium on Information Theory (ISIT), Nice, France, 24–29 June 2007
59. Grassl, M., Rötteler, M.: Non-catastrophic encoders and encoder inverses for quantum convolutional codes. In: International Symposium on Information Theory (ISIT), Seattle, WA, USA, 9–14 July 2006
60. Guenda, K.: Quantum duadic and affine invariant codes. Int. J. Quantum Inf. **7**, 373–384 (2009)
61. Guenda, K., La Guardia, G.G., Gulliver, T.A.: Algebraic quantum synchronizable codes. J. Appl. Math. Comput. **55**, 393–407 (2017)
62. Hamada, M.: Concatenated quantum codes constructible in polynomial time: efficient decoding and error correction. IEEE Trans. Inf. Theory **54**, 5689–5704 (2008)
63. Hocquenghem, A.: Codes correcteurs d'erreurs. Chiffres **2**, 147–156 (1959)
64. Hoffman, K., Kunze, R.: Linear Algebra. Prentice-Hall Inc., New Jersey (1971)
65. Hole, K.J.: On classes of convolutional codes that are not asymptotically catastrophic. IEEE Trans. Inf. Theory **46**, 663–669 (2000)
66. Houshmand, M., Hosseini-Khayat, S., Wilde, M.M.: Minimal-memory, non-catastrophic, polynomial-depth quantum convolutional encoders. IEEE Trans. Inf. Theory **59**, 1198–1210 (2013)
67. Huffman, W.C., Pless, V.: Fundamentals of Error-Correcting Codes. Cambridge University Press, Cambridge (2003)
68. Hutchinson, R., Rosenthal, R., Smarandache, R.: Convolutional codes with maximum distance profile. Syst. Control Lett. **54**, 53–63 (2005)
69. Iglesias-Curto, J.I.: Generalized AG convolutional codes. Adv. Math. Commun. **3**, 317–328 (2009)

70. Ioffe, L., Mezard, M.: Asymmetric quantum error-correcting codes. Phys. Rev. A **75**, 1–4 (2007)
71. Jin, L.: Quantum stabilizer codes from maximal curves. IEEE Trans. Inf. Theory **60**, 313–316 (2014)
72. Jin, L., Ling, S., Luo, J., Xing, C.: Application of classical Hermitian self-orthogonal MDS codes to quantum MDS codes. IEEE Trans. Inf. Theory **56**, 4735–4740 (2010)
73. Jin, L., Xing, C.: Euclidean and Hermitian self-orthogonal algebraic geometry codes and their application to quantum codes. IEEE Trans. Inf. Theory **58**, 5484–5489 (2012)
74. Jin, L., Xing, C.: A construction of new quantum MDS codes. IEEE Trans. Inf. Theory **60**, 2921–2925 (2014)
75. Johannesson, R., Zigangirov, K.S.: Fundamentals of Convolutional Coding. Digital and Mobile Communication. Wiley-IEEE Press, New Jersey (1999)
76. Kai, X., Zhu, S.: Quantum negacyclic codes. Phys. Rev. A **88**, 012326(1–5) (2013)
77. Kai, X., Zhu, S.: New quantum MDS codes from negacyclic codes. IEEE Trans. Inf. Theory **59**, 1193–1197 (2013)
78. Kai, X., Zhu, S., Li, P.: Constacyclic codes and some new quantum MDS codes. IEEE Trans. Inf. Theory **60**, 2080–2086 (2014)
79. Kai, X., Zhu, S., Li, P.: Some new classes of quantum MDS codes from constacyclic codes. IEEE Trans. Inf. Theory **61**, 5224–5228 (2015)
80. Ketkar, A., Klappenecker, A., Kumar, S., Sarvepalli, P.K.: Nonbinary stabilizer codes over finite fields. IEEE Trans. Inf. Theory **52**, 4892–4914 (2006)
81. Kim, J.-L., Walker, J.: Nonbinary quantum error-correcting codes from algebraic curves. Discret. Math. **308**, 3115–3124 (2008)
82. Kitaev, A.Yu.: Fault-tolerant quantum computation by anyons. Ann. Phys. **303**, 2–30 (2003)
83. Klappenecker, A., Rötteler, M.: Beyond stabilizer codes I: nice error bases. IEEE Trans. Inf. Theory **48**, 2392–2395 (2002)
84. Klappenecker, A., Rötteler, M.: Beyond stabilizer codes II: Clifford codes. IEEE Trans. Inf. Theory **48**, 2396–2399 (2002)
85. Knill, E.: Non-binary unitary error bases and quantum codes. Los Alamos National Laboratory Report LAUR-96-2717 (1996)
86. Knill, E., Laflamme, R.: Theory of quantum error-correcting codes. Phys. Rev. A **55**, 900–911 (1997)
87. Krishna, A., Sarwate, D.V.: Pseudocyclic maximum-distance-separable codes. IEEE Trans. Inf. Theory **36**, 880–884 (1990)
88. Laflamme, R., Miquel, C., Paz, J.P., Zurek, W.H.: Perfect quantum error correction code. Phys. Rev. Lett. **77**, 198–201 (1996)
89. La Guardia, G.G.: Constructions of new families of nonbinary quantum codes. Phys. Rev. A. **80**(042331), 1–11 (2009)
90. La Guardia, G.G.: New families of asymmetric quantum BCH codes. Quantum Inf. Comput. **11**, 239–252 (2011)
91. La Guardia, G.G.: New quantum MDS codes. IEEE Trans. Inf. Theory **57**, 5551–5554 (2011)
92. La Guardia, G.G.: New families of asymmetric quantum BCH codes. Quantum Inf. Comput. **11**, 239–252 (2011)
93. La Guardia, G.G.: On nonbinary quantum convolutional BCH codes. Quantum Inf. Comput. **12**, 820–842 (2012)
94. La Guardia, G.G.: Asymmetric quantum Reed-Solomon and generalized Reed-Solomon codes. Quantum Inf. Process. **11**, 591–604 (2012)
95. La Guardia, G.G.: Asymmetric quantum product codes. Int. J. Quantum Inf. **10**, 1250005(1–11) (2012)
96. La Guardia, G.G.: Nonbinary convolutional codes derived from group character codes. Discret. Math. **313**, 2730–2736 (2013)
97. La Guardia, G.G.: Asymmetric quantum codes: new codes from old. Quantum Inf. Process. **12**, 2771–2790 (2013)

98. La Guardia, G.G.: On classical and quantum MDS-convolutional BCH codes. IEEE Trans. Inf. Theory **60**, 304–312 (2014)
99. La Guardia, G.G.: On negacyclic MDS-convolutional codes. Linear Algebra Appl. **448**, 85–96 (2014)
100. La Guardia, G.G.: On the construction of nonbinary quantum BCH codes. IEEE Trans. Inf. Theory **60**, 1528–1535 (2014)
101. La Guardia, G.G.: On optimal constacyclic codes. Linear Algebra Appl. **496**, 594–610 (2016)
102. La Guardia, G.G.: Asymmetric quantum convolutional codes. Quantum Inf. Process. **15**, 167–183 (2016)
103. La Guardia, G.G.: Quantum codes derived from cyclic codes. Int. J. Theor. Phys. **56**, 2479–2484 (2017)
104. La Guardia, G.G., Alves, M.M.S.: On cyclotomic cosets and code constructions. Linear Algebra Appl. **488**, 302–319 (2016)
105. La Guardia, G.G., Palazzo Jr., R.: Constructions of new families of nonbinary CSS codes. Discret. Math. **310**, 2935–2945 (2010)
106. La Guardia, G.G., Pereira, F.R.F.: Good and asymptotically good quantum codes derived from algebraic geometry. Quantum Inf. Process. **16**, 1–12 (2017)
107. Lee, L.N.: Short unit-memory byte-oriented binary convolutional codes having maximum free distance. IEEE Trans. Inf. Theory **22**, 349–352 (1976)
108. Li, R., Li, X.: Binary construction of quantum codes of minimum distances five and six. Discret. Math. **308**, 1603–1611 (2008)
109. Li, S., Xiong, M., Ge, G.: Pseudo-cyclic codes and the construction of quantum MDS codes. IEEE Trans. Inf. Theory **62**, 1703–1710 (2016)
110. Lidl, R., Niederreiter, H.: Finite Fields. Cambridge University Press, Cambridge (1997)
111. Ling, S., Luo, J., Xing, C.: Generalization of Steane's enlargement construction of quantum codes and applications. IEEE Trans. Inf. Theory **56**, 4080–4084 (2010)
112. Lipschutz, S., Lipson, M.L.: Linear Algebra. McGraw-Hill Companies, New York (2009)
113. Ma, Z., Lu, X., Feng, K., Feng, D.: On non-binary quantum BCH codes. Lect. Notes Comput. Sci. **3959**, 675–683 (2006)
114. MacWilliams, F.J., Sloane, N.J.A.: The Theory of Error-Correcting Codes. North-Holland (1977)
115. Mandelbaum, D.M.: Two applications of cyclotomic cosets to certain BCH codes. IEEE Trans. Inf. Theory **26**, 737–738 (1980)
116. Martin, F.J.P., Curto, J.I., Sotelo, G.S.: On the construction of 1-D MDS convolutional Goppa codes. IEEE Trans. Inf. Theory **59**, 4615–4625 (2013)
117. Matsumoto, R.: Improvement of Ashikhmin-Litsyn-Tsfasman bound for quantum codes. IEEE Trans. Inf. Theory **48**, 2122–2124 (2002)
118. Munkres, J.R.: Topology. Prentice Hall Inc., New Jersey (2000)
119. Munuera, C., Tenorio, W., Torres, F.: Quantum error-correcting codes from algebraic geometry codes of castle type. Quantum Inf. Process. **15**, 4071–4088 (2016)
120. Niederreiter, H., Xing, C.: Algebraic Geometry in Coding Theory and Cryptography. Princeton University Press, Princeton, New Jersey (2009)
121. Nielsen, M.A., Chuang, I.L.: Quantum Computation and Quantum Information. Cambridge University Press, Cambridge (2000)
122. Nigg, D., Müller, M., Martinez, E.A., Schindler, P., Hennrich, M., Monz, T., Martin-Delgado, M.A., Blatt, R.: Quantum computations on a topologically encoded qubit. Science **345**, 302–305 (2014)
123. Ollivier, H., Tillich, J.-P.: Description of a quantum convolutional code. Phys. Rev. Lett. **91**, 1–4 (2003)
124. Ollivier H., Tillich J.-P. (2004) Quantum convolutional codes: fundamentals. In: arXiv. https://arxiv.org/pdf/quant-ph/0401134.pdf. Accesses 20 March 2018
125. Ozbudak, F., Stichtenoth, H.: Constructing codes from algebraic curves. IEEE Trans. Inf. Theory **45**, 2502–2505 (1999)

126. Pellikaan, R., Shen, B.-Z., van Wee, G.J.M.: Which linear codes are algebraic-geometric? IEEE Trans. Inf. Theory **37**, 583–602 (1991)

127. Pereira, F.R.F., La Guardia, G.G., de Assis, F.M.: Classical and quantum convolutional codes derived from algebraic geometry codes. IEEE Trans. Commun. **67**, 73–82 (2019)

128. Pérez, J.A.D., Porras, J.M., Sotelo, G.S.: Convolutional codes of Goppa type. A.A.E.C.C. **91**, 51–61 (2004)

129. Peterson, W.W., Weldon Jr., E.J.: Error-Correcting Codes. MIT Press, Cambridge (1972)

130. Piret, Ph.: Convolutional Codes: An Algebraic Approach. The MIT Press, Cambridge (1988)

131. Piret, Ph.: A convolutional equivalent to Reed-Solomon codes. Philips J. Res. **43**, 441–458 (1988)

132. Rains, E.M.: Quantum codes of minimum distance two. IEEE Trans. Inf. Theory **45**, 266–271 (1999)

133. Rains, E.M.: Nonbinary quantum codes. IEEE Trans. Inf. Theory **45**, 1827–1832 (1999)

134. Rains, E.M.: Quantum shadow enumerators. IEEE Trans. Inf. Theory **45**, 2361–2366 (1999)

135. Rosenthal, J., Smarandache, R.: Maximum distance separable convolutional codes. Appl. Algebra Eng. Commun. Comput. **10**, 15–32 (1998)

136. Rosenthal, J., York, E.V.: BCH convolutional codes. IEEE Trans. Inf. Theory **45**, 1833–1844 (1999)

137. Rotman, J.J.: Advanced Modern Algebra. Prentice Hall (2003)

138. Sarvepalli, P.K., Klappenecker, A.: Nonbinary quantum Reed-Muller codes. In: International Symposium on Information Theory (ISIT), Adelaide, Australia, 4–9 Sept 2005

139. Sarvepalli, P.K., Klappenecker, A., Rötteler, M.: Asymmetric quantum LDPC codes. In: International Symposium on Information Theory (ISIT), Toronto, Canada, 6–11 July 2008

140. Sarvepalli, P.K., Klappenecker, A., Rötteler, M.: Asymmetric quantum codes: constructions, bounds and performance. Proc. R. Soc. A **465**, 1645–1672 (2009)

141. Sharma, A., Bakshi, G.K., Dumir, V.C., Raka, M.: Cyclotomic number and primitive idempotents in the ring $GF(q)/(x^{p^n} - 1)$. Finite Fields Appl. **10**, 653–673 (2004)

142. Sharma, A., Bakshi, G.K., Raka, M.: The weight distributions of irreducible cyclic codes of length 2^m. Finite Fields Appl. **13**, 1086–1095 (2007)

143. Sharma, A., Bakshi, G.K., Raka, M.: Polyadic codes of prime power length. Finite Fields Appl. **13**, 1071–1085 (2007)

144. Shor, P.: Scheme for reducing decoherence in quantum computer memory. Phys. Rev. A **54**, R2493–R2496 (1995)

145. Smarandache, R., Gluesing-Luerssen, H., Rosenthal, J.: Constructions of MDS-convolutional codes. IEEE Trans. Inf. Theory **47**, 2045–2049 (2001)

146. Soares Jr., W.S., Silva, E.B.: Hyperbolic quantum color codes. Quantum Inf. Comput. **18**, 0307–0381 (2018)

147. Steane, A.M.: Simple quantum error correcting-codes. Phys. Rev. A **54**, 4741–4751 (1996)

148. Steane, A.: Enlargement of Calderbank-Shor-Steane quantum codes. IEEE Trans. Inf. Theory **45**, 2492–2495 (1999)

149. Stephens, A.M., Evans, Z.W.E., Devitt, S.J., Hollenberg, L.C.L.: Asymmetric quantum error correction via code conversion. Phys. Rev. A **77**, 062335(1–5) (2008)

150. Stichtenoth, H.: Self-dual Goppa codes. J. Pure Appl. Algebra **55**, 199–211 (1988)

151. Stichtenoth, H.: Transitive and self-dual codes attaining the Tsfasman-Vladut-Zink bound. IEEE Trans. Inf. Theory **52**, 2218–2224 (2006)

152. Stichtenoth, H.: Algebraic Function Fields and Codes. Springer, Berlim (2009)

153. Sundeep, B., Thangaraj, A.: Self-orthogonality of q-ary images of q^m-ary codes and quantum code construction. IEEE Trans. Inf. Theory **53**, 2480–2489 (2007)

154. Tan, P., Li, J.: Efficient quantum stabilizer codes: LDPC and LDPC-convolutional constructions. IEEE Trans. Inf. Theory **56**, 476–491 (2010)

155. Thangaraj, A., McLaughlin, S.: Quantum codes from cyclic codes over $GF(4^m)$. IEEE Trans. Inf. Theory **47**, 1176–1178 (2001)

156. Tonchev, V.D.: Quantum codes from caps. Discret. Math. **308**, 6368–6372 (2008)

157. Wang, L., Feng, K., Ling, S., Xing, C.: Asymmetric quantum codes: characterization and constructions. IEEE Trans. Inf. Theory **56**, 2938–2945 (2010)
158. Weinstein, Y.S.: Syndrome measurement strategies for the [[7, 1, 3]] code. Quantum Inf. Process. **14**, 1841–1854 (2015)
159. Weinstein, Y.S.: Syndrome measurement order for the [[7, 1, 3]] quantum error correction code. Quantum Inf. Process. **15**, 1263–1271 (2016)
160. Wilde, M.M., Brun, T.A.: Unified quantum convolutional coding. In: International Symposium on Information Theory (ISIT), Toronto, Canada, 6–11 July 2008
161. Wilde, M.M., Brun, T.A.: Entanglement-assisted quantum convolutional coding. Phys. Rev. A **81**, 042333 (2010)
162. Xiaoyan, L.: Quantum cyclic and constacyclic codes. IEEE Trans. Inf. Theory **50**, 547–549 (2004)
163. Xu, Y., Ma, Z., Zhang, C.: On classical BCH codes and quantum BCH codes. Chin. J. Electron. **26**, 64–70 (2009)
164. Yixuan, X., Yuan, J., Fujiwara, Y.: Quantum synchronizable codes from quadratic residue codes and their supercodes. In: Proceedings of IEEE Information Theory Workshop, Hobart, TAS, Australia 172–176 Nov 2014
165. Yue, D.W., Feng, G.Z.: Minimal cyclotomic coset representatives and their applications to BCH codes and Goppa codes. IEEE Trans. Inf. Theory **46**, 2625–2628 (2000)
166. Yue, D.W., Hu, Z.M.: On the dimension and minimum distance of BCH codes over $GF(q)$. Chin. J. Electron. **18**, 263–269 (1996)
167. Zhang, T., Ge, G.: Quantum block and synchronizable codes derived from certain classes of polynomials. https://arxiv.org/abs/quant-ph/1508.00974 (2015). Accessed 15 May 2019
168. Zhang, T., Ge, G.: Some new classes of quantum MDS codes from constacyclic codes. IEEE Trans. Inf. Theory **61**, 5224–5228 (2015)
169. Zhu, S., Kai, X.: A class of constacyclic codes over $\mathbb{Z}_{p^m}$. Finite Fields Appl. **16**, 243–254 (2010)
170. Zhu, S., Wang, L., Kai, X.: New optimal quantum convolutional codes. Int. J. Quantum Inf. **13**, 1550019–14 (2015)

Index

© Springer Nature Switzerland AG 2020   225
G. G. La Guardia, *Quantum Error Correction*, Quantum Science and Technology,
https://doi.org/10.1007/978-3-030-48551-1

Printed in the United States
by Baker & Taylor Publisher Services